Conservation Biology

Conservation Biology

THE THEORY AND PRACTICE OF NATURE CONSERVATION PRESERVATION AND MANAGEMENT

● ● ●

PEGGY L. FIEDLER AND SUBODH K. JAIN, EDITORS

CHAPMAN AND HALL
NEW YORK AND LONDON

First published in 1992 by
Chapman and Hall
an imprint of
Routledge, Chapman & Hall, Inc.
29 West 35 Street
New York, NY 10001-2291

Published in Great Britain by
Chapman and Hall
2-6 Boundary Row
London SE1 8HN

Library of Congress Cataloging-in-Publication Data

Conservation biology : the theory and practice of nature conservation,
    preservation, and management / editors, Peggy L. Fiedler and Subodh
    K. Jain.
        p. cm.
    Includes bibliographical references and index.
    (hb) ISBN 0-412-01951-5   (pb) ISBN 0-412-01961-2
    1. Nature conservation.   2. Plant conservation.   3. Biological
diversity conservation.   I. Fiedler, Peggy L.   II. Jain, Subodh K.,
1934–
QH75.C662   1992   92-30942
333.95'16—dc20   CIP

NWst
IAEB7690

British Library Cataloguing-in-Publication Data also available.

TO TIMOTHY CHARLES PLOWMAN
1944-1989
INTREPID BIOLOGIST, MENTOR, FRIEND
AND
FOR MY PARENTS, WHO TAUGHT ME
THE JAINISM VIEW OF CONSERVATION ETHICS

● ● ●

CONTRIBUTORS

• • •

Jeremy J. Ahouse
Department of Biology
Bassine 235
Brandeis University
Waltham, MA 02254

Miguel Altieri
Division of Biological Control
University of California, Berkeley
1020 San Pablo Ave
Albany, CA 94706

M. Kat Anderson
Department of Forestry and Resource
  Management
University of California
Berkeley, CA 94720

Peter S. Ashton
Harvard Herbaria
Harvard University
22 Divinity Avenue
Cambridge, MA 02138

Douglas T. Bolger
Department of Biology C-0116
University of California
La Jolla, CA 92093

Marybeth Buechner
Institute of Ecology
Wickson Hall, University of California
Davis, CA 95616

C. Ron Carroll
Institute of Ecology
University of Georgia
Athens, GA 30602

Ted J. Case
Department of Biology C-0116
University of California
La Jolla, CA 92093

Donald Falk
Center for Plant Conservation
Missouri Botanical Garden
St. Louis, MO 63166

Peggy L. Fiedler
Department of Biology
San Francisco State University
1600 Holloway Avenue
San Francisco, CA 94132

Edward O. Guerrant, Jr.
The Berry Botanic Garden
11505 SW Somerville Avenue
Portland, OR 97219

John E. Hafernik, Jr.
Department of Biology
San Francisco State University
1600 Holloway Avenue
San Francisco, CA 94132

John L. Harper
School of Plant Biology
University College of North Wales
Bangor, Gwynnedd, Wales

Larry D. Harris
Department of Wildlife and Range
  Sciences
118 Newins-Ziegler Hall
University of Florida
Gainesville, FL 32611-0301

Laura L. Jackson
The Desert Botanical Garden
1201 N. Galvin Parkway
Phoenix, AZ 85008

Subodh K. Jain
Department of Agronomy and Range
  Science
University of California
Davis, CA 95616

Robert C. Lacy
Chicago Zoological Society
Brookfield Zoo
3300 S. Golf Road
Brookfield, IL 60513

Robert A. Leidy
U.S. Environmental Protection Agency
Region IX
75 Hawthorne Street (W-7-2)
San Francisco, CA 94105

Eric S. Menges
Archbold Biological Station
P.O. Box 2057
Lake Placid, FL 33852

Peter B. Moyle
Department of Wildlife and Fisheries
  Biology
University of California, Davis
Davis, CA 95616

Reed F. Noss
925 NW 31st Street
Corvallis, OR 97330

V. Thomas Parker
Department of Biology
San Francisco State University
1600 Holloway Avenue
San Francisco, CA 94132

Mary C. Pearl
Wildlife Conservation International
New York Zoological Society
Bronx, NY 10460

Steward T.A. Pickett
Institute for Ecosystem Studies
New York Botanical Garden
Mary Flager Cary Arboretum, Box AB
Millbrook, NY 12545

Adam D. Richman
Department of Biology C-0116
University of California
La Jolla, CA 92093

Christine Schonewald-Cox
National Park Service Cooperative
  Studies Unit
Institute of Ecology, and Division of
  Environmental Studies
Wickson Hall, University of California
Davis, CA 95616

Gilberto Silva-Lopez
Department of Wildlife and Range
  Sciences
118 Newins-Ziegler Hall
University of Florida
Gainesville, FL 32611-0301

G. Ledyard Stebbins
Department of Genetics
University of California
Davis, CA 95616

CONTENTS

• • •

Part III Population Biology and Genetics

Part IV The Practice of Conservation, Preservation, and Management

FOREWORD

• • •

John Harper

● ● ●

Nature conservation has changed from an idealistic philosophy to a serious technology. Ecology, the science that underpins the technology of conservation, is still too immature to provide all the wisdom that it must. It is arguable that the desire to conserve nature will in itself force the discipline of ecology to identify fundamental problems in its scientific goals and methods. In return, ecologists may be able to offer some insights that make conservation more practicable (Harper 1987).

The idea that nature (species or communities) is worth preserving rests on several fundamental arguments, particularly the argument of nostalgia and the argument of human benefit and need. Nostalgia, of course, is a powerful emotion. With some notable exceptions, there is usually a feeling of dismay at a change in the status quo, whether it be the loss of a place in the country for walking or rambling, the loss of a painting or architectural monument, or that one will never again have the chance to see a particular species of bird or plant. But nostalgia is selective, as it rarely applies to poison ivy, venomous snakes, electricity pylons, and the smallpox virus or to species of microscopic algae, pathogenic fungi, soil mites, and collembola!

Part of the nostalgic argument for nature conservation is the feeling that we have the duty to allow our children to experience what has given us great pleasure or piqued our curiosity. Thus the nostalgia argument dominates the universal concern with colorful birds, furry mammals, and beautiful plants, although there is no codified scientific argument for concentration of interest on these special groups. Nostalgia for them is, of course, a crucial element determining the funding of much conservation research, based largely on the appeal to our cultivated biological interests.

The argument of human benefit and need is more powerful in terms of the Descartian scientific tradition. In essence this is the argument that unless there is nature conservation, we may lose something of direct value to humans that has not yet been discovered. Every species (and variants within species) might conceivably be developed as a food crop or contain a valuable drug or potential pesticide. Even if the species itself is not directly useful, it might contain genetic material that one day might be used transgenically in genetic engineering. The argument of benefit and need places no barriers on what should be conserved, be it a bat or a rare pathogenic fungus. However, the argument does not imply that the conservation of all species must be accomplished in natural communities. In some cases, particularly many plant species and some lower invertebrates, their conservation inevitably will be satisfied by gene banks, zoos, or botanic gardens.

It is doubtful whether there is any community of organisms on the face of the earth that has not been disturbed by human activities, even if only through the long-distance aerial drift of pesticides or changes in the composition of the atmosphere. If nature has been disturbed by human technologies, it is not surprising that these technologies also have been called into play to minimize their deleterious effects. As just one example, the survival of naturally regenerating woodlands in Ireland presently depends on the control of invading *Rhododendron ponticum,* and this has called for the modern technology of selective control by chemical herbicides. Thus, although many conservationists have resisted the notion that nature reserves require active management, it is now widely accepted that conservation is not just leaving well enough alone.

Fundamental ecological questions arise out of the need to manage nature. What constitutes an "adequate" size and shape of a reserve? What is a "sustainable" population size for a species? Are there general rules for conservation, or is every species and community a special case? Could we more easily conserve some species if we eliminated some others, for example, their pests or competitors? Should our concern be with particular species that are at present rare or endangered, or are conservation aims best realized by the conservation of the essence of whole communities, even if this involves losing a few rarities?

Perhaps the most fundamental scientific question that now concerns the conservationist is what factors determine population sizes of species. All species, perhaps with the exception of a few species of bacteria, are absent from most of the world for most of geologic time. Absence is the normal condition in nature! Thus it is perhaps particularly unfortunate that so much of ecologists' energy has been devoted to discovering what it is about particular organisms that allows them to live where they do, while we remain so ignorant of what excludes them from areas where they do not live.

Scarcely have ecologists asked what happens to organisms that are *deliberately* introduced to places or communities in which they are not typically found. The exceptions are, of course, the purposeful introduction of crops to new agricultural regions and ornamental plants to private and public gardens. But in both cases, usually there is deliberate amelioration of the environment to which they are introduced and, in agriculture, manipulated genetic change making existence possible in the new environment. It is still true that our most thorough knowledge of the forces that control species abundance comes from accidental introductions of noxious pests and weeds, as well as from studies of their biological control. No comprehensive science of introduced species that might reveal the kinds of forces leading to local extinctions exists yet.

Some of the arguments for the conservation of species in nature are arguments for maintaining a status quo, and the very idea of experiments involving deliberate introduction of exotic species to new places is anathema. Indeed, nostalgia can be used to justify the maintenance of rare species in their rare or local condition, because rare species lose some of their nostalgic value if they become more abundant or widespread!

Today the conservation biologist needs the expertise that will allow the control and regulation of individuals of particular species or communities. For this he or she needs to understand the species' population biology—in particular, what controls their numbers where they live and, perhaps even more important, what forces exclude them from where they are absent. This question was elegantly posed by Charles Darwin in *The Origin of Species,* as were so many critical questions in biology that we still cannot answer. "Look at a plant in the midst of its range, why does it not double or quadruple its numbers?" Darwin thought the answer lay in resource competition and predation offered by other species. We need still to identify these controlling and regulating factors so that we may deliberately adjust their intensity in habitats or ecosystems of concern. Only experiments, specifically field experiments, can provide this information.

Ecologists have outgrown the simplistic view that communities can be regarded as homogeneous assemblages. Largely due to the pioneering work of Alex Watt (1947), it has been recognized that communities are patchworks. Moreover, the patches commonly represent "internal successions" in which particular species persist only temporarily (Connell and Slatyer 1977). Species persistence depends on continued dispersal or migration, "discovering" and colonizing one patch while disappearing from another. Abundance of a species may then depend on the (i) abundance and size of its specific patch, (ii) time for which the patch is colonizable and habitable, (iii) distance between such patches, and (iv) dispersal properties of the species (see, e.g., Gadgil 1971).

At least in theory any one of these intrinsic features may restrict the abundance of a species and cause it to be rare or to go extinct. And in theory, every one of these features is modifiable experimentally, or as part of a deliberate conservation practice. This is perhaps the most conspicuous area of modern ecology in which the study of population dynamics merges with the study of plant and animal physiology, as well as with community, ecosystem, and landscape ecology. This merger does not often happen but is required for a predictive science of nature conservation. The physiologist may determine what patch conditions are needed for the growth of a species or genotype, the population biologist studies how patches are colonized and how numbers change within them, and the community ecologist studies the causes and distribution of the patchwork.

One of the consequences of nature being a dynamic patchwork is that so many empirical demographic studies document histories of declines in natural populations. We need far more studies of the dynamics of expanding populations, especially the initial stage of patch colonization. Obviously it is much easier to find an established population to investigate than to initiate a study to detect and document colonization where a particular species is absent! It is almost certainly more important for the conservationist to discover *why* communities and patches fail to be colonized than to interpret the much rarer occasions when they succeed.

The great body of effort devoted to the conservation of nature by governments and private organizations is concerned with conserving organisms and communities *where they currently exist*. The problems involved in this accomplishment are enormous, but conservation problems produced by climatic change may be much greater. Predictions of climatic change as part of the greenhouse effect suggest that many, perhaps most, natural communities are doomed in their present state and spatial position. Predictions of an increase in mean temperatures of 2° to 5° C within 50 years, associated with increased variance in climatic factors and perhaps major changes in water balance, suggest that most natural reserve areas would be in the "wrong" places. Temperature rises of the predicted order would produce, within 50 years, hotter climates than have been experienced since the evolution of *Homo sapiens*. Such changes will occur within the lifetimes of many individuals, perhaps even within the juvenile stage in the life of some forest trees.

The fossil and historic records provide clues to the forces that may have led to species extinctions and also to those forces that have allowed species to survive. When major climatic changes have occurred as part of ice age cycles, populations and portions of whole communities apparently migrated great longitudinal distances. However, the areas over which future migrations might occur as a result of our present global warming are now fragmented: The vegetation continuum no longer exists, and species migrations will be from islands of "natural" vegetation across an ocean of managed vegetation. I find it hard to believe that there will not be a massive loss of species and communities as a consequence of the greenhouse effect unless there are dramatic changes in conservation practices.

There appear to be just two ways in which biodiversity might be conserved in the face of rapid climatic change—*either* by artificially maintaining climates of existing nature reserves in their present state, which is theoretically conceivable but impossibly costly, *or* by moving and reestablishing whole communities to regions where the changed environment is compatible with their survival. I see no other solutions. For this reason, I believe that "restoration ecology" is an emerging science of the most profound importance in conservation biology.

A science of restoration ecology has developed mainly because of a perceived need to recreate whole communities of organisms on land, or in waters or wetlands that have been damaged by (usually anthropogenic) disturbance (e.g., mining spoils, oil spills, agricultural reclamation, deforestation). If predictions of climatic change are realistic and if climates are as important for organisms' and communities' viability as most ecologists believe, nostalgic laissez-faire is suicidal for conservation. Restoration ecology will naturally extend to the recreation of assemblages of species in a way that keeps pace with climatic change.

I have argued elsewhere (Harper 1987) that there are fundamental ecological questions that require answers if we are to recreate communities. These same questions (the list is far from exclusive) must be answered if we are to try to conserve nature where it is at present. And answering the same questions seems to me to be necessary if climatic change requires us to recreate communities elsewhere.

1. Is the full species diversity of a recreated community necessary for its sustainability?
2. Is genetic diversity of an introduced species a necessary condition for its persistence?
3. Should an introduced species complement already have coevolved in the same habitat?

4. Does the age structure of the dominant species need to be controlled deliberately by phased introductions to establish appropriate patch dynamics and to avoid the ecological monotony of even-aged stands?
5. Can succession be bypassed? This question is, in essence, the same as one posed in the now-classic paper by Connell and Slatyer (1977).
6. Does the re-creation of vegetation come first and the introduction of the animal component follow, or are animals such a vital part of the community structure that their presence must be incorporated in the early stages?
7. How important are mutualists? Is it perhaps important to envisage effective conservation practices for the maintenance (or recreation) of tight species complexes represented by mutualisms and specialized food chains, rather than concentrating on the conservation of single species or of whole natural communities?

In partial answer, this book brings together a diversity of science and its practitioners—ecological empiricists and theorists—appropriate for this enormously daunting task. There are few branches of biology that will need not to contribute, and those that are critical to the evolution of conservation science are well represented in this volume. As conservation biology develops, it will become, like agriculture, a broad branch of applied ecology. It will depend on an underpinning by fundamental ecological science (Begon, Harper, and Townsend, 1986) and, it is hoped, give a new urgency to this parent discipline. Already, as the contents of this book demonstrate, the ecological foundation is still relatively too embryonic to offer generalizations that a more mature science will ultimately provide. Conservation biologists tend to treat their problems as situational, specialized, and often species-specific. Unifying principles are beginning to emerge, and this book goes far in starting to assemble them. Predictive generalizations are needed beyond just the case-by-case local solutions to local problems. It is for the theoretician and, above all, the experimentalist to produce these ground rules for conservation science. The aim should be to develop a predictive science of ecological management. This might enable us to conserve some of the diversity of nature where we have it now and, perhaps now even more necessary, provide the principles and technology to reassemble it, like London Bridge, elsewhere.

John Harper
University College of North Wales
Bangor, Gwynnedd, Wales

LITERATURE CITED

Begon, M., J.L. Harper, and C.R. Townsend. 1986. *Ecology: Individuals, populations, and communities.* Sunderland, Mass.: Sinauer.

Connell, J.H., and R.O. Slatyer. 1977. Mechanisms of succession in natural communities and their role in community stability and organization. *Amer. Nat.* 111:1119–44.

Gadgil, M. 1971. Dispersal: Population consequences and evolution. *Ecology* 52:253–61.

Harper, J.L. 1987. The heuristic value of ecological restoration. In *Restoration Ecology*, ed. W.R. Jordan, M.E. Gilpin, and J.D. Aber, 35–45. Cambridge: Cambridge University Press.

Watt, A.S. 1947. Pattern and process in the plant community. *J. Ecol.* 35:1–22.

PROLOGUE

• • •

Peggy L. Fiedler and Subodh K. Jain

• • •

The frontiers are not east or west, north or south, but wherever man fronts a fact.

Henry David Thoreau
Thursday, *A Week on the Concord and Merrimack Rivers*
1849

This is an age of much knowledge, little wisdom and even less ethics.
Jack Harlan quoting from Albert Szent-Györgi

Frontiers in science can be the fulcrums of change, and the frontier of nature conservation, preservation, and management is no exception. The newly coalesced discipline, known widely as "conservation biology," represents such a frontier in the theory and practice of an unparalleled amalgam of traditional and innovative, pure and applied sciences—wildlife management, ecology, evolution, systematics, population genetics, molecular genetics, restoration ecology, atmospheric science, agricultural economics, and environmental and resource policy, to name only a few. Conservation biology's fulcrum of change and decision making balances the conservation, preservation, and management of nature with the inevitable homocentric development of our natural resources and the exploitation of nature. In all future events, it behooves academic biologists, natural historians, and conservation practitioners to study and teach the scientific basis of our understanding and management of the planet's resources and biota.

But why a fulcrum of *change*? Sobering evidence is accumulating rapidly that the resource-based calculus of local and global carrying capacities is being violated to the detriment and potential collapse of our ecosystems. Landscapes are being altered by and are responding to anthropogenic influences at rates and in directions not easy, perhaps impossible, to predict. Thus *Homo sapiens* is confronting this uneasy fact. In response, conservation biology is a discipline that attempts, during the process of its crystallization, to develop a new body of theory that will prescribe a course of action for the reversal of environmental degradation, accelerated species extinctions and the consequent loss of biodiversity, and for the maintenance of ecosystem integrity.

Conservation biology is considered by some writers a faddish new science. However, much of it is not new, nor has it ever been terribly fashionable in academe. Roots of this enormously synthetic discipline penetrate through the early works of wildlife managers and land stewards who devoted their professional and personal lives to the maintenance of, for example, genetically distinct fish stocks or the restoration of a small patch of mixed-grass prairie. Thus in this claim of infancy is the neglect of a large and unappreciated cadre of scientists who have being doing "conservation biology" for decades under the broad aegis of "resource management" (see Aplet, Laven, and Fiedler 1992). More recently, environmental concerns about industrial waste, pollution, and rapid urbanization focused the attention of many serious writers and activists during the 1960s and 1970s. These dedicated but often less visible applied scientists and students recently have been passed over by the more glamorous doomspeakers, but their works in the *Journal of Wildlife Management, Journal of Forestry, Biological Conservation, Journal of Applied Ecology, Journal of Range Management*, and *Transactions of the American Fisheries Society*, for example, should not be ignored. In fact, belief in the integrity and intrinsic value of all living things as a professional and personal ethic, as well as a source of scientific inspiration, dates to centuries past. We feel that this belief will continue to be a primary, motivating ethic for biologists and biologically literate politicians who, in the relatively near future, will make crucial, perhaps irreversible decisions affecting the future health of our planet.

The history of conservation science involves many continents and cultures, with the works of a vast spectrum of land managers, scientists, politicians, ethicists, and artisans who are the past and present heroes of our science. It is not possible or appropriate here to document their contributions, and such summaries can be found elsewhere. We have chosen instead to provide a concise précis only of the significant historical legislative and intellectual landmarks in the United States, beginning in 1872 with the designation of approximately 2 million acres in Wyoming and Montana as Yellowstone National Park, our nation's first national park. Although many wildlands like Yellowstone that received federal protection status initially were considered "useless" for resource exploitation, the dedication of federally protected lands was a crucial beginning in the preservation of the nation's natural resources and, by extension, its natural areas. Two additional legislative actions around the turn of the century are of note: the 1891 Forest Reserve Act that established the first federal forest reserve and authorized the president to designate additional federal lands for timber and watershed protection, and the Lacey Act of 1900 that made it illegal to transport either live or dead animals or their parts across state borders without a permit. The latter act was prompted by a perceived need among the nation's citizenry to regulate the previously unregulated hunting of the nation's wildlife.

The first international conservation-related legislation occurred in 1911, with the passage of the Fur Seal Treaty. Shortly thereafter, the U.S. Congress ratified the 1918 Migratory Bird Act, an agreement between Great Britain and the U.S. to protect migratory birds in common. Following these early international agreements, the U.S. Congress passed a long series of federal acts to protect the nation's wildlands, wildlife, and natural resources. Notable among these are the Taylor Grazing Act (1934), Federal Bird Hunting Stamp Act (1934), Wilderness Act (1964), Clean Air Act (1965), National Environmental Policy Act (1969), Federal Water Pollution Control Act [Clean Water Act] (1972), Endangered Species Act (1973), Forest and Rangelands Renewable Resources Planning Act (1974), National Forest Management Act (1976), Federal Land Policy and Management Act (1976), and the Resource Conservation and Recovery Act (1976). Taken together, these (and other lesser-known federal laws enacted by the U.S. Congress, representing the people of the United States) embody the most far-reaching attempts by any public body to preserve, conserve, and manage its wildlands and natural resources.

The intellectual history of nature conservation, preservation, and management in great part parallels the development of the science of ecology, including the fields of wildlife and wildland management. Foremost among the writings of Western plant and animal ecologists are F. Clements' classic text on plant ecology (1916), later followed by Allee et al.'s classic text on animal ecology (1949); A. Leopold's *Game Management* (1933) and *A Sand County Almanac* (1949); Andrewartha and Birch's (1954) and Lack's (1954) landmark works on the dynamics and regulation of animal populations; and the writings of H. Gleason (1926), A.G. Tansley (1920, 1935) and R. Lindeman (1941, 1942) on ecosystem structure and processes. These scholars in particular laid the intellectual foundations for the emergence of conservation science as we know it today.

Several decades later, a variety of influential books and papers became available for the teaching of conservation biology and environmental science during the 1960s and 1970s. David Ehrenfeld's (1970) *Biological Conservation*, a very forceful introductory text, provided a succinct review of the field's main concepts at the level of species and communities and listed important sources of readings in conservation. Even twenty years ago, *ex situ* and *in situ* approaches to conservation and the fact that conservation meant more than just the preservation of flora and fauna were strongly emphasized. Also in 1970, O.H. Frankel and E. Bennett published a noteworthy volume on plant genetic resources documenting the outcome of the first major conference on the topic. Eric

Duffey and A.S. Watt's (1971) *The Scientific Management of Animal and Plant Communities for Conservation* appeared one year later as a bench mark for developing the clearly numerous links between theory and practice.

In 1974, R. Miller and D. Botkin published an extraordinarily influential paper dealing with the modeling of endangered species. Norman Myers, a longtime international champion of conservation, also published an important text, *The Sinking Ark* (1979), projecting the grim future of rare and endangered species. Similar concerns were later expressed in the significant publication by Paul and Anne Ehrlich (1981), *Extinction: The Causes and Consequences of the Disappearance of Species.* We may also mention and strongly recommend the insightful writings of B. Commoner (1968), G. Hardin (1968), M.W. Holdgate (Holdgate, Kassas, and White, 1982), R. Dasmann (1964, 1965, 1968), J. Passmore (1974), N.W. Moore (1962; Moore and Hooper 1976), M.J. Usher (1975), and P. Sears (1980) during this period of heightened concern and initiation of many ecological studies in conservation biology. And, never to be overlooked, MacArthur and Wilson's 1967 classic *Island Biogeography* overwhelmingly influences to this day quantitative thinking about extinction probabilities, population ecology analyses of species diversity, and the design of nature reserves (e.g., Diamond 1975).

The concept of minimal required areas or habitat for species survival also has long been widely recognized since the 1960s (e.g., see S. Cain's marvelous chapter in the classic work edited by F.F. Darling and J.P. Milton, *Future Environments of North America* [1966]). Further readings on specific population biology issues and environmental concerns during this time include G.W. Cox's (1969) *Readings in Conservation Ecology*, T.R. Detwyler's (1971) *Man's Impact on Environment*, the Ehrlichs' *The Population Bomb* (1968), and the proceedings from the 1972 United Nations Conference on the Human Environment, held in Stockholm (UNESCO 1973). During the same period, several biome projects provided excellent status reports emphasizing ecosystem analysis and resource management.

Conservation needs and programs are truly global, however, and indeed, one finds in this brief historical survey numerous national and regional contributions. The role of the International Union of the Conservation of Nature and Natural Resources (IUCN) is particularly noteworthy; and the international journal *Biological Conservation*, established in 1969, has provided from its inception an excellent medium for conservation research and issues. Early issues describe, for example, the biology of North America's endangered trumpeter swan (*Olor buccinator*) and the management approaches for Great Britain's rare chalk endemics. In addition, many wildlife and environmental management publications have been successful worldwide in continuing applied ecology interests in these areas. It is notable, however, that *evolutionary* and *population genetic* aspects of wildlife management or nature reserve establishment were treated only sketchily until recently. O.H. Frankel and M.E. Soulé's 1981 book *Conservation and Evolution,* M.E. Soulé and B.A. Wilcox's 1980 book *Conservation Biology: An Evolutionary-Ecological Perspective,* and C. Schonewald-Cox et al.'s 1983 book *Genetics and Conservation* clearly filled this need. Soon after, it became apparent how certain ideas on genetic bottlenecks, extinction proneness of small populations, the role of inbreeding, and diversity of gene pools found potential uses in the ecological framework for conservation projects.

We recognize that scientific "revolutions" such as that occurring in ecology and evolutionary biology today frequently represent a long gestation as well as a period of synergistic commingling of both scientific and sociopolitical interests. Thus it is imperative that we mention the important and incisive critiques of our homocentric exploitation of nature, beginning with Rachel Carson's *Silent Spring* (1962). The writings of E. Abbey (1968), particularly his "subversive" *Monkey Wrench Gang* (1975), were influential during the late 1970s, while more recently, M. Reisner's *Cadillac Desert* (1986) and A. Chase's *Playing God in Yellowstone* (1986) should be required reading for all those interested in the outcome of twentieth-century environmental policy.

Thus, in the 1980s, conservation biology in the United States emerged not only as a legitimate scientific endeavor for a broad spectrum of scientists and practitioners, but as the only conscionable one. In particular, great recognition is due to Michael Soulé (Soulé 1986, 1987) for providing the critical spark that recently coalesced so many theorists and practitioners into the main encampment of conservation biologists. These newly directed scientists, initially population geneticists and zoo biologists, joined a larger ground swell of applied ecologists and land managers interested in the immediate application and relevancy of their own research. All those now committed to the conservation and preservation of nature and natural resources must recognize and credit the redirected problem solving that Mike Soulé cogently brought into focus in the United States.

Nearly fifty years of ecological thinking and activism have just now begun to convince us about the relevancy of this "crisis discipline." Conservation biology as defined today encompasses the works of a large segment of biological practitioners—government scientists, research biologists, private environmental consultants, and ecology teachers. It is a discipline, rather, an umbrella of science, art, and politics, that addresses our sustainable needs as well as the integrity of all life on earth. But because our knowledge of life's networks and linkages is still uneasily trifling, we must admit to the premature application of conservation theory, and we must accept the consequences of its application.

To this end, we have assembled a collection of papers that deal with many issues in the theory and application of conservation biology as it emerges in the final decade of the twentieth century. Our anthology is divided into four main parts that represent a natural harmony. John Harper's foreword commences with his characteristically elegant commentary on the need and scope for conservation efforts. Part I provides several chapters outlining the natural order of the biological world. Peter Ashton (chapter 1) begins with a brief overview of plant species richness in the context of understanding the dynamics controlling species assemblages that are of concern in conservation biology. Peggy Fiedler and Jeremy Ahouse (chapter 2) seek to clarify, at the level of individual taxa, what it means to be rare. They illustrate that rarity is a muddled concept, typically not explicitly defined, and one that has looming ramifications for the preservation of species and ecosystems alike. Miguel Altieri and M. Kat Anderson (chapter 3) provide a glimpse of the conservation of landraces in traditional agricultural systems and the need for their *in situ* continuance. Their chapter serves to emphasize that we should not place man apart from nature or from the life-sustaining products of human civilization. Together these chapters describe three levels (communities, species, genes) of biodiversity and our concerns for their understanding and description.

Part II illustrates the processes and effects of change as discussed in these introductory comments. Ted Case and his colleagues (chapter 5) document the extinctions of reptiles within the last 10,000 years and comment in particular on the role of humans in the increase of reptilian extinction rates on islands. Peter Moyle and Robert Leidy (chapter 6) present a review of the worldwide loss of biodiversity in aquatic systems as represented by fishes, and these trends are mirrored in John Hafernik's (chapter 7) discussion of the loss of invertebrate biodiversity in the United States. This part also includes a discussion of habitat fragmentation by Larry Harris and Gilberto Silva-Lopez (chapter 8), who clarify the various kinds of fragmentation that occur in forested ecosystems. Five types of fragmentation are defined and illustrated by the region surrounding Florida's Ocala National Forest. This collection of chapters provides a forceful message about the role of habitat fragmentation among other potential threats to ecosystems and rare species.

Because conservation biology has been termed informally "the biology of small populations," Part III provides background and insight into this field. Eric Menges (chapter 10) defines a demographic-based version of the minimum viable population concept using stochastic modeling of plant populations. Such modeling of population behavior has enormous promise for managing endangered populations using empirically derived demographic data. Robert Lacy (chapter 11) examines the minimum size of populations from a geneticist's point of view, providing portentous questions concerning current theory of inbreeding depression in small populations. Mary Pearl (chapter 12) discusses the behavioral aspects of Asian primates and illustrates the complex role of behavior in addition to genetics in maintaining viable populations of complex organisms. Thus genetics, demography, and social behavior clearly represent an integrated triad of conservation science research and application. The ill-perceived gulf between the theoretical and applied fields clearly has begun to disappear.

Finally, Part IV provides examples of conservation biology in practice—that is, management of preserves and natural areas and *ex situ* conservation. Ron Carroll (chapter 14) offers a taxonomy of natural areas that are particularly sensitive to anthropogenic influences and that require active management for their maintenance. Christine Schonewald-Cox and Marybeth Buechner (chapter 15) provide examples of the seemingly endless fragmentation by public roads within the U.S. National Park system and illustrate the conflict between public access to natural areas and their preservation. Donald Falk (chapter 16) outlines the role of propagation, breeding, and research in botanical gardens (and by analogy, zoological parks) in *ex situ* conservation of plant species. He suggests that, at least for plants, an integrated strategy of conservation that includes off-site research and propagation, habitat restoration, and *in situ* management may be the most effective strategy for the preservation of rare plants and animals.

We have also included five essays throughout the book that represent viewpoints, sometimes rather personal and informal, of practicing conservation biologists. Steward Pickett and his colleagues (chapter 4) write of a palette of concerns of the community ecologist, while Reed Noss (chapter 9) reminds us of the importance of scale in the definition of conservation problems and in the seeking of their answers. Edward Guerrant, Jr., (chapter 13) discusses the importance of empirically derived demographic and genetic data in the management of plant populations, arguing that both can be critically important to ensure population persistence. Laura Jackson (chapter 17) speaks to those who recognize that conservation biology, specifically restoration ecology, is as much art as it is science, and such distinctions are, in fact, detrimental to the renewal and revival of degraded landscapes. Finally, a legendary evolutionary biologist and conservationist, G. Ledyard Stebbins (chapter 18), provides a personal ethic for why he has been a champion in the conservation of California's biota for nearly sixty years.

This volume speaks to the young scientists of conservation biology as well as to those interested in the newly revived directions of resource management, for they are really the same society. Designing such an anthology is tremendously difficult, however, because of the disparate yet synthetic nature of this beast. Initial ideas and inspiration for the preparation of this text sprang from a graduate seminar series in conservation biology, organized by one of us (PLF) and held in 1988 at San Francisco State University. Eighteen students and fourteen speakers, in essence, outlined the structure of our book, and to the insight and enthusiasm of all those involved we give due acknowledgment. We also wish to thank our colleagues C.R. Carroll, T.C. Foin, J. Hafernik, R.F. Holland, R.D. Laven, R.A. Leidy, J. Major, V.T. Parker, R. Patterson, and G.L. Stebbins for their considerable patience and encouragement that began long before the inception of this book, and who have taught us much in different forums. Gregory Payne of Chapman and Hall is an editor of superb skill and insight, and we have only the highest praise for his unflagging assistance in the preparation of this volume. Finally, but not only because it is customary to do so in this fashion, we thank our families for all that goes unspoken.

LITERATURE CITED

Abbey, E. 1968. *Desert solitaire.* New York: McGraw-Hill.

———. 1975. *The monkey wrench gang.* New York: Avon Books.

Allee, W.C., A.E. Emerson, O. Park, T. Park, and K.P. Schmidt. 1949. *Principles of animal ecology.* Philadelphia: W.B. Saunders.

Andrewartha, H.G., and L.C. Birch. 1954. *The distribution and abundance of animals.* Chicago: University of Chicago Press.

Aplet, G.H., R.D. Laven, and P.L. Fiedler. 1992. The relevance of conservation biology to natural resource management. *J. Cons. Biol.* In press.

Cain, S. 1966. Biotope and habitat. In *Future environments of North America,* ed. F.F. Darling and J.P. Milton, 38–54. Garden City, N.Y.: The Natural History Press.

Carson, R. 1962. *Silent spring.* Boston: Houghton Mifflin.

Chase, A. 1986. *Playing God in Yellowstone: The destruction of America's first national park.* New York: Atlantic Monthly Press.

Clements, F.E. 1916. *Plant succession: An analysis of the development of vegetation.* Carnegie Institution of Washington Publ. 242. Washington, D.C.: Carnegie Institution.

Commoner, B. 1968. *The closing circle.* London: Lowe and Brydone.

Cox, G.W., ed. 1969. *Readings in conservation ecology.* New York: Appleton-Century-Crofts.

Dasmann, R.F. 1964. *African game ranching.* London: Pergamon Press.

———. 1965. *The destruction of California.* New York: Macmillan.

————. 1968. *Environmental conservation.* New York: John Wiley & Sons.

Detwyler, T.R. 1971. *Man's impact on environment.* New York: McGraw-Hill.

Diamond, J. 1975. The island dilemma: Lessons of modern biogeographic studies for the design of natural reserves. *Biol. Cons.* 7:129–46.

Duffey, E., and A.S. Watt. 1971. *Scientific management of animal and plant communities for conservation.* London: Blackwell Scientific Publications.

Ehrenfeld, D. 1970. *Biological conservation.* New York: Holt, Rinehart and Winston.

Ehrlich, P.R., and A.H. Ehrlich. 1968. *The population bomb.* New York: Ballantine.

————. 1981. *Extinction: The causes and consequences of the disappearance of species.* New York: Random House.

Frankel, O.H., and E. Bennett, eds. 1970. *Genetic resources in plants—their exploration and conservation.* IBP Handbook no. 11. Oxford: Blackwell Scientific Publications.

Frankel, O.H., and M.E. Soulé. 1981. *Conservation and evolution.* Cambridge: Cambridge University Press.

Gleason, H.A. 1926. The individualistic concept of the plant association. *Bull. Torrey Club* 53:7–26.

Hardin, G. 1968. The tragedy of the commons. *Science* 162:1243–48.

Holdgate, M.W., M. Kassas, and G.F. White, eds. 1982. *The world environment 1972–1982.* Dublin: UNEP/Tycolly International Publishing.

Lack, D. 1954. *The natural regulation of animal numbers.* Oxford: Clarendon.

Leopold, A. 1933. *Game management.* New York: Scribners.

————. 1949. *A Sand County almanac and sketches here and there.* New York: Oxford University Press.

Lindeman, R.L. 1941. Seasonal food cycle dynamics in a senescent lake. *Amer. Midl. Nat.* 26:636–73.

————. 1942. The trophic-dynamic aspect of ecology. *Ecology* 23:399–408.

MacArthur, R.H., and E.O. Wilson. 1967. *The theory of island biogeography.* Princeton, N.J.: Princeton University Press.

Miller, R.S., and D.B. Botkin. 1974. Endangered species: Models and predictions. *Amer. Sci.* 62:172–81.

Moore, N.W. 1962. The heaths of Dorset and their conservation. *J. Ecol.* 50:369–91.

Moore, N.W., and Hooper, M.D. 1975. On the number of bird species in British woods. *Biol. Conserv.* 8:239–50.

Myers, N. 1979. *The sinking ark.* London: Pergamon.

Passmore, J. 1974. *Man's responsibility for nature.* New York: Charles Scribner's Sons.

Reisner, M. 1986. *Cadillac desert.* New York: Viking Press.

Schonewald-Cox, C.M., S.M. Chambers, B. MacBryde, and L. Thomas, eds. 1983. *Genetics and conservation: A reference for managing wild animal and plant populations.* Menlo Park, Calif.: Benjamin/Cummings.

Sears, P.B. 1980. *Deserts on the march.* Norman: University of Oklahoma Press.

Soulé, M.E. 1986. *Conservation biology: The science of scarcity and diversity.* Sunderland, Mass.: Sinauer.

————. 1987. *Viable populations for conservation.* Cambridge University Press, New York.

Soulé, M.E., and B. Wilcox, eds. 1980. *Conservation biology: An evolutionary-ecological perspective.* Sunderland, Mass.: Sinauer.

Tansley, A.G. 1920. The classification of vegetation and the concepts of development. *J. Ecol.* 8:118–44.

————. 1935. The use and abuse of vegetational concepts and terms. *Ecology* 16:284–307.

UNESCO. 1973. Expert panel on Project 8: Conservation of natural areas and the genetic material they contain. In *MAB report series no. 12*, Paris: UNESCO.

Usher, M.B. 1975. *Biological management and conservation: Ecological theory, application and planning.* London: Chapman and Hall.

PART
I
• • •

THE NATURAL ORDER

• • •

CHAPTER
1

Species Richness in Plant Communities

PETER S. ASHTON

ABSTRACT
Knowledge of species richness is important for establishing priorities in conservation planning and management. Patterns of species richness in four exceptionally species-rich plant communities and correlations between species richness patterns, physical habitat, and historical biogeographic factors are examined here. I conclude that several of the apparently conflicting hypotheses for the maintenance of species richness in plant communities are reconcilable under different ecological circumstances.

INTRODUCTION

Knowledge of the general spatial patterns of species richness is important in conservation science because it allows one to identify geographic regions of exceptional richness. Knowledge of how species richness varies in relation to physical site characteristics also can provide a means to predict where, in unknown terrain, such areas of exceptional conservation value might lie, such as in relation to local topography or geology. This is critical, because preservation of these sites may be the most economical way to conserve the largest number of species. In addition, knowledge of how species richness is maintained may be particularly important, because the way in which it is maintained may modify the rate at which extinctions increase when a community becomes restricted in its range, fragmented, or isolated by human activity. As we shall see, if species richness is maintained solely through the balance between stochastic events of immigration and extinction, then any reduction in the area may increase extinction rates and, consequently, the total species complement. If, however, species accumulate through competition, leading to specialization determined by the resources available within a particular site, then change in the areal extent of the habitat within certain bounds may have relatively little influence on extinction rates.

Vegetation can be rich in species either because individual plant communities are intrinsically rich in total number of species (α, alpha diversity), or because the spatial variability of the physical habitat influences species composition so that a mosaic of communities exists whose total species complement is high (β, beta diversity). Although it is frequently difficult to disentangle the second from the first, this chapter will concentrate on the likely causes of the first—that is, intrinsic plant species richness. The term richness will be used to connote total number of species (S) in preference to species diversity that could equally mean diversity of plant structure—ecophysiology or chemistry, for instance.

Variation in species richness among plant communities requires explanation in the context of two processes, origin and maintenance. Origin is inherently unpredictable and, in some measure, uninterpretable. This is because the origin of species richness is dependent on, in part, (1) opportunities for sporadic immigration of new species through habitat corridors in the more or less unknowable past; (2) the chance arrival of new floras through the collision of drifting continents; and (3) the geographical extent of the habitat, itself fortuitous because it will determine opportunities for new species to evolve locally and rare species to become extinct. All else being equal, the greater the areal extent of a community, the greater the chance that local specialization among populations within it can lead to speciation. The smaller the areal extent, the greater the likelihood that chance fluctuations within populations of the rarest species will cause their extinction (MacArthur and Wilson 1967). Therefore, large expanses of a single flora or plant community have an inherently greater likelihood of producing greater species richness than small ones.

Nevertheless, the applicability of MacArthur and Wilson's theory of island biogeography seldom has been demonstrated convincingly for plants. Data remain inadequate to test rigorously the respective influences of area, topographic and edaphic variation, and opportunities for immigration on species richness within plant communities. The data that do exist suggest that the influence of area alone on regional species diversity is generally obscured by other factors. This may be in part because sessile organisms, such as plants, may differentiate locally within an island more rapidly than vagile organisms, such as, animals. On the other hand, island biogeography theory may indeed be valid for tropical rain forest floras where long-lived woody plants predominate. The poor dispersal abilities that overall characterizes this flora may mean that an equilibrium between immigration and extinction rates rarely may be achieved before climatic (or other) perturbations intervene.

For example, in relation to the large continental island of Madagascar, the island of Reunion in the Indian Ocean is larger (251,000 ha) and closer (780 km) than the nearby oceanic island of Mauritius (186,500 ha, 830 km). Both Reunion and Mauritius are composed of similar basic volcanic rocks whose fertile soils may be inimical to edaphic differentiation of the flora. However, Reunion has greater altitudinal diversity than Mauritius, with mountains rising to 3000 m. Reunion also is under the influence of trade winds that produce greater variability in the lowland rainfall seasonality. Nevertheless, Reunion Island has an angiosperm flora of approximately 500 indigenous species, 30 percent of which are endemic. Mauritius has approximately 870 indigenous species, about 35 percent of which are endemic. The only available explanation for these differences appears to be age of the flora. Mauritius is seven million years old, whereas Reunion is four million, so, until the decimating influence of human settlement intervened, the flora of Reunion was still diversifying relative to that of Mauritius (see chapter 2; also Fiedler [1986] for discussion of the "age and area hypothesis").

Maintenance of species richness also may involve chance events on a local scale over relatively short periods of time (Hubbell and Foster 1983, 1985, 1986). Importantly, species richness also can involve predictable processes of natural selection leading to communities with fine levels of organization through evolution of mutualistic interdependencies, particularly between plants and insects or microorganisms (Ashton 1969; Janzen 1970). Communities therefore can either be randomly varying assemblages of ecologically "equivalent" species that have convened at one site due to historical exigencies, or they can made up of a parsimonious grouping of specialists, each of which must be "fitter" in at least one respect than all the others to avoid extirpation through competitive exclusion. Of course, species composition of communities is, in fact, determined by several kinds of processes, and the extent to which each is important will vary (see Diamond and Case 1986 for numerous discussions of this dichotomy and for consensus on pluralistic views).

The implications of these alternative community processes described previously for the management of nature reserves are important. If species composition as measured by species richness (S) is due *solely* to events of historical biogeography (the Mauritius/Reunion example), then any reduction in the areal extent of a reserve will require active management due to the consequent increased probability of chance extinctions (MacArthur and Wilson 1967). If, however, community composition of a nature reserve is largely a consequence of natural selection, then a species complement may possibly be retained intact, with minimal intercession, in areas far smaller than occupied in nature *provided* the area remains large enough for the survival of every interdependent component, animal and plant as well as microorganism.

In this chapter, the likely causes of species richness in plant communities will be discussed in the context of the variation among those kinds of vegetation that are the richest in plant species. I will focus on tree communities in the tropical evergreen lowland forests of Borneo, and comment more briefly on three others (tropical montane epiphyte communities, mediterranean shrub communities, and temperate short grasslands). These communities represent dramatically different plant life forms and physical structures. Our challenge will be to seek the shared factors that may underlie their comparable measures of species richness.

The maintenance of species richness has been examined in detail only in tropical lowland forests and in short grasslands where in the latter, the comparatively smaller scale and shorter life cycle of the grassland components have allowed experimentation (Grubb 1977; Mitchley and

Grubb 1986). Recent hypotheses regarding the origin and maintenance of the patterns of variation in the species richness will serve as the starting point for discussion.

HYPOTHESES OF SPECIES RICHNESS

As we will see shortly, Borneo's rain forests exhibit a pattern of species richness very similar to that predicted in part by a model for variation in species richness put forward by Tilman (1982, 1986; Figure 1.1a).

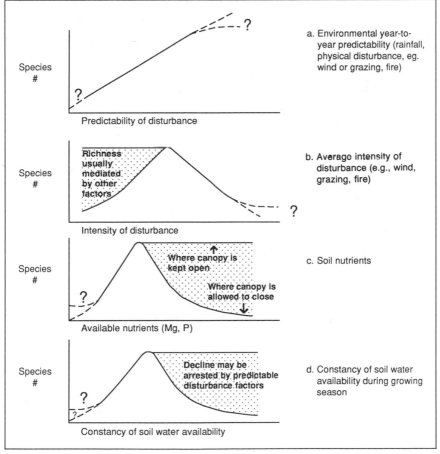

Figure 1.1
Environmental correlates of species richness.

Tilman reasoned that opportunities for species to specialize in their exploitation of a heterogenous resource such as soil are limited where the availability of major constituents of the resource is limited. Species richness is therefore low. As resource availability increases, opportunities for specialization increase, particularly when several different elements of the resource increase in approximate parallel. Tilman argued that slight differences in spatial distribution of soil resources will impose an increasingly marked intrinsic heterogeneity on species composition and hence enhance overall species richness.

Tilman also reasoned that resource availability frequently can exceed the threshold above which resources are no longer limiting to plant growth. He suggested that after this happens, competition is purely for space; under such circumstances one or a few dense-crowned species will outcompete most others for light. These tree species will come to dominate the canopy, creating a uniform shade that depresses species richness in the understory by limiting light and reducing its spatial heterogeneity.

Tilman's hypothesis has been criticized, however, because it assumes that vegetation is in a steady state at ecosystem as well as biogeographic levels, which, critics claim, is unlikely to occur in nature. Earlier work by Grubb (1977) considered that periodic disturbance occurs frequently enough in terrestrial vegetation to prevent assemblages of fine-tuned resource specialists, as assumed by Tilman, from coexisting in competitive equilibrium. Thus, species survival is determined early during seedling establishment. Species richness is, in turn, determined in part by the diversity of microhabitats into which species can establish, and in part by the diversity of regenerative characteristics manifested by the flora available at a site.

Grubb's view of the "regenerative niche" encompasses not only heterogeneity in the soil surface, but importantly includes the frequency, scale, and intensity of disturbance as a major cause of habitat heterogeneity. In this, his views are compatible with those of Connell (1971, 1978), who likened rain forest dynamics to the effects of disturbance caused by wave action on coral reefs. Connell proposed an "intermediate disturbance hypothesis" for optimization of species richness. He argued that very frequent, large-scale, or intense disturbances will depress species richness by eliminating the mature old-growth denizens (Figure 1.1b). Conversely, infrequent, slight, or small-scale disturbances will reduce the total number of species in gaps by encouraging the gradual increase of areal dominance by a few species in old-growth forest. Note that Connell's predictions of competitive exclusion at low disturbance levels are therefore in agreement with (and in some sense equivalent to) Tilman's predictions for sites where soil nutrients are not limiting.

TROPICAL LOWLAND MOIST FORESTS

Borneo

The lowland forests of the Sunda Shelf—that is, the Southeast Asian continental shelf region that currently includes the Malaysian peninsula and the giant islands of Sumatra and Borneo—contain as many tree species as (or more than) any comparable area of the world. One 50 ha block of forest on uniform undulating land, recently censused in peninsular Malaysia, contains over 800 vascular plant species, or about one-third of the lowland tree flora of the peninsula (Ashton n.d.). This is an example of intrinsic species richness par excellence, apparently rivaled only by the forests of the northern Andean foothills, some of which seem to be of strikingly similar richness (Gentry 1986; Figure 1.2).

The richest forests in the Far East actually are in northwest Borneo. There, the region north of the Kapuas River in the west, and northeast throughout the Malaysian state of Sarawak and the Sultanate of Brunei Darussalam to the extreme southwest lowlands of Sabah, is markedly richer in total number of species and in local endemics than is the rest of the island (Ashton 1984). This region almost certainly is richer in vascular plant species than the whole of the Old World. It is tempting to conclude that this region was a "refugium" where present-day rain forest species persisted during the Pleistocene. During those times when a more seasonal continental climate over much of the area presumably prevailed, Borneo, Java, Sumatra, and peninsular Malaysia were joined into a continent not much smaller than northern South America. Although currently there is little conclusive evidence to support a historical explanation for species richness, we do know that northwest Borneo is more rugged topographically than the rest of the island—that is, it is geologically relatively young and consists mostly of sandstones yielding infertile shallow soils. Geologic age and poor edaphic conditions are associated with very great local differentiation of the flora and, in some forest communities, with truly exceptional tree species richness (Ashton 1989a). This has been documented in a series of several hundred sample plots spanning the range of lowland habitats, within which both the tree floras and soils have been recorded and analyzed. The results reveal a distinctive, predictable pattern of species richness that bespeaks a major role for the forces of natural selection (Ashton 1989a).

Borneo rain forests on very fertile soils are relatively poor in species. Sometimes one upper-canopy species is much more abundant than all others and may occasionally even be the exclusive dominant on these soils. Curiously, then, the forests with the greatest number of species are

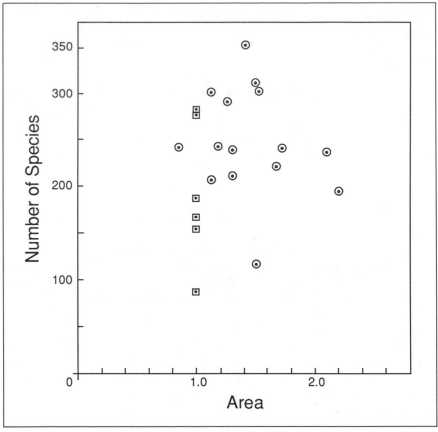

Figure 1.2
Tree species richness in tropical lowland forests (squares: Upper Amazonia, after Gentry 1982a; circles: Sarawak, Malaysia, Northwest Borneo; from Ashton n.d.).

not found on the most fertile soils. Species richness is variable on these soils. For example, on mesic lower slopes (well-drained valley soils in particular) there is a marked depression in species richness at all structural levels, from emergent canopy to understory. The highest values of S are found near the low end of the soil fertility range, apparently at or near the point in the lowland humid tropics where the soil surface is acid enough for partially decomposed raw humus to accumulate.

In fact, further examination of the evidence from Borneo forests lends support to Tilman's resource hypothesis (Figure 1.1c). The nutrients most closely correlated with species richness on less-fertile soils in Borneo are magnesium and phosphorous. Magnesium mediates phosphorous uptake by roots (Baillie et al. 1987; Garraway and Evans 1986), and symbiotic mycorrhizal fungi play an important part in phosphorous uptake in less-fertile soils in the tropics as well as temperate regions (Smits 1983). Ectotrophic basidiomycete mycorrhizae can be highly host-specific, and some evidence for this specificity exists among closely related tree hosts in Borneo (Smits 1983). Several mycorrhizal basidiomycete genera also are exceptionally high in number of species in the humid tropics. For example, although still poorly known, there are some 130 described species of *Boletus* alone in peninsular Malaysia (Corner 1972). Though individual tree families differ in the conformity of their patterns of species richness with Tilman's model, those of ectotrophic mycorrhizal families such as Dipterocarpaceae and Myrtaceae fit closely (Ashton, pers. ob.). This suggests that ectomycorrhizal fungi may specialize with their hosts in their uptake of minerals; however, this has yet to be demonstrated experimentally.

Although species dominance is not correlated with species richness on the comparatively *less*-fertile soils in the mixed forests of lowland Borneo, the cumulative dominance of the five leading tree species is correlated closely with species richness on the more-fertile soils on mesic sites. The fact that this rain forest example remains the only demonstration of Tilman's model in natural terrestrial vegetation speaks strongly for the climatic equability of northwest Borneo.

However, on drought-prone ridgetops and on steep slopes where disturbance is more frequent and lateral light penetrates, understory species richness is relatively low on fertile soils. Partially, then, this topographic pattern likely is attributable to disturbance, as proposed in Connell's intermediate disturbance hypothesis, or at least to conditions that maintain the forest canopy in a partially open condition. Interestingly, just across the hills on the east coast of Borneo, intense droughts at intervals of perhaps one per century may devastate the emergent canopy and curtail regeneration. A severe drought happened most recently in 1982–1983. Its effects were recorded by Leighton and Wirawan (1988). Thus it is possible that droughts as frequent extinction events may explain the comparatively low species richness and endemism on ecological islands of freely draining, sandy, white podsolic soils in eastern Borneo (Ashton 1984). This contrasts strongly with the comparatively high regional endemism and very high species richness in islands of similar habitat in the northwest. There, areas of distinctive podsolic soils as small as 30 km^2 exhibit no measurable loss of species richness.

But a third pattern, a correlation between increasing species richness and decreasing photosynthetic area, suggests that competitive exclusion by canopy dominance is most effective where soil water stress is least severe (Figure 1.1d). Narrow leaves evaporate less water and potentially may be better adapted to water stress. In addition, narrow, diffusely arranged leaves cast less shadow and thus permit more, and more variable, light to penetrate beneath the forest canopy than do broad, densely arranged leaves. Thus, in this instance, Grubb's emphasis on the importance of a diversity of microsite and regenerative options in the establishment phase is perhaps supported.

In summary, patterns of species richness in the forests of northwest Borneo can be described by three hypotheses of species richness as proposed by Tilman (resources), by Connell (disturbance), and by Grubb (regeneration niche). The large plant families with speciose genera, such as the Dipterocarpaceae and Myrtaceae, may conform to the soil fertility patterns and mycorrhizal specializations predicted by Tilman. The forests of the ridgetops and steeper slopes are subject to the intermediate forms of disturbance predicted by Connell and thus demonstrate high species richness. Forests dominated by narrow, diffusely leaved adults may provide confirmation for Grubb's hypothesis concerning the role of the regeneration niche in the evolution of species-rich communities.

Comparison with South and Central American Tropical Forests

Studies of species richness in the forests of Central and South America interestingly have failed to show any correlation between species richness and available soil nutrients (Gentry 1982a, 1982b). Instead, in a regional study of lowland rain forest samples from Central and South America, Gentry (1982a) found that mean annual rainfall explained more of the variation in total floristic richness (i.e., all life forms) than any other variable measured. Such vegetation differences might merely reflect analytical procedural differences between the two studies, but a real difference in environmental influences also is possible. For example, wind squalls in the New World Tropics are more frequent than in Borneo. Thus, it is likely that the reduction in species number and increase of individual species dominance observed on mesic sites in Borneo do not occur as a result of the severity and frequency of climatic disturbance. However, the correlation between infertile soil and certain plant families and their mycorrhizal specialists as observed in Borneo might be overridden in the Neotropics by the effects of differentials in disturbance regimes.

In addition, the consideration of the gap phase flora is crucial to the consideration of species richness of lowland tropical forests. In Borneo, the pioneer tree flora of light gaps is a small component of the total tree flora. It is possible that relative gap size would have to remain consistently large over evolutionary time before the number of species in gap phase flora would grow large relative to the number of species in mature phase flora for the gap phase flora to influence patterns of richness in the manner that Connell predicted.

One way in which this could occur in South America would be through the influence of the abundant epiphyte flora, particularly the members of the Bromeliaceae. Bromeliads are common at all altitudes of the New World tropics but are unknown in Asia. Strong (1977) has suggested that their mass might cause branch and tree breakage, and consequently increase the frequency of gap formation. This hypothesis needs to be field-tested because, if correct, it could provide the mechanism required to explain forest tree species richness in the New World tropics according to the intermediate disturbance hypothesis (Connell 1978).

As mentioned previously, Gentry (1982a) found that mean annual rainfall, not soil fertility, explained the majority of floristic richness variation in Central and South American lowland moist forests. The correlation was lost, however, in the areas of highest rainfall and low seasonality, as in Borneo. In Asia, where forest samples from such a wide range of rainfall seasonality have yet to be collected, it is probable that close correlations exist between tree and total species richness and the mean length of the dry season.

It is unclear how either of the two rainfall measures (i.e., amount and distribution of rainfall) could influence species richness directly, particularly because dry seasons of the higher-latitude tropics occur during cooler times of the year when their influence is less. A more likely cause of species richness is a related rainfall phenomenon—that is, *between-year variability* (Figure 1.3). Between-year variability in rainfall increases from the aseasonal wet equatorial regions to the desert margins of the tropics (Von Riehl 1979). Seed fall and early establishment occur mainly at the beginning of the wet season so that variability in the onset of the rains could act as an increasingly severe constraint on increasing specialization within the regeneration niche (see also Ashton 1989b).

This interpretation harkens to Stevens' (1989) observations that the global pattern of increasing species richness towards lower latitudes parallels Rapoport's rule—that is, species have increasingly narrower geographical ranges toward low latitudes.[1] Stevens inferred from Rapoport's rule that the average environmental tolerances of species may be narrower at low latitudes than at high latitudes, and he assumed that no parallel

---

[1]In fact, the increase is not continuous in that there is a surge of species number in Mediterranean climates and latitudes and a dramatic fall in species numbers in the Horse latitudes.

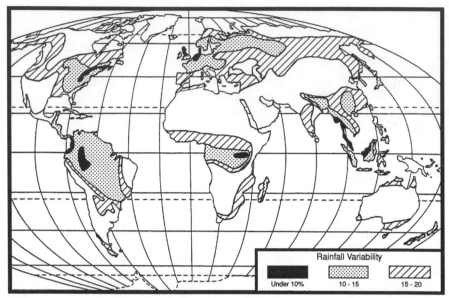

Figure 1.3
Annual rainfall variability of the earth (percentage departures from average; modified after V. Riehl 1979).

trend in seed dispersal ability with latitude exists. He also speculated that the increase in species richness could be caused by the greater opportunity for individuals to periodically and temporarily enrich adjacent communities outside their optimal habitat because species' habitat ranges are narrow and typically fragmented. However, tree species of the mature phase rain forest where species richness is concentrated usually have relatively poor dispersal abilities and low seed dormancy. Thus, latitudinal and climatic patterns of species richness, as opposed to local and edaphic, may perhaps be explained more plausibly in terms of the large-scale global trend toward supra-annual climatic variability with increasing latitude.

Climatic Variability and Species Richness

It is appropriate to ask how long climatic predictability needs to persist before local site-related patterns, such as those related to soil, come to control the variation in patterns of species richness. In northwest Borneo as in other regions of extremely rich vegetation, the observed variation in species richness is almost entirely attributable to variation in the number of sympatric species of a few speciose genera. Many of the species confined to the richest community types are endemic to northwest Borneo. This implies either that climate predictability has persisted long enough for speciation to occur, or that it has persisted longer than elsewhere so that northwest Borneo has been a refuge for a vast number of species.

Yet tropical plant communities of great species richness are not as a rule characterized by high endemicity at generic or higher systematic levels. We may conclude, then, that their climates must be as old as it takes for the life form in question to speciate, but not as long as is required for genera to differentiate. Thus, the high tree species endemicity of the ecological islands of sandstone-derived soils in northwest Borneo, compared with those of the rest of the island (see, e.g., Ashton 1984), must date to the Pliocene. This is further supported by the relationship between local endemicity and the known history of geomorphological change in the region (Ashton 1973).[2]

In certain circumstances, however, there are dramatic differences in species richness between plant communities in apparently identical habitats sharing a common flora and climatic history. A striking Asian example is the lowland rain forests of Sri Lanka where the range of soils and climate are very similar to those of northwest Borneo. The flora at the familial and generic level is remarkably similar, too, although there is much higher species endemicity of Sri Lanka (ca. 70% in the wet forests alone) than in northwest Borneo (ca. 55% endemism for the lowland forests of the whole island, ca. 65% for the northwest alone). Floristic differentiation on less-fertile soils follows the same pattern as on Borneo (Gunatilleke and Ashton 1987). But whereas a 4 ha forest sample in northwest Borneo may contain 335 species exceeding 20 cm in diameter, in Sri Lanka, a sample from a similar habitat yields only 62 tree species. The most abundant species on the two islands differ only slightly in population density, so the observed difference in species richness is due entirely to differences in the number of rare species. The most likely explanation for this difference is the very small area of wet lowlands in Sri Lanka (ca. 1,500,000 ha) and the long isolation from forests in similar aseasonal wet climates of the Far East (although climatic change during the Pleistocene may have occurred in Sri Lanka). The physical isolation and small areal extent of Sri Lanka's wet lowland forest likely have nearly eliminated immigration, restricted opportunities for speciation, and,

---

[2]Interestingly, however, there is evidence that the lowland tree communities of New Guinea are as rich as those of Borneo (Paijmans 1970). Because this lowland flora is overwhelmingly of Asiatic origin, we can be reasonably certain that it arrived in New Guinea (or speciated within it) since the 15 million years following the collision of the Australasian and Laurasian plates.

most of all, increased the likelihood of extinctions among rare taxa. The Sri Lankan forests therefore are a clear example of the influence of chance historical factors on species richness.

It is moot whether a ceiling is ever reached for species richness in plant communities occupying extensive continental areas of uniform climate, such as in northwest Borneo. Niche specialization in species-rich plant communities obviously cannot be the result of interspecific competition for soil resources, except perhaps among the few most abundant species. As species richness increases, the predictability of direct species-to-species competition among large sessile organisms for common resources must decrease, such that a limit of such differentiation must exist (although it may differ according to the soil type). Janzen (1970) pointed out, however, that niche differentiation in species-rich plant communities is more likely to be mediated by relatively mobile organisms (e.g., insect seed predators, herbivores, and pathogenic organisms) than directly through competition between plant species. Although Janzen's hypothesis is not yet confirmed, the high specificity of potentially defensive secondary chemicals that characterizes tropical plant tissues suggests that this may be so. However, Janzen indicated that these interactions also are eventually density-dependent, because herbivores and pathogens have increasingly greater difficulty in tracking their hosts as distance between host conspecifics increases. Low plant population density also will affect pollination success and hence may depress fecundity among outbreeders. In support of this, Hubbell and Foster (1986) have found that dioecious species are relatively poorly represented among rain forest tree populations consisting of sparsely distributed individuals. Nevertheless, plants have evolved various means of reproduction, including self-pollination and apomixis, that can overcome this problem. It may be significant that apomixis appears to be widespread among rain forest trees in regions of high species richness (Kaur et al. 1978), and that it appears to prevail in those large genera whose species numbers dominate the variation in overall forest species richness (Ashton 1989a). If a substantial number of species can exist in population densities below the threshold at which species-specific competition operates *and* without significant reduction of fecundity, then there may be no limit to accumulation of species richness in widespread continental plant communities. At the same time, an increasing proportion of the species complement of such communities would be dependent on the exigencies of immigration and extinction (Ashton 1984). The fact that tree species richness in tropical forests is remarkably predictable within a biogeographic region, irrespective of the areal extent, suggests that this is not accurate but serves to emphasize that our knowledge of the means by which rare species persist in these communities is meager.

SPECIES RICHNESS IN OTHER
PLANT COMMUNITIES

Epiphyte Species Richness

It would be difficult to think of a plant guild that contrasts more strongly, particularly in the nature and diversity of the regeneration niche, from those of rain forest trees than that of their epiphytes. Whereas tree species richness is greatest at low altitudes, epiphyte communities reach their zenith within the cloud layer, at 1,000 to 3,500 m, in the mountains of the humid tropics. Although comparative quantitative data do not exist, the richest epiphyte communities seem to be in the foothills of the northern Andes between 1,200 to 2,500 m (Gentry 1982b; Gentry and Dodson 1987), and on several large tropical mountain massifs, such as Kinabalu in northern Borneo.

Historical biogeography plays its part. The highly speciose epiphytic Bromeliaceae are found largely in mountain forests in the Neotropics. Also, there is a vastly greater area of emergent montane habitat represented in the Andes than elsewhere. But the concentration of species-rich epiphyte floras on big massifs in the equatorial tropics in the lower portions of the cloud forest zone also correlates with the most reliably constant cloudiness of tropical climates. Some observational evidence suggests that bark roughness and, interestingly, nutrient status of the host's soil may influence epiphyte richness and abundance. Compared with that of a rain forest tree, the environment for establishment of an epiphyte has (1) a water regime at least as variable and more continuously limiting; (2) a similarly variable light regime; (3) a nutrient regime less variable but universally limiting; and (4) a substrate that, at least in dense epiphyte communities, experiences frequent sudden destruction and continuous renewal.

The plants themselves generally do not form closed communities, and compared with their tree hosts, epiphytes have short life cycles and remarkably effective dispersal of generally small seeds that lack a carbohydrate store but often possess dormancy mechanisms. Also unlike rain forest trees, very local endemism is common, and hybridization seems to play a major part in local speciation processes. In general, speciation in the epiphyte flora is likely to be much more rapid than it is for tropical tree species.

In summary, species-rich epiphyte guilds share in common with species-rich Bornean rain forest tree communities a physical environment of limiting nutrients and water, yet an equable and predictable moist climate permitting fine niche differentiation. But although there is evidence of differentiation in relation to light (Sandford 1969), other differentiation appears to be in relation to competition for biotic resources, notably pollinators, as in the outstanding case of the Orchidaceae. It is difficult to avoid the conclusion that the hazards of and opportunities for immigration and extinction play the major role in determining establishment and hence species richness in this open, dynamic, yet climatically dependable environment.

Mediterranean Shrub Communities

Following rain forests, mediterranean shrub communities contain the greatest number of plant species, making instructive a brief comparison with both epiphyte and terrestrial rain forest communities. Also more or less open or with a diffuse canopy of small leaves, plants in mediterranean shrub communities are limited in growth by drought during hot summers and by temperature during moist but cool winters. Species richness reaches its peak on low nutrient soils as it does in rain forests (see, e.g., Hopper 1979; Beard 1983), and the community is, as a rule, shaded out by one or a few taller tree species on mesic sites. As, too, with rain forests, many mediterranean shrub species are mycorrhizal, and their tissues contain highly specific chemical defenses. But mediterranean shrublands are subject to frequent fires in nature. Many species are adapted to a specific fire regime, as plants regenerate from fire-resistant root crowns and dormant seed. Much of the species richness must therefore be attributed (as in epiphyte communities) to factors of chance and opportunity in an open environment rather than to site diversity at early establishment. Mediterranean shrub floras also share with epiphytes high levels of biotic differentiation in relation to pollinators, interspecific hybridization, and local endemism. Perhaps most telling is the fact that mediterranean shrub communities have originated independently in distantly isolated regions. They vary greatly in the maximum species richness within communities, but the richest, which are in the Cape Province of South Africa and in Southwestern Australia, in that order, persist in climates of less annual rainfall variability, particularly in the hot summer months, than the others (Figure 1.3).

Temperate Grasslands

Finally, the short grasslands of temperate mountains and limestone hills of the British Isles make a third useful comparison. Although their species richness is comparatively modest, much is known of the mechanisms that maintain species richness due to the work of Grubb and his colleagues (e.g., Grubb 1977; Mitchley and Grubb 1986). Temperate grassland communities form closed-canopy seres whose succession is deflected by a highly uniform, predictable grazing regime (nowadays by sheep and other domestic livestock) during the growing season. The role of nutrients in sustaining high species richness is *not* completely understood, but species richness appears not to decline at high soil nutrient levels provided the grassland is kept short by grazing. This situation is reminiscent of the difference in S between tropical moist forests of Borneo and the New World. Mitchley and Grubb (1986) have shown, however, that the rank order of species abundance within these temperate grasslands is predictable and returns to the same order after manipulation. This implies that the role of rare as well as abundant species is controlled by directional selection. Grubb (1977) has emphasized the importance of disturbance at various intensities and scales in providing a diversity of habitats for early seedling establishment in the maintenance of species richness. In the most obvious case, exclusion of grazing as a highly predictable disturbance in temperate grasslands leads to suppression of species richness and a rise in dominance of a few species in a pattern that is compatible with Connell's predictions of intermediate disturbance.

CONCLUSIONS

Can a general theory for observed patterns of species richness emerge from such diverse systems? I think so, but species richness certainly is not mediated by a single factor. Rather, several current hypotheses are reconcilable because the conditions required by each apply under different circumstances, even perhaps within the same vegetation type!

There seems little doubt that the geographical factors of isolation and small area do suppress species richness in plant communities. They can represent the main determinant of observed differences in species richness where habitats are comparable and not in themselves limiting richness or where differences in area and isolation are sufficiently extreme.

High predictability—that is, low year-to-year variation in the pattern and intensity of stress or disturbance, whether it be drought or storms, grazing or fire (Table 1.1)—is the necessary and primary condition for the accumulation of a large number of species within plant communities. This is in part because high environmental predictability provides the conditions for narrow niche differentiation to develop, and in part because it represents reduction to intermediate levels of frequency and intensity of those stochastic events that can cause extinction in small populations. Also, random perturbations may prevent the community from reaching a potential steady state where one or a few species would preempt others from obtaining the available resources.

In rain forests the limiting factor appears to be either soil nutrient availability, water stress, or frequency of canopy disturbance, according

Table 1.1
An overview of suggested causes of species richness.

			Condition		
Community	**Nutrients**	**Water**	**Canopy Disturbance**	**Historical Biogeographic Sectors**	**Climatic Predictability**
Mixed rain forest tree communities, N.W. Borneo	Weakly limiting	Weakly limiting (predictable)	Little	High age relative to rates of evolution	High. Effectively no rainfall seasonality
Mixed rain forest, Peruvian Amazon	Not limiting	Not limiting	Moderate (constant)	High effective age (as above)	High (as above)
Montane epiphyte communities, N. Andes	Limiting	Limiting but predictable	Moderate (constant)	High effective age (as above); large area	High (as above)
Mediterranean shrub communities, Cape Province, S. Africa	Weakly limiting	Predictably limiting	Moderate, predictable (fire)	High effective age	High
Short, grazed temperate grassland communities	Not limiting	Weakly limiting	Moderate, predictable (grazing)	High effective age	Moderate

to local circumstances. In temperate grasslands the perturbation is principally grazing, whereas in epiphyte communities and mediterranean shrublands it seems to be small- to medium-scale catastrophic events, such as branch break and fire, respectively, that keep the communities permanently in cycles of succession. In conclusion, environmental predictability at all time scales seems to be the universal requirement of high species richness in plant communities.

LITERATURE CITED

Ashton, P.S. 1969. Speciation among tropical forest trees: Some deductions in the light of recent evidence. *Bio. J. Linn. Soc.* 1:155–96.

———. 1973. The quaternary geomorphological history of western Malesia and lowland forest phytogeography. In *The quaternary era in Malesia*, ed. P. Ashton and M. Ashton, 35–62. University of Hull Deptartment of Geography.

————. 1984. Biosystematics of tropical forest plants: A problem of rare species. In *Plant biosystematics,* ed. W.F. Grant, 497–518. New York: Academic Press.

————. 1989a. Species richness in tropical forests. In *Tropical forests,* ed. L.B. Holm-Nielsen, I.C. Nielsen, and H. Balslev, 239–51. New York: Academic Press.

————. 1989b. Dipterocarp reproductive biology. In *Ecosystems of the world. 14B. Tropical rain forest ecosystems,* ed. H. Lieth and M.J.A. Werger, 219–40. New York: Elsevier.

Baillie, I.C., P.S. Ashton, M.N. Court, J.A.R. Anderson, E.A. Fitzpatrick, and J. Tinsley. 1987. Site characteristics and the distribution of the species in mixed Dipterocarp forests on tertiary sediments in central Sarawak, Malaysia. *J. Trop. Ecol.* 3:201–20.

Beard, J.S. 1983. Ecological control of the vegetation of Southwestern Australia: Moisture versus nutrients. In *Mediterranean-type ecosystems,* ed. F.J. Kruger, D.T. Mitchell, and J.V.M. Jarvis, 66–73. NY: Springer-Verlag.

Connell, J.H. 1971. On the role of natural enemies in preventing competitive exclusion in some marine animals and rain forest trees. In *Dynamics of numbers in populations,* Proceedings of the Advanced Study Institute, ed. P.J. den Boer and G.R. Gradwell, 298–312. Osterbeek, Wageningen, Netherlands, 1970.

————. 1978. Diversity in tropical rain forests and coral reefs. *Science* 199:1302–10.

Corner, E.J.H. 1972. Boletus *in Malaysia.* Singapore: Government Printing Office.

Diamond, J., and T.J. Case, eds. 1986. *Community ecology.* New York: Harper & Row.

Fiedler, P.L. 1986. Concepts of rarity in vascular plant species, with special reference to the genus *Calochortus* Pursh (Liliaceae). *Taxon* 35:502–18.

Garraway, M.O., and R.C. Evans. 1986. *Fungal nutrition and physiology.* New York: John Wiley & Sons.

Gentry, A.H. 1982a. Patterns of neotropical plant species diversity. *Evol. Biol.* 5:1–84.

————. 1982b. Neotropical floristic diversity: Phytogeographical connections between Central and South America, Pleistocene climatic fluctuations, or an accident of Andean orogeny? *Ann. Missouri Bot. Gard.* 69:557–93.

————. 1986. Endemism in tropical versus temperate plant communities. In *Conservation biology: The science of scarcity and diversity,* ed. M. Soulé, 153–81. Sunderland, Mass.: Sinauer.

Gentry, A.H., and C.H. Dodson. 1987. Diversity and biogeography of neotropical vascular epiphytes. *Ann. Missouri Bot. Gard.* 74:205–33.

Grubb, P.J. 1977. The maintenance of species richness in plant communities: The importance of the regeneration niche. *Biol. Rev.* 52:107–45.

Gunatilleke, C.V.S., and P.S. Ashton. 1987. New light on the plant geography of Ceylon II. The ecological biogeography of the lowland endemic flora. *J. Biogeogr.* 14:295–327.

Hopper, S.D. 1979. Biogeographical aspects of speciation in the Southwestern Australian flora. *Ann. Rev. Ecol. Syst.* 10:399–422.

Hubbell, S.P., and R.B. Foster. 1983. Diversity of canopy trees in a neotropical forest and implications for conservation. In *Tropical rain forest: Ecology and management*, ed. S.L. Sutton, T.C. Whitmore, and A.C. Chadwick, 25–42. Oxford: Blackwell.

———. 1985. Biology, chance, and history, and the structure of tropical rain forest tree communities. In *Community Ecology*, ed. J. Diamond and T.J. Case, 314–29. New York: Harper & Row.

———. 1986. Commonness and rarity in a tropical forest: Implications for tropical tree conservation. In *Conservation biology: The science of scarcity and diversity*, ed. M.E. Soulé, 205–31. Sunderland, Mass.: Sinauer.

Janzen, D.H. 1970. Herbivores and the number of tree species in tropical forests. *Amer. Nat.* 104:501–28.

Kaur, A., C.O. Ha, K. Jong, V.E. Sands, H.T. Chan, E. Soepadmo, and P.S. Ashton. 1978. Apomixis may be widespread among trees of the climax rain forest. *Nature* 271:440–41.

Leighton, M., and N. Wirawan. 1988. Catastrophic drought and fire in Borneo tropical rain forest associated with the 1982–83 El Niño southern oscillation event. In *Tropical forests and the world atmosphere*, ed. G.T. Prance, 75–102. American Association for the Advancement of Science Selected Symposium 101.

MacArthur, R., and E.O. Wilson. 1967. *The theory of island biogeography*. Princeton, N.J.: Princeton University Press.

Mitchley, J., and P.J. Grubb. 1986. Control of relative abundance of perennials in a chalk grassland in southern England. Constancy of rank order and results of pot and field experiments on the role of interference. *J. Ecol.* 74:1139–66.

Paijmans, K. 1970. An analysis of four tropical rain forest sites in New Guinea. *J. Ecol.* 58:77–101.

Riehl, H. von. 1979. *Climate and weather in the tropics*. New York: Academic Press.

Sandford, W.W. 1969. The distribution of epiphytic orchids in Nigeria in relation to each other and to geographic location and climate, type of vegetation and tree species. *Biol. J. Linn. Soc.* 1:247–85.

Shmida, A., and M.V. Wilson. 1985. Biological determinants of species diversity. *J. Biogeogr.* 12:1–20.

Smits, W.Th.M. 1983. Dipterocarps and mycorrhiza. An ecological adaptation and a factor in forest regeneration. *Fl. Mal. Bull.* 36:3926–37.

Stevens, G.C. 1989. The latitudinal gradient in geographical range: How so many species coexist in the tropics. *Amer. Nat.* 133:240–56.

Strong, D.R., Jr. 1977. Epiphyte loads, tree falls and perennial disruption: A mechanism for maintaining higher tree species richness in the tropics without animals. *J. Biogeogr.* 14:215–18.

Tilman, D. 1982. *Resource competition and community structure*. Princeton N.J.: Princeton University Press.

———. 1986. Evolution and differentiation in terrestrial plant communities. In ed. J. Diamond and T.J. Case, 359–80. *Community Ecology*. New York: Harper & Row.

Hierarchies of Cause:
Toward an Understanding of
Rarity in Vascular Plant Species

PEGGY L. FIEDLER

and JEREMY J. AHOUSE

ABSTRACT

*Four classes of naturally rare vascular plant species are described
and classified, based on parameters of spatial distribution and longevity. Properties intrinsic to these time/space parameters are explored and an importance hierarchy of causes of rarity is proposed
for each class. These hierarchies serve as the basis for a predictive
classification. Human causes of rarity such as habitat destruction
and taxonomic difficulties are not considered in detail here but are
discussed as confounding factors in the elucidation of rarity in vascular plants. Several examples are provided to illustrate this classification and provide testable hypotheses concerning the origins of
natural rarity in plant species.*

INTRODUCTION

Rarity in vascular plant species is a topic replete with anecdotes. A great deal of technical and nontechnical literature currently exists concerning the biology and management of intrinsically rare (i.e., naturally occurring) plant species, but a general theory regarding their evolution and persistence is still far from our grasp. One of the difficulties in understanding the reasons why plant species are rare is that practitioners who study rare life forms often confuse cause and consequence. This confusion follows from the neglect of a broader historical (i.e., evolutionary) perspective of rarity. A short-term ecological time scale commonly is the perspective taken by those who work with rare plant species, often because of the immediacy of management objectives. However, given that virtually all plant taxa are characterized by histories marked in evolutionary time, it seems appropriate to examine the causes of vascular plant rarity from such a perspective.

Also appropriate to the management of naturally rare species is that now we no longer assign all patterns of rarity to one phenomonological level. It is generally agreed that plants may be rare for an astonishingly large number of reasons; some but not all of which work in concert to maintain a rare taxon's distribution, abundance, or both. Equally important, it is necessary to distinguish causes of rarity that relate to the geologic and evolutionary history of the taxon in question (e.g., glaciation and vicariance) from those causes that are the result of human intervention, such as certain land-use policies, habitat conversion, or taxonomic classification. Different causes of rarity are the result of phenomena that are distinguished by, at minimum, different temporal and spatial scales.

In this chapter we describe factors influencing rarity in vascular plants in the context of four general combinations of temporal persistence and spatial distribution. In conjunction with describing these four classes, we have assigned hierarchies of probable cause of rarity that relate to their general biology. These "hierarchies of cause" represent an attempt to elucidate the most probable and most important factors explaining rarity with different time/space histories. Our motive in describing rarity in this fashion is to clarify what it means to be intrinsically rare for vascular plant species, as well as to provide a framework for developing testable hypotheses about causes of rarity and the persistence of vascular plants. Examples of hierarchical classifications of importance factors also can be found in paleontology (Gould 1985) and ecology (Pickett, Collins, and Armesto 1987). Therefore, motivated by the recognition that it is time to move the study of the theory of rare organisms from a descriptive into a testable discipline, we present a classification that allows the ordering and comparison of competing hypotheses. At minimum, the classification outlines some of the necessary and sufficient data needed to address management problems of intrinsically rare plant species.

One result of developing these hierarchies is that the focus of study is expanded, from organism or population to species (and higher taxonomic categories). In a very tangible sense, naturally rare plant species provide unparalleled experiments by nature, admittedly imperfectly controlled, upon which to test current theories of distribution and abundance of all grades of organisms, from species up the taxonomic hierarchy, that we typically describe as rare. This inherent quality of rare taxa continues to hold the focus of evolutionary biologists.

Theoretical Considerations

Rarity is of both theoretical and applied concern, and has been so for several centuries. Although it is not our purpose to review the concepts of rarity here (see Fiedler 1986), a brief summary is appropriate because it provides a historical context in which to view our classification. We begin with Darwin, not because he was the first to write about rarity, but because in the scientific literature, rarity became a central focus of discussion for the mid- to late nineteenth-century field naturalists and systematists as a result of his work.

Darwin wrote that rarity was a necessary precursor to extinction and that until the reasons for rarity were established, we could not explain extinction (Darwin 1872). Although today we are not a great deal closer in our understanding of rarity, Darwin's sentiments are perhaps verified, as causes of extinction are intensively documented in concert with causes of rarity (see Nitecki 1984, Myers 1989). In addition, Stebbins' (1980) relatively recent call for a "synthetic" approach is predated by nearly a century in Darwin's concept of rarity in that he concluded that causes of rarity were typically multifactorial (Darwin 1872).

In the early twentieth century, however, scientists began to look for singular causes of rarity. Willis (1922) claimed that rare species, specifically endemics, were new species, while Fernald (e.g., 1918, 1929, 1943) wrote extensively that rare species (i.e., endemics) were relictual. Gleason (1924) expanded Willis's age-and-area hypothesis and argued that species could be either new or old and that the geographic extent of a taxon is at least in part a function of its age. Today, however, it is generally accepted that naturally rare species may be new, old, or of intermediate age.

The twentieth century also saw the establishment of a mathematical basis for describing the variable abundance of species in plant communities. Raunkiaer (1918), Gleason (1929), and Preston (1948, 1962a, 1962b) are perhaps most well known in this endeavor, as it became evident that within any given community of plants or animals, there tends to be only a few relatively common species and *many* rare ones. Finally, this century also witnessed the rise of the "modern synthesis," in which geneticists published several variations on the genetic theme of rarity. Stebbins (1942) argued that rare taxa were rare, genetically homogeneous "biotypes" that could not spread because of a genetic "conservatism." Wright (1956) and later Huxley (1963) felt that rare species had lower genetic heterozygosity than common ones and that the smaller absolute number of individuals in rare species allowed the spread of deleterious genes. These ideas are still hotly debated today, with no conclusive evidence for any generalizations about the genetics of rare plant species (e.g., Levin, Ritter and Ellestrand 1979; Warwick and Gottlieb 1985; Karon 1987; Waller, O'Malley and Gawler 1991; Lesica et al. 1988; but see Hamrick et al. in press). Finally, the notion of "competitive incompetancy" was advanced by Griggs (1940) to explain why some rare plants appear to be poor competitors. This has been echoed by Kruckeberg (1951) and others to explain the restriction of some endemics to serpentine and other ultramafic substrates.

Current Thinking

Stebbins (1980) called for a synthetic approach to the understanding of rare plants and proposed the gene pool–niche interaction theory for the evolution of rare taxa. This theory explains rarity as the interaction of three factors peculiar to each rare taxon—a specific evolutionary history, a unique and localized environment, and a specific genetic structure. Although still not a complete description of the causes of rarity, when proposed it was a marked improvement over the singular explanations of rarity put forth in the previous decades because of its multifactored approach. No other comprehensive theory concerning the *causes* of rarity exists currently.

Perhaps the most persuasive classification of rarity in vascular plants is that proposed by Rabinowitz (1981), who defined seven possible forms of rarity on the basis of distribution, abundance, and habitat specificity. Such a categorization ties a plant species to its habitat and thus can be used to classify rarities for floristic regions (e.g., Rabinowitz, Cairns, and Dillon 1986). Useful in a heuristic sense, such a classification has only moderate predictive value and does not clarify whether the classes are causes or consequences of rarity.

An important contribution to the understanding of patterns of tropical plant species abundance and rarity is that of Hubbell and Foster (1986a, 1986b), who examined the possibilities of ecological specialization and nonequilibrium forest dynamics to explain patterns of commonness and rarity on Barro Colorado Island in Panama. They proposed that chance (stochasticity) and evolutionary history are the two forces dominating tropical forest dynamics. As such, the species-rich tropical forest consists primarily of "generalists" and a few rare "specialists" (e.g., heliophiles, riparian and other wetland species). Their data are circumstantial, but illustrate that species-rich communities may require different explanations for causes of rarity in sparsely distributed vascular plants. Grubb (1986) has presented a similar argument for the role of sparsely and patchily distributed species in species-rich environments in general.

Recently Schoener (1987) defined two ways that abundances are misjudged due to fluctuations in local abundances. He called these suffuse and diffuse rarity. Suffuse species are those that appear rare in censuses and are rare throughout their ranges, whereas diffuse species are rare in censuses but are common in the majority of their ranges elsewhere. This classification, though not bearing directly on actual causes of rarity, is noteworthy because it points to difficulties in working with rare entities in general, and rare organisms in "the field" in particular. Although the classification is more immediately relevant to highly vagile organisms such as birds, it provokes interesting questions concerning the validity of relative abundances of rare species in plant inventories.

THE CLASSIFICATION

Formal Definition and Partitioning of the Problem

Using the word "rare" in population biology is really a statement about the distribution and abundance of a particular species or population. Thus "rarity" describes three different situations: (1) taxa whose distribution is broad but population sizes are never large where found; (2) those whose distribution is clumped or narrow yet whose populations are represented by many individuals where they are found; and finally, (3) those taxa whose distribution is clumped and whose individual abundance is low where found (Figure 2.1). Even though this definition of rarity has been described previously (Mayr 1963; Drury 1974, 1980), the geographic form of rarity frequently is not made explicit.

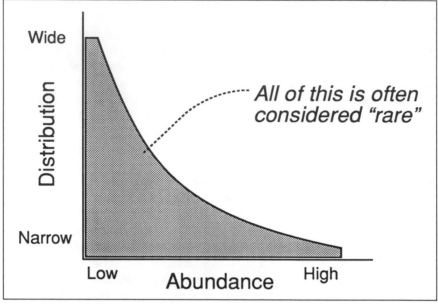

Figure 2.1
The term "rarity" is used to describe a spectrum of combinations of abundance and distribution values, from very narrow distributions with great abundance (i.e., large population sizes) to very broad distributions with population composed of a few individuals.

Forms of Rarity in Time and Space

The geographic definitions of rarity can be extended by including a temporal component. Figure 2.2 represents a three-dimensional space of abundance, distribution, and time. Time in this case is a measure of temporal persistence—that is, taxon age). Our classification of rarity is based on these three parameters. Thus, we are interested in distinguishing rare distributions that are followed or preceded by very different relative abundances from those that persist as rare over a very long period of time. Note that the classification also distinguishes between those rarities that are highly localized (endemics) and those that are broadly distributed. For the sake of brevity, we will refer to the categories as "short" and "long" when referring to temporal persistence and "narrow" and "wide" when describing spatial distribution. This leaves the following categories: short/wide, long/wide, short/narrow, and long/narrow. Four combinations of temporal persistence and spatial distribution are shown in Figure 2.3.

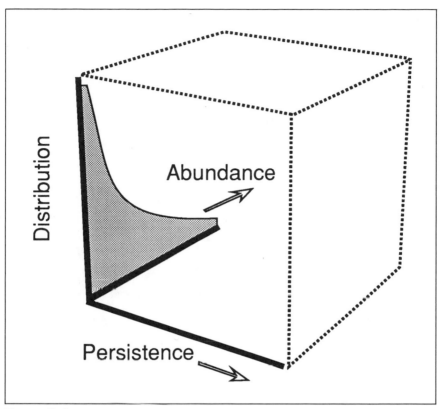

Figure 2.2
Three primary dimensions are used to classify rarity. All taxa can be placed into this distribution/abundance/persistence cube. Rare taxa are restricted to certain values of distribution and abundance as indicated in Figure 2.1.

It is important to recognize that temporal persistence and spatial distribution, though often measured in absolute terms, will need to be sensitive to the biology of the organism if we are to compare long-lived, slow-emigrating taxa with short-lived ones that disperse more quickly. Thus, although some reasonable temporal and spatial limits for categorizing taxa are suggested, it is assumed that factors such as generation time and ability to disperse will influence the categorization of any given taxon. It also is better to define temporal persistence in terms of generation time and spatial distribution in the context of gene flow, because these definitions allow a more meaningful comparison among taxa. However, rarely does sufficient data exist to allow such comparisons. Therefore, we suggest time in years and areas in hectares.

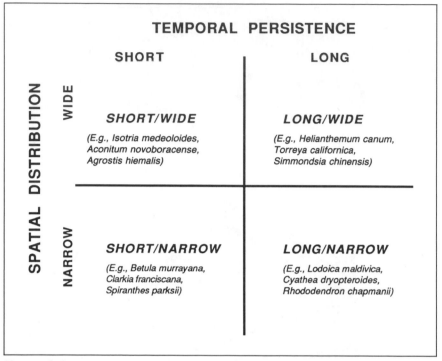

Figure 2.3
Four categories of rarity in vascular plants, as defined by temporal persistence and spatial distribution.

Long time is defined here as greater than 10^4 to 10^5 years (approximately 500 generations [see Brubaker 1988]). Short time is within 10^0 to 10^3 years. Notice that for long-lived woody species a single generation may require 500 years. Intermediate time scales range from five to twenty-five generations, or from 10^3 to 10^4 years, depending on the plant life form. A wide geographic distribution is considered arbitrarily to be >500,000 km², whereas a narrow area constitutes <10,000 km². Intermediate distributions range from 10,000 to 500,000 km². It should be remembered, however, that restrictions to gene flow (and thus one potential cause of rarity) may be a problem for outcrossing plants in any arbitrarily defined measure of area.

Class Characteristics

The short/wide class describes those rare taxa that are rare over a large area but have been in existence for a relatively short time period. These taxa are, by definition, never abundant where found, most are herbaceous, many are "neoendemics" or representatives of post-Pleistocene floras (thus the fossil record is of little assistance), most of their environments tend to be species-rich, and they are usually found in temperate climates. Examples of this form of rarity include several North American orchids, such as the small whorled pogonia (*Isotria medeoloides*) that is distributed throughout the temperate forests of the northeastern United States. Populations consist of few individuals, often less than ten (Mehrhoff 1983). Another set of examples are the North American tall-grass prairie grasses, such as *Sphenopholis obtusa, Agrostis hiemalis,* and *Festuca paradoxa,* species that Rabinowitz has convincingly discussed as widely but sparsely distributed (Rabinowitz 1978; Rabinowitz, Bassett, and Renfro 1979; Rabinowitz and Rapp 1981). This category, however, represents comparatively few taxa (see Cook and Dixon 1989).

As with those species in the short/wide class, rare plants described by the long/wide class are not abundant where found. Most are woody trees or shrubs, some are "paleoendemics" (see Stebbins and Major 1965), and their environments tend to be species-rich. One example of this form of rarity is *Welwitchia mirabilis,* a relictual dioecious gymnosperm distributed throughout the Karoo-Namib floristic region of Africa, one of the most species-rich regions of vascular plants in the temperate zone. A second example, the California nutmeg (*Torreya californica),* is a gymnosperm widely distributed throughout a variety of forest communities in California's North Coast Ranges and the western slope of the Sierra Nevada. Interestingly, the forest communities in which it is found are not considered "relictual" habitats (Raven and Axelrod 1978). Finally, many tropical moist forest trees that are rare by temperate standards (Hubbell and Foster 1986a, 1986b) are described by this category.

The third category, short/narrow, is occupied by rare plants that are found in a small area that maintain their narrow distribution for a short period of time. By definition, population size or abundance is not a descriptive feature because they are narrowly distributed. Most are ecological "specialists" such as edaphic endemics or narrow hydrophytes; many are herbaceous and are often annuals in temperate climates. California has many rare plants that fit into this categroy. Examples include *Clarkia franciscana,* an ultramafic endemic of the San Francisco Bay Area; Raven's manzanita (*Arctostaphylos hookeri* var. *ravenii*); and many dicotyledonous species of California's vernal pool systems, such as *Limnanthes bakeri, Navarettia plieantha,* and *Downingia concolor* var. *brevior.* Interestingly, endemic vernal pool grass genera, such as *Tuctoria* and *Orcuttia,* are considered to be relatively old genera, whereas the rare species may be quite recent in origin (Stebbins 1974). Although many species in this category occur in large population numbers, a few species in this category, such as Murray birch (*Betula murrayana*) and Graves' beach plum (*Prunus maritima* var. *gravesii*), represent notable woody species described from individuals.

The last category, long/narrow, describes those taxa distributed over a small area (or that are highly clumped) for a long period of time. Population abundance is not of descriptive importance. Typically these rare taxa are woody shrubs or trees and are members of a pre-Pleistocene flora (i.e., the fossil record often is useful to determine the distributional history of the taxa). Many long/narrow species are regarded as "paleoendemics." Again, notable California representatives can be included here, including the Macnab cypress (*Cupressus macnabiana*), Catalina ironwood (*Lyonthamnus floribundus*), and several species of manzanita, such as the Ione manzanita (*Arctostaphylos myrtifolia*), a species restricted to Eocene sediments in the Sierra Nevada foothills. More distant examples are also familiar, such as the double coconut restricted to Seychelles (*Lodoica maldivica*) and Madagascarian baobabs (e.g., *Adansonia madagascariensis, A. gandidieri*).

This category, however, is problematic because several well-known rare plants that historically were not rare, but are so today, fit into this category. The California redwood (*Sequoia sempervirens*) and giant sequoia (*Sequoiadendron gigateum*), both very widely distributed before the Pleistocene glaciations, are now restricted to relictual habitats along the fog belt of the California coast and unglaciated valleys of the Sierra Nevada, respectively. Therefore, this category holds those taxa that *presently* are rare with typically large population sizes, but may or may not have been more widely distributed during an earlier time in their taxon history.

Hierarchies of Cause

Recently Fiedler (1986) proposed a list of factors as necessary and sufficient to the understanding of the causes of rarity in vascular plant species. We have modified this list somewhat (Table 2.1) and propose that four causal hierarchies, derived from these factors, be assigned to the four categories of rarity as defined previously (Table 2.2). Thus these hierarchies serve as hypotheses for the causes of rarity for four combinations of spatial distribution and temporal persistence. By assigning causes of rarity in three discrete tiers to each class, we imply that specific causal factors are of relative importance in "explaining" rarity for each category. The hypotheses are testable and thus are proposed so as to determine both the validity and the utility of this classification. However, the four hierarchies of probable cause ignore those causes of rarity that find their origin in human intervention, such as habitat conversion or

Table 2.1
List of Probable Causes of Rarity in Vascular Plant Species (Modified from Fiedler 1986).

I. AGE OF TAXON
  1. Old and senescent
  2. Young and incipient
  3. Intermediate age
II. COEVOLUTION
III. EARTH HISTORY
  1. Plate techtonics
  2. Vicariance
IV. ECOLOGY
  1. Habitat specificity
  2. Effects of present climate
  3. Effects of edaphic conditions
  4. Effects of specific predators
  5. Effects of generalist predators
  6. Effects of specific pathogens
  7. Effects of generalist pathogens
  8. Competitive ability
V. EVOLUTIONARY HISTORY
  1. Effects of past climatic changes or stasis
  2. Mode of origin of species
  3. Phyletic momentum or inertia
VI. LAND-USE HISTORY
  1. Effects of land management
  2. Habitat conversion
  3. Interactions with exotic plant competitors
  4. Interactions with exotic flower visitors that do not enact pollination
  5. Interactions with exotic herbivores
  6. Effects of fire suppression and wildfires
VII. LIFE HISTORY "STRATEGIES"
  1. Herbivory escape
  2. Allelopathy
VIII. POPULATION DYNAMICS
  1. Schedule of births and factors influencing recruitment
  2. Schedule of deaths and factors influencing mortality
  3. Assessment of status of populations (e.g., e^r declining, stable, increasing)

IX. HUMAN USES
  1. Horticultural trade
  2. Aboriginal uses
  3. Role in ancient and/or modern medicine
  4. Role in past and/or current industry
X. REPRODUCTIVE BIOLOGY
  1. Average number and range of flower production per reproductive individual
  2. Average number and range of fruit production per reproductive individual
  3. Average number and range of fruit set per reproductive individual
  4. Average number and range of seed produced per reproductive individual
  5. Average number and range of seed set per fruit
  6. Average number and range of seed set per reproductive individual
  7. Pollination biology
  8. Seed dispersal methods, agents, and distance
  9. Seed germination dynamics
  10. Seedling establishment
XI. STOCHASTICITY
  1. Demographic
  2. Environmental
XII. TAXON GENETICS
  1. Individual genetics (e.g., depauperate/depleted genotype, detectable heterogeneity comparable to closely related common taxon)
  2. Population genetics
  3. Product of hybrid speciation
XIII. TAXONOMIC HISTORY
  1. Meaningful taxonomic level (i.e., is rarity a taxonomic artifact?)

Table 2.2

Hierarchical Ranking of Most Probable Causes of Rarity for the Four Major Categories: Short Persistence/Wide Distribution, Long Persistence/ Wide Distribution, Short Persistence/Narrow Distribution, and Long Persistence/Narrow Distribution

Short/Wide	Long/Wide	Short/Narrow	Long/Narrow
1st Tier	*1st Tier*	*1st Tier*	*1st Tier*
Stochasticity	Earth history	Taxon ecology	Evolutionary history
Population biology	Evolutionary history	Life history strategies	Earth history
Reproductive biology	Taxon age	Stochasticity	Population biology
2nd Tier	*2nd Tier*	*2nd Tier*	*2nd Tier*
Taxon ecology	Population dynamics	Reproductive biology	Taxon age
Coevolution	Taxon ecology	Taxon genetics	Stochasticity
Taxon genetics	Reproductive biology	Population dynamics	Taxon ecology
3rd Tier	*3rd Tier*	*3rd Tier*	*3rd Tier*
Taxon age	Coevolution	Coevolution	Reproductive biology
Evolutionary history	Stochasticity	Taxon age	Taxon genetics
Earth history	Taxon genetics	Earth history	Life history strategies

overcollection. These human-induced causes apply to a rather small subset of the time/space parameters and thus are discussed elsewhere in this chapter. We recognize, however, that evolutionary consequences of human intervention are clearly possible and in some cases probable.

Any discussion of the factors that cause and maintain rarity must address the issue of proximate and ultimate causes, proximate referring to those factors that created the current situation and ultimate being those that set the stage. Simberloff (1986), in discussing causes of extinction, has suggested that ultimate causes of extinction can be manifold. This is true for rarity as well, as they are not unrelated phenomena. Both proximate and ultimate causes of rarity can be found in the causal hierarchies (Table 2.2).

Factors related to the immediate environment and biology of the rare taxon are proposed as most likely to be involved in the ultimate causes of rarity for plants in the short/wide class. The second order of likelihood and importance involves the taxon's autecology, coevolutionary relationships, and genetics of both individuals and populations. The third tier in the hierarchy of causes of rarity for this group of vascular plants is large-scale phenomena (e.g., earth and evolutionary history). The first tier in the hierarchy of causes of rarity for species of long/wide class includes those characteristics that relate to evolutionary time, while the second order of causes of rarity are those that relate to the intrinsic biology of the rare plant species. The third level in the hierarchy includes proximate factors least likely to be of consequence in causing rarity, such as environmental or demographic stochasticity and coevolution.

The hierarchy of causes in the short/narrow class focuses primarily on small-scale events. The second tier in the causes of rarity includes the reproductive biology, genetics, and population dynamics of the taxon, while the third tier involves long-term processes—coevolution, taxon age, and earth history. Probable causes of rarity for the long/narrow class begin with the ultimate first-order causes that are largely long-term processes. The second order of probable causes involves taxon age, stochasticity, and taxon ecology. The third order includes reproductive biology, taxon genetics, and life history strategies.

In summary, the hierarchies of causes of rarity are polythetic when taken as a whole, but when viewed individually for each class of natural rarity, the hierarchies are discrete, monothetic units. In other words, each of our groups (short/wide . . . long/narrow) has its own individual hierarchy of causes of rarity, regardless of considerable overlap in the causes among the categories. The rank order, however, differs for each one and reflects a predicted hierarchy of importance of each cause of rarity.

Human Intervention and Causes of Rarity

Three major causes of rarity as listed in Table 2.1, land-use history (VI), human uses (IX), and taxonomic history (XIII), often can explain the rarity of a taxon without invoking an intrinsic feature about its biology, habitat requirements, or evolutionary history. Thus our classification does not address these *extrinsic* (i.e., anthropogenic) rarities, and generalizations about *intrinsic* causes of rarity are not necessarily applicable. However, some factors are important enough to interact with and modify the devastating effects of "anthropogenic" rarities. Two obvious examples are demographic stochastic factors and intrinsic biology of the species. These are particularly important for those taxa that are "recent" (whether narrowly or widely distributed) and for those species that are "old" and are narrowly distributed. These taxa can be expected to be affected negatively by human intervention.

A great many species are, however, "anthropogenic" rarities, simply through the species' negative interaction with human populations. For example, Cook and Dixon (1989), in a recent review of federally listed endangered and threatened plant species in the United States, report that

81 percent of the listed species are threatened by some form of human intervention. In particular, land uses such as agriculture, mining, urban and suburban development, and the introduction of exotic species have contributed to regional lists of rare, threatened, or endangered species. Habitat conversion, disturbance, and degradation are contributing significantly to the growing number of rarities and extinctions in the tropical and subtropical regions. Some moderately long-term land-use practices, such as fire repression and water diversion, likely result in more subtle but negative effects on rare taxa.

A nearly intractable effect of human intervention, largely in temperate regions, is the "overdescription" of taxa. This may result in the description of taxonomic entities that are not discrete evolutionary units. Examples from the United States include *Mahonia sonnei* (Truckee barberry), *Hudsonia caerulea* var. *faxonorum* (alpine bluet), *Achyranthes splendens* var. *reflexa, Priva portoricencis,* and many species and infraspecific taxa of *Lipochaeta* and *Nototrichium.*

For tropical floras, however, many taxa remain uncollected, undescribed, or, as noted by Gentry (1986), are described on the superficial comparison of herbarium specimens rather than on a thorough taxonomic revision. Thus, rare tropical taxa by and large remain unknown. Of the estimated 10,000 species in the Neotropics remaining to be described (Gentry 1982; Raven in Gentry 1986), many of these doubtless are rare endemics. As an example, the ninety new species known only from the extraordinary species-rich cloud forests of Centinela, the first Andean foothill ridge in western Ecuador, are now extinct as a result of deforestation (Gentry 1986). Clearly, the effects of human intervention are experienced across all forms of rarity without regard to their evolutionary history or ecological specialization. The most immediate management objective for anthropogenic rarities must be habitat protection.

APPLICATION OF THE CLASSIFICATION

Example 1: Furbish's Lousewort (Pedicularis furbishiae S. Wats.)

Furbish's lousewort is a perennial herb of considerable notoriety in the snapdragon family (Scrophulariaceae). It was thought to be extinct (Smithsonian 1975) for nearly thirty years until the Dickey-Lincoln Dam was proposed for construction on the St. John River in Maine. Furbish's lousewort was rediscovered prior to dam construction, however, and thus became the botanical "snail darter." Although the dam was not built due to economic reasons, a detailed study of the biology of this rare lousewort was initiated (see Menges 1988 for summary). As a result, Furbish's lousewort may represent the best studied of all rare plant species in the United States.

In our classification, Furbish's lousewort is best described by short/ narrow, a young taxon endemic to a small area. The species is restricted to several localized populations within a 140-mile reach of the St. John River in the state of Maine and adjacent New Brunswick, Canada, a region notably depauperate of endemic species. It is believed to be a recently evolved species whose closest relatives are found in boreal habitats and in the Rocky Mountains (Macior 1978). It is listed by the U.S. Fish and Wildlife Service (1978) as an endangered species. Assuming that renewed attempts at dam construction and other human activities, such as timber harvesting within the St. John watershed, are curtailed, effects of human intervention presently are unlikely to be a significant cause of rarity. If, however, as predicted by the classification, taxon ecology, life history strategies, and stochasticity represent the most probable causes of rarity, we should be able to discern ultimate first-order causes for Furbish's lousewort.

Details are known about all three predicted first-order causes of rarity for Furbish's lousewort. The first, taxon autecology, is perhaps the most important factor in the rarity of this herb. It is restricted to a riverine habitat dominated by the disturbance regime of the St. John River (e.g., yearly flooding, ice scouring, bank slumping). Specifically, it prefers mesic, calcium-rich soils in shrub-dominated openings within an acidic, closed-forest riparian community (Gawler 1985; Menges, Gawler, and Waller, 1985; Menges 1990). The specialized yet ephemeral nature of this habitat must severely limit the long-term viability of this species (see Menges 1988, 1990). In addition, the severe disturbances of the St. John River that extinguish populations and open new habitats for the lousewort reflect the precarious ecological regime to which it is adapted.

The second cause of rarity in the first tier of probable cause is life history strategy. As with many species in the Scrophulariaceae, Furbish's lousewort is a non–host-specific hemiparasite (Macior 1978, 1980). Once established, seedlings grow comparatively slowly, possibly due to the hemiparasite nature of the species (Gawler, Waller, and Menges 1987). It is likely that this life history character may prevent it from invading unvegetated, disturbed riverbanks and thus limit its ability to colonize new habitats (Menges 1988). The third probable cause in tier 1 is stochasticity, and environmental (vs. demographic) stochasticity is the more likely form affecting populations of Furbish's lousewort. Menges (1988, 1990, chapter 10) modeled catastrophic extinction probabilities for this species and found that catastrophic extinctions in the form of natural disasters associated with the disturbance regime of the St. John River may doom many of the populations to extinction regardless of any other feature of the species biology. Thus, environmental stochasticity coupled with taxon autecology may be the single most important factor in the cause of rarity for tier 1. Long-term species viability may largely be a function of metapopulation dynamics (Menges 1988, 1990).

The second tier of probable causes of rarity for the short/narrow category includes reproductive biology, taxon genetics, and population dynamics. Several aspects of the reproductive biology of Furbish's lousewort may be, in part, secondarily responsible for its rare status. The lousewort does not reproduce vegetatively and thus must rely exclusively on sexual reproduction. There is, however, great variability in number of seed set, although inefficient pollination does not appear responsible for this variation in reproductive output (Menges, Waller, and Gawler 1986; Menges 1988). In addition, seed predation by both specific and generalist herbivores can be significant (Menges 1988), and may be particularly so when coupled with low seedling recruitment. Also, despite an earlier study that suggested Furbish's lousewort was self-incompatible (Macior 1978), more recent work indicates that it is autogamous (Menges 1988). Selfing potentially could counteract any depression in seed set through outcrossing.

The genetics of Furbish's lousewort recently has been documented. In an electrophoretic analysis, Waller et al. (1987) found no allelic variation at any of twenty-two loci. They postulated that the complete lack of detectable genetic variation in this species is the result of repeated founder effects, local inbreeding, and repeated population extinctions through environmental stochasticity. There is no evidence at this time, however, of any negative effects of homozygosity (i.e., inbreeding depression) in this species. Finally, coupled with genetics and reproductive biology in the second tier of rarity is population dynamics. Using a transition matrix theory to model the population growth rates for Furbish's lousewort, Menges (1988, 1990; chapter 10) found that most of the undisturbed populations appeared stable or increasing slightly in size. Therefore, although taxon genetics and reproductive biology may prove important in the cause of rarity for Furbish's lousewort, the current population dynamics of this species do not appear to be a significant factor.

The third tier of rarity predicted for short/narrow rare plant species involves earth history, coevolution, and taxon age. With regard to earth history and taxon age, Furbish's lousewort is a narrow endemic restricted to the post-Pleistocene habitats of less than 12,000 years of age (Macior 1982; Menges 1988). Thus it is most likely a "neo-endemic," a notion that is correlated with its lack of genetic variability and its narrow distribution. Coevolution with any specific pollinator or disperser, although predicted by the classification as a lower order cause of rarity, does not appear important for this species at this time. However, there is some circumstantial evidence regarding coevolution. Pollination success for Furbish's lousewort is highest when bumblebees (*Bombus vagans*) are the pollinators (Menges 1988).

In conclusion, for this first example, the first tier of rarity for Furbish's lousewort (and thus the most probable cause of rarity) is the precarious autecology of the species—that is, its confinement to a limited, unstable and ephemeral riverine habitat that frequently is subjected to stochastic, catastrophic natural disturbances. Other primary causes of rarity that might continue to influence its rare status include variable reproductive output, a hemiparasitic seedling establishment requirement, its apparent youth, and possibly its genetic homogeneity.

Example 2: California Fan Palm (Washingtonia filifera (Lindl.) Wendl.)

The California fan palm appears to be a relictual endemic in southwestern United States and northern Baja California, Mexico (Munz 1974). Fossils of this fan palm date to the Miocene and Pliocene and are found from the California coast to the Mojave Desert (Bailey 1936; Vogl and McHargue 1966). In our classification this long-lived monocot belongs in the long/narrow category. It represents one of those "problematic" genera—that is, a taxon formerly widespread and common but presently locally restricted.

As described by the classification, first-order causes of rarity for this category include evolutionary history, earth history, and population biology. The evolutionary history of the California fan palm can be reconstructed from the fossil record and, as mentioned previously, provides evidence that this species is a relictual one. Therefore, it possibly is a "senescent" species, with extinction impending. Earth history of this species suggests that both geologic and climatic changes eliminated the Mojave and coastal California populations (Axelrod 1950). Existing populations represent both its relictual distribution and more recent colonization events. Establishment of the recent populations was possible as periodic incursions of the Gulf of California inundated the eastern portion of southern California and adjacent Arizona (Jaeger 1955). More recently, diversion of the Colorado River into the Salton Basin has created a suitable habitat for this palm (Oakeshott 1978), although it is primarily restricted to oases associated with the San Andreas Fault (Vogl and McHargue 1966). Population biology of this species, the third probable cause of rarity, is unknown but likely is significant. Currently disjunct populations range in size from 1 to approximately 3,000 individuals

(Smith 1958; Moran 1977), and the smaller populations (i.e., fewer than 10 individuals) are not likely to sustain long-term viability (see Menges 1990, chapter 10).

The second tier of probable cause of rarity for the California fan palm includes taxon age, stochasticity, and taxon ecology. Taxon age already has been implicated in the evolutionary history of this palm, and again, it may represent a senescent species. Demographic stochasticity is a probable cause of rarity for this species due in large part to its variable population sizes, and thus its susceptibility to chance events of individual death, reproductive variability, and population extirpation. Finally, this palm is restricted to the desert springs, seeps, and canyon bottoms of the Colorado Desert. This physiological restriction greatly limits its potential for colonizing new habitats because of the paucity of springs and oases in the Colorado Desert.

The third tier of probable cause of rarity for this species includes reproductive biology, genetics, and life history strategy. Again, very little is known about its reproduction biology, with the exception that it does not reproduce vegetatively, and its pollination is typically accomplished by generalist insects (Cornett in McClenaghan and Beauchamp 1986). Although vegetative reproduction could insure long-term *in situ* survival for a particular genotype, its sexual reproduction, specifically pollination ecology, may not be of serious concern in establishing the causes of its rarity. In addition, seed dispersal is accomplished by a variety of wide-ranging species, including birds, coyotes, and humans (McClintock 1978; Bullock 1980). Therefore, gene flow may be considerable.

The genetics of this palm is relatively well studied, however, and may represent the most important proximate cause of rarity in the third tier. McClenaghan and Beauchamp (1986) examined sixteen loci for sixteen populations of the California fan palm and found low levels of within-population genetic variability. In addition, they found that population size was positively correlated with the proportion of polymorphic loci. McClenaghan and Beauchamp (1986) suggest that the climatic and geological changes (ultimate cause of rarity, tier 1) that have occurred within the region have reduced the levels of genetic variability in refugium populations (proximate cause of rarity, tier 3). However, low genetic heterozygosity or former bottlenecks does not de facto cause rarity in all vascular plants.

Potential human causes of rarity are substantial. Smith and Berg (1988) suggest that many populations are threatened by fire and vandalism. Perhaps more significant is the severe groundwater pumping and cooption of water resources that occur in the Colorado Desert. Such human activities not only threaten the health of existing populations but may prevent the recruitment of new cohorts.

In summary of the second example, the California fan palm is rare because of the rapid and severe climatic changes that have occurred in the southwestern United States since the Miocene, and because of its current physiological restriction to the oases associated with the San Andreas Fault zone (tiers 1 and 2, ultimate causes). It is also likely that many of the populations are too small to be viable. But because it is long-lived, populations that currently appear stable may not be, and thus we may see a continued contraction of its range as local extirpations occur. Other less-significant causes of rarity for this monocot may include its low genetic heterozygosity and current forms of human disturbance (tier 3, proximate causes).

Example 3: Rarity on a Community Basis—The Ring Mountain Preserve

One outcome of this classification of rarity is that areas of notable endemism containing relatively large numbers of rare species can be examined to determine whether the rare taxa present are the result of different causes of rarity, or whether the rare species simply survive in an area of refuge or active speciation. As an example, the Ring Mountain Preserve is a 148 ha remnant serpentine bunchgrass coastal prairie (Fiedler and Leidy 1987) on the Tiburon peninsula of the San Francisco Bay. It harbors a rather large number (seven) of rare, locally restricted plant taxa. Three of these taxa (Tiburon mariposa [*Calochortus tiburonensis*], Tiburon buckwheat [*Eriogonum caninum*], and Marin dwarf flax [*Hesperolinon congestum*]) may be considered ultramafic neoendemics based on the recent time of exposure of most serpentine substrates in the Coast Ranges (Raven and Axelrod 1978). These species are habitat specialists on three microhabitats of the serpentine substrate (Fiedler, pers. obs.). Tiburon mariposa prefers the large boulders and rock outcrops found on the deeper soils of the upper slopes. Tiburon buckwheat is restricted to the steep, serpentine skrees that experience regular disturbance through slumping and surface runoff. Marin dwarf flax is found predominately on the barren ultramafic outcrops characterized by high light, little plant competition, and little if any soil development.

As predicted by the first tier of probable cause in our classification, species of short persistence and narrow distribution should be habitat specialists whose rarity is determined primarily by their narrow ecologic latitude, including edaphic restriction and competitive ability. Additional causes of rarity both predicted and verified by the classification (e.g., reproductive biology, population dynamics, herbivory) have been documented only for the Tiburon mariposa (Fiedler 1987). Fiedler (1987) established that, in comparison with a common species, the Tiburon mariposa exhibited smaller sexual and asexual reproductive output, notably low seed survivorship and seedling establishment, and slower growth rates. In addition, transition matrix analyses established that the population dynamics of one population of this species exhibit remarkable inertia to small changes in life history characteristics (Fiedler 1987). If, however, as suggested by a phylogenetic analysis (Fiedler n.d.) and geologic evidence specifically for the Tiburon peninsula (Taliafero 1943; Dudley 1972), the Tiburon mariposa is an older species only recently restricted to a serpentine substrate (long/narrow rarity), then evolutionary and earth history, as well as population biology, are the most probable causes of rarity for this species.

A fourth rare species, serpentine reed grass (*Calamagrostis ophitidis*) is a member of a monocot genus with several rare species in California (Smith and Berg 1988). Nygren (1954) proposed that many North American species of *Calamagrostis* represent old ("expiring") species. This reed grass, like the Tiburon mariposa, may represent a species of long persistence and narrow distribution, with a similar hierarchy of causes of rarity. Again, we hypothesize that the ultimate factors in the rarity of the serpentine reed grass are evolutionary history, earth history, and population dynamics, with taxon age, autecology, and stochasticity of proximate importance.

A fifth species, Tiburon Indian paintbrush (*Castilleja neglecta*), is marked by a complex taxonomic history. Tiburon Indian paintbrush no longer is recognized as a distinct species (Heckert 1993), but as an ill-defined subspecies of the widespread *Castilleja affinis*. In our classification, this Indian paintbrush belongs to the short/wide category, a species of moderate to young age, restricted to a relatively small area. It is a local endemic in the San Francisco Bay Area, protected on the Ring Mountain Preserve, with the largest populations threatened by mining and development elsewhere (Smith and Berg 1988). The Tiburon paintbrush therefore likely is an anthropogenic rarity that is considered rare because of several unrelated effects of human intervention (e.g., taxonomic history, mining), and, like the remaining two species discussed next, constitute local endemics that are in large part anthropogenic rarities and not appropriately described in our classification.

The sixth and seventh rare species on Ring Mountain Preserve, seaside tarplant (*Hemizonia multicaulis* ssp. *vernalis*) and Oakland star tulip (*Calochortus umbellatus*), are questionably rare. Both are local endemics of the grasslands and forest edges in the San Francisco Bay region. As taxa of relatively young age, they may not have had time to spread before their potential habitats were coopted by historic and recent human settlement. Population numbers for both species remain relatively high

(Fiedler, pers. obs.), although they are increasingly vulnerable to rarity through the effects of land management. For example, the urban parks and preserves in which they are most common are presently succeeding from grassland to dense scrub as a result of several decades of fire suppression. This land-use policy may soon be the single most important proximate cause of rarity for these taxa. In summary, the seven rare species of this nature preserve are rare for different reasons, some of which are the result of the biology and age of the species, and others because of various and unrelated effects of human intervention, and still others that are the result of both.

CONCLUSION

The classification of rarity presented here is offered to managers and researchers of rare vascular plants to provide a starting point for research into why a particular species is rare. We recognize the financial and time constraints typically imposed on those who manage rare plants, as well as those who do not have the resources for the necessary but extensive documentation of the causes of rarity in specific taxa of concern. Given that a species is not exclusively an "anthropogenic" rarity, with a modicum of knowledge about its abundance, geographic distribution, and evolutionary history, one can examine tiers of likelihood of rarity using our classification. Only in understanding causes of rarity for a specific taxon can one move toward developing conservation strategies that reduce its likelihood of extinction, as Darwin (1872) so cogently explained over a century ago. Our classification also allows those interested in rarity to sort the many probable causes into a hierarchy of likelihood, not just on a species-by-species basis, but also on a community (or landscape) level that can provide insight into the role of space and time in the evolution of present-day community organization.

Demonstration of the utility of the classification will come only when much more is known about the biology of rare species, as we move from an anecdotal discipline to one of precision. The model is not proposed as a substitute for data collection, but instead should provide a starting point for those with limited time, money, and personnel in understanding why a taxon is rare and subsequently how to manage it. With increasing knowledge about rare plant biology, our predictions of the probable causes, ultimate and proximate, will be verified, negated, or modified through additional hypothesis testing. It is our hope, therefore, that those who are mandated to mitigate or manage against population extirpation and species extinction will find this classification useful in their efforts.

ACKNOWLEDGMENTS

We wish to thank Sybil Lockhardt, E.O. Guerrant, Jr., S.K. Jain, R.A. Leidy, and R. Patterson for their many thoughtful discussions and editorial comments that served to greatly enhance this chapter.

LITERATURE CITED

Axelrod, D.I. 1950. Evolution of desert vegetation in western North America. *Carnegie Inst. Wash. Publ.* 590:215–306.

Bailey, L.H. 1936. *Washingtonia. Gentes Herbarium* 4:53–82.

Brubaker, L.B. 1988. Vegetation history and anticipating future vegetation change. In *Ecosystem management for parks and wilderness*, ed. J.K. Agee and D.R. Johnson, 41–61. Seattle: University of Washington Press.

Cook, R.E., and P. Dixon. 1989. A review of recovery plans for threatened and endangered plant species. Unpublished report, World Wildlife Fund, Washington, D.C.

Bullock, S.H. 1980. Dispersal of a desert palm by opportunistic frugivores. *Principes* 24:29–32.

Darwin, C. 1872. *On the origin of species.* 6th edition. USA: Mentor.

Drury, W.H. 1974. Rare species. *Biol. Conserv.* 6:162–69.

———. 1980. Rare species of plants. *Rhodora* 82:3–48.

Dudley, P. 1972. Comments on the distribution and age of high-grade blueschists, associated eclogites, and amphibolites from the Tiburon Peninsula, California. *Geol. Soc. Amer. Bull.* 83:3497–500.

Federal Register. 1985. 50 CFR Part 17. Endangered and threatened wildlife and plants; review of plant taxa for listing as endangered or threatened species; Notice of review. Department of Interior, U.S. Fish and Wildlife Service, 27 September 1985.

Fernald, M.L. 1918. The geographic affinities of the vascular floras of New England, the Maritime Provinces, and Newfoundland. *Am. J. Bot.* 5:219–47.

———. 1929. Some relationships of the floras of the Northern Hemisphere. *Proc. Intern. Congr. Plant Sciences* 2:1487–1507.

———. 1943. *Scirpus longii* in North Carolina. *Rhodora* 45:55–65.

Fiedler, P.L. 1986. Concepts of rarity in vascular plant species, with special reference to the genus *Calochortus* Pursh (Liliaceae). *Taxon* 35:502–18.

———. 1987. Life history and population dynamics of rare and common mariposa lilies (*Calochortus* Pursh: Liliaceae). *J. Ecol.* 75:977–95.

Fiedler, P.L., and R.A. Leidy. 1987. Plant communities of Ring Mountain Preserve, Marin County, California. *Madrono* 34:173–92.

Gawler, S.C. 1985. Characteristics of the riparian plant community surrounding the endemic Furbish's lousewort (*Pedicularis furbishiae* S. Wats.) on the St. John River, Maine (Abstract). *Bull. Ecol. Soc. Amer.* 66:177.

Gawler, S.C., D.M. Waller, and E.M. Menges. 1987. Environmental factors affecting establishment and growth of *Pedicularis furbishiae*, a rare endemic of the St. John River Valley, Maine. *Bull. Torrey Bot. Club* 114:280–92.

Gentry, A.H. 1982. Neotropical floristic diversity: Phytogeographic connections between Central and South America, Pleistocene climatic fluctuations, or an accident of Andean orogeny? *Ann. Missouri Bot. Gard.* 69:557–93.

———. 1986. Endemism in tropical versus temperate plant communities. In *Conservation biology: The science of scarcity and diversity*, ed. M. E. Soulé, 153–81. Sunderland, Mass.: Sinauer.

Gleason, H.A. 1924. Age and area from the viewpoint of phytogeography. *Am. J. Bot.* 4:541–46.

———. 1929. The significance of Raunkiaer's law of frequency. *Ecology* 10:406–8.

Gould, S.J. 1985. The paradox of the first tier: An agenda for paleobiology. *Paleobiology* 11:2–12.

Griggs, R.F. 1940. The ecology of rare plants. *Bull. Torrey Bot. Club* 67:575–94.

Grubb, P.J. 1986. Problems posed by sparse and patchily distributed species in species-rich plant communities. In *Community ecology*, ed. J. Diamond and T.J. Case, 207–25. New York: Harper & Row.

Hamrick, J.L., M.J. Godt, D.A. Murawski, and M.D. Loveless. In press. Relationships between species characteristics and the distribution of allozyme variation. In *The biology and conservation of rare plants*, ed. D.A. Falk and K.E. Holsinger. 75–86, New York: Oxford University Press.

Heckert, L.R. 1993. *Castilleja*. In *The Jepson Manual of California Plants*, ed. J. Hickman and D. Wilken. Berkeley: University of California Press.

Hubbell, S. P., and R. B. Foster. 1986a. Commonness and rarity in a neotropical forest: Implications for tropical tree conservation. In *Conservation biology: The science of scarcity and diversity,* ed. M. E. Soulé, 205–31. Sunderland, Mass.: Sinauer.

———. 1986b. Biology, chance, and the structure of tropical rain forest tree communities. In *Community ecology*, ed. J. Diamond and T. Case, 314–29. New York: Harper & Row.

Huxley, J. 1963. *The modern synthesis.* London: George Allen and Unwin.

Jaeger, E.C. 1955. *The California deserts.* Stanford, Calif.: Stanford University Press.

Karon, J.D. 1987. A comparison of levels of genetic polymorphism and self-compatibility in geographically restricted and widespread plant congeners. *Evol. Ecol.* 1:47–58.

Kruckeberg, A.R. 1951. Intraspecific variability in the response of certain native plants to serpentine soils. *Am. J. Bot.* 38:408–19.

Kruckeberg, A. R., and D. Rabinowitz. 1985. Biological aspects of endemism in higher plants. *Ann. Rev. Ecol. Syst.* 16:447–79.

Lesica, P., F.W. Allendorf, R.F. Leary, and D.E. Bilderback. 1988. Lack of genetic diversity within and among populations of an endangered plant, *Howellia aquatilis. Cons. Biol.* 2:275–82.

Levin, D.A., K. Ritter, and N.C. Ellestrand. 1979. Protein polymorphism in the narrow endemic *Oenothera organensis. Evol.* 33:534–42.

Macior, L.W. 1978. The pollination ecology and endemic adaptation of *Pedicularis furbishiae* S. Wats. *Bull. Torrey Bot. Club* 105:268–77.

———. 1980. The population biology of Furbish's lousewort (*Pedicularis furbishiae* S. Wats.). *Rhodora* 82:105–11.

————. 1982. Plant community and pollinator dynamics in the evolution of pollination mechanisms in *Pedicularis* (Scrophulariaceae). In *Pollination and evolution*, ed. J.A. Armstrong, J.M. Powell, and A.J. Richards, 29–45. Sydney: Royal Botanic Gardens.

Mayr, E. 1963. *Animal species and evolution.* Cambridge: The Belknap Press of Harvard University Press.

McClenaghan, L.R. Jr. and A.C. Beauchamp. 1986. Low genic differentiation among isolated populations of the California fan palm (*Washingtonia filifera). Evol.* 40:315–22.

McClintock, E. 1978. The Washington fan palm. *Fremontia* 6:3–5.

Mehrhoff, L.A. III. 1983. Pollination in the genus *Isotria* (Orchidaceae). *Amer. J. Bot.* 70:1444–53.

Menges, E.S. 1988. Conservation biology of Furbish's lousewort. Report no. 126. Indianapolis: Holcomb Research Institute.

Menges, E.S. 1990. Population viability analysis for an endangered plant. *Cons. Biol.* 4:52–62.

Menges, E.S., S.C. Gawler, and D.M. Waller. 1985. Population biology of the endemic plant, Furbish's lousewort (*Pedicularis furbishiae*), 1984 research. Report to the U.S. Fish and Wildlife Service. Holcomb Research Institute Report no. 40. Indianapolis: Butler University.

Menges, E.S., D.M. Waller, and S.C. Gawler. 1986. Seed set and seed predation in *Pedicularis furbishiae*, a rare endemic of the St. John River, Maine. *Amer. J. Bot.* 73:1168–77.

Moran, R. 1977. Palms of Baja California. *Env. Southwest* 478:10–14.

Munz, P.A. 1974. *A flora of southern California.* Berkeley: University of California Press.

Myers, N. 1989. A major extinction spasm: Predictable and inevitable? In *Conservation for the twenty-first century*, ed. D. Western and M. Pearl, 42–53. New York: Oxford.

Nesom, G.L. 1983. New species of *Calochortus* (Liliaceae) and *Linum* (Linaceae) from northern Mexico. *Madroño* 30:250–54.

Nitecki, M.H., ed. 1984. *Extinctions.* Chicago: University of Chicago Press.

Nygren, A. 1954. Investigations on North American *Calamagrostis. I. Hereditas* 40:377–97.

Oakeshott, G.B. 1978. *California's changing landscapes.* New York: McGraw-Hill.

Pickett, S.T.A., S.L. Collins, and J.J. Armesto. 1987. A hierarchical consideration of causes and mechanisms of succession. *Vegetatio* 69:109–14.

Preston, F.W. 1948. The commonness and rarity of species. *Ecology* 29:95–116.

————. 1962a. The canonical distribution of commonness and rarity. *Ecology* 43:185–215.

————. 1962b. The canonical distribution of commonness and rarity. *Ecology* 43: 410–32.

Rabinowitz, D. 1978. Abundance and diaspore weight. *Oecologia* 37:213–19.

————. 1981. Seven forms of rarity. In *The biological aspects of rare plant conservation*, ed. H. Synge, 205–17. New York: John Wiley & Sons.

Rabinowitz, D., B. K. Bassett, and G. E. Renfro. 1979. Abundance and neighborhood structure for sparse and common prairie grasses in a Missouri prairie. *Am. J. Bot.* 66:867–69.

Rabinowitz, D., and J. K. Rapp. 1981. Dispersal abilities of seven sparse and common grasses from a Missouri prairie. *Am. J. Bot.* 68:616–24.

Rabinowitz, D., S. Cairns, and T. Dillon. 1986. Seven forms of rarity and their frequency in the flora of the British Isles. In *Conservation biology: The science of scarcity and diversity,* ed. M. E. Soulé, 182–204. Sunderland, Mass: Sinauer.

Raunkiaer, C. 1918. Recherches statistiques sur les formations végétales. *Biol. Meddr.* 1:1–80.

Raven, P. H., and D. I. Axelrod. 1978. Origin and relationships of the California flora. *Univ. Calif. Publ. Bot.* 72:1–134.

Schoener, T. W. 1987. The geographic distribution of rarity. *Oecologia* 74:161–73.

Simberloff, D. 1986. The proximate causes of extinction. In *Patterns and processes in the history of life,* ed. D.M. Raup and D. Jablonski, 259–76. Berlin: Springer-Verlag.

Smith, D. 1958. The California habitat of *Washingtonia filifera*. *Principes* 2:41–51.

Smith, J.P. Jr., and K. Berg. 1988. *California Native Plant Society's inventory of rare and endangered vascular plants of California.* Special Publication no. 1, 4th Ed. Berkeley: California Native Plant Society.

Smithsonian Institution. 1975. Report on endangered and threatened plant species of the United States. House Document 94–51. Washington, D.C.: U.S. Government Printing Office.

Stebbins, G.L. 1942. The genetic approach to problems of rare and endemic species. *Madrono* 6:241–72.

———. 1974. *Flowering plants, evolution above the species level.* Cambridge, Mass.: Harvard University Press.

———. 1980. Rarity of plant species: A synthetic viewpoint. *Rhodora* 82:77–86.

Stebbins, G.L., and J. Major. 1965. Endemism and speciation in the California flora. *Ecol. Monogr.* 35:1–35.

Taliafero, N.L. 1943. Franciscan-Knoxville problem. *Bull. Amer. Assoc. Petroleum Geol.* 27:109–219.

U.S. Fish and Wildlife Service. 1978. Determination that various plant taxa are endangered or threatened species. *Federal Register* 43(81):17910–16.

Vogl, R.J., and L.T. McHargue. 1966. Vegetation of California fan palm oases on the San Andreas Fault. *Ecol.* 47:532–40.

Waller, D.M., D.M. O'Malley, and S.C. Gawler. 1987. Genetic variation in the extreme endemic *Pedicularis furbishiae* (Scrophulariaceae). *Cons. Biol.* 1:335–40.

Warwick, S.I., and L.D. Gottlieb. 1985. Genetic divergence and geographic speciation in *Layia* (Compositae). *Evol.* 39:1236–41.

Willis, J.C. 1922. *Age and Area.* Cambridge: Cambridge University Press.

Wright, S. 1956. Modes of selection. *Am. Nat.* 90:5–24.

CHAPTER

3

Peasant Farming Systems, Agricultural Modernization, and the Conservation of Crop Genetic Resources in Latin America

MIGUEL A. ALTIERI

and M. KAT ANDERSON

ABSTRACT

Many traditional agroecosystems found in Latin America constitute major in situ repositories of crop genetic diversity. This native germplasm is crucial to developing countries and industrialized nations alike. Native varieties expand and renew the crop genetic resources of developed countries while also performing well under the ecological and economic conditions of the traditional farms where they are grown. With agricultural modernization and environmental degradation, crop genetic diversity is decreasing in peasant agricultural systems. Research is urgently needed to document rates and causes of genetic erosion in these systems and the role that peasants play in maintenance of crop genetic diversity. It is proposed that multidisciplinary teams that work under the paradigms of ethnoecology and agroecology be assembled to integrate farmers' knowledge with Western scientific approaches to design meaningful in situ crop genetic conservation strategies.

INTRODUCTION

The earth's major crops on which the world's population relies have their geographic centers of genetic diversity in the Third World (Hawkes 1983). Much of this crop germplasm can be found in small fields of peasants who historically have grown landraces of ancestral crop species and traditional varieties. Among the most genetically diverse systems are the farms within areas of crop differentiation, as is the case with maize diversity in Mexican agroecosystems (Wilkes 1979) and with the potato in traditional agroecosystems of the Andes (Jackson, Hawkes, and Rowe 1980). These *in situ* repositories of crop genetic diversity are important to world agriculture because native germplasm is critical for the expansion and renewal of crop genetic resources. In fact, the "health" of agriculture in industrial countries depends to a great extent on the "genetic wealth" of developing countries.

On the other hand, native crop diversity is crucial to developing countries for several reasons:

1. Native varieties are highly adapted to the ecological heterogeneity that characterizes agricultural landscapes of the Third World.
2. Native varieties offer risk-averting advantages to resource-poor farmers confined to marginal lands and they outperform improved varieties under environmentally stressing conditions and/or low-input management.
3. Native crops in these traditional farming systems make a substantial contribution to the food self-sufficiency of rural and urban-poor populations of the Third World.

Unfortunately, native crop genetic diversity is decreasing in peasant agricultural systems. Agricultural development and environmental degradation have been linked to this decrease in diversity, especially in areas where traditional farming systems have been modified through the introduction of modern crop varieties and/or high-input technologies (Mooney 1983).

Most farmers throughout the developing world grow a mix of traditional and improved varieties for subsistence and commercial purposes that results in some cases in enhanced diversity, whereas in others decreased diversity. These patterns change across agroecological zones, ethnic groups, socioeconomic conditions, and technological domains, with local varieties predominating in traditional agroecosystems located in ecologically and economically marginal areas. Understanding the dynamics of these genetic changes in traditional agroecosystems, as well as quantifying the rates and extent of genetic erosion in these systems, is crucial. Much research remains to be done to determine the factors affecting rates of genetic erosion and the ways in which peasants adapt to modernization by either abandoning the cultivation of landraces or maintaining genetic diversity in the midst of technological innovation.

Landraces of crops such as maize, potatoes, peppers, beans, tomatoes, and squash are still prevalent in the hillsides of Mesoamerica, the Andes, and the lowlands of South America (Ford-Lloyd and Jackson 1986). In many cases their maintenance is dependent on local ecosystem diversity, farmers' knowledge about the environment, modes of production, seed selection criteria, and cultural factors. Therefore, the role that peasants play in the maintenance of crop genetic diversity must be assessed before any meaningful *in situ* conservation strategy can be developed.

Recent work has begun to focus on the levels of crop genetic diversity intrinsic to peasant agroecosystems and on the cultural and ecological factors that allow such diversity to persist. Studies conducted on maize in Mexico (Beadle 1980), cassava in tropical South America (Jennings 1976) and potato in Peru (Brush 1982) point to the following factors to account for the maintenance of racial diversity of Latin American staple crops:

1. Cultivation of primitive races by traditional methods
2. Small-scale farming systems
3. Environmental diversity within fields
4. Local adaptation to environmental and biological stress
5. Conscious seed selection and maintenance by farmers
6. Geographic fragmentation that creates isolating mechanisms conducive to rapid differentiation
7. Seed networks and exchanges among farmers within and between villages and cultural groups
8. Ethnic diversity leading to various classifications, uses, and management of distinct crop varieties
9. Tolerance of weedy relatives within and around fields promoting hybridization
10. Ecological exchanges between the vegetation in farmers' fields and the surrounding wild ecosystems

Each of these factors impinges on socioeconomic, agroecological, and ethnoecological matters and are therefore herein discussed accordingly in light of the realities of Latin American peasant agriculture.

PEASANT AGRICULTURE IN LATIN AMERICA

Socioeconomic Features

In Latin America there are about nine million peasant units with an average farm size of 2.1 ha. These peasant farmers constitute about 80 percent of the total number of farmers, holding 18 percent of the total agricultural land and only 7 percent of the arable land. It is in this sector where 51 percent, 77 percent, and 61 percent of the output of corn, potatoes, and beans, respectively, originate for domestic consumption (Ortega 1986). About 60 percent of the total peasant families live under conditions of impoverishment, devoting their agricultural activities mostly to subsistence and the selling of their labor off-farm for income. Lack of access to land and low productivity in marginal lands are important factors explaining the poverty of this sector. Poverty has forced peasants to cultivate an important proportion of the total land in steep slopes. Fifty to 75 percent of the small farms of each country are concentrated in these areas, totaling about eight million farms (Posner and McPherson 1982). Upland and highland cropping systems are therefore of utmost importance to peasants in many Latin American countries, not only in terms of the total arable area devoted to these systems, but also in terms of their contribution to the total agricultural production of crops such as corn, potatoes, beans, and so on (Table 3.1). Most of these hillside agroecosystems are confined to ecologically marginal areas prone to problems associated with erratic rainfall, low soil fertility, pest problems, and erosion. Peasants hedge against these risks by enhancing the genetic diversity of their crop mixtures by planting several varieties of various crops. Paralleling species diversity with genetic diversity is perhaps the most effective long-lasting mechanism of stabilizing yields under conditions of environmental and ecological uncertainty.

New technology has not yet reached this large group of impoverished hillside farmers. Many peasants insist on preserving traditional varieties and technologies even when other alternatives, including new varieties better suited for monoculture cropping, become available. For example, in Mexico, between 10 percent and 25 percent of the total peasants have adopted improved seed fertilizers, pesticides, and machinery. On Colombian hillsides, only 15 percent of the peasants have adopted new maize varieties, whereas 65 percent of the producers have adopted them in the lowland valleys (Ortega 1986).

Three main factors may explain why most peasants resist or have difficulty in adopting new varieties and technologies:

Table 3.1
Estimated Arable Land and Population on Steep Slopes of Selected Latin American Countries and Their Contribution to Total Agricultural Output.[1]

Country	Arable Land (%)	Agricultural Population (%)	Percent Contribution to Agricultural Output (Including Coffee)	Contribution to Country's Total Agricultural Production	
				Corn (%)	Potato (%)
Ecuador	25	40	33	50	70
Colombia	25	50	26	50	70
Peru	25	50	21	20	50
Guatemala	75	65	25	50	75
El Salvador	75	50	18	50	—
Honduras	80	20	19	40	100
Haiti	80	65	30	70	70
Dominican Republic	80	30	31	40	50

[1]Modified after Posner and McPherson (1982).

1. Modern varieties and global technological recommendations have proved to be seriously unfit to the tremendous ecological and socio-economic heterogeneity that characterizes peasant farms and households (de Janvry 1981).
2. The physical and socioeconomic conditions of agricultural experiment stations where technologies are generated sharply differ from those of peasants. Therefore, the technology generated at these research experiment stations usually does not fit peasants' needs or their ecological and socioeconomic conditions (Chambers and Ghildyal 1985).
3. Many peasants have a resistance to adopting technologies (including improved varieties) that they perceive as increasing risks by further monetarizing their economies and making them more dependent on the market.

Agroecological Features

In Latin America peasants manage a diversity of agricultural systems distributed along a range of ecological conditions. A salient feature of most systems is their degree of spatial and temporal plant diversity in the form of rotation, polycultures, and/or agroforestry systems. An important proportion of the staple crops such as maize, beans, cassava, and rice are grown in polycultures (Table 3.2).

Table 3.2
Prevalence of Polycultures in Latin American Countries.[1]

Country	Dominant Crop	Percentage of the Crop Grown in Polyculture
Brazil	Maize	11
Colombia	Rice	6
Dominican Republic	Maize	40
Guatemala	Beans	73
Mexico	Maize	20
Paraguay	Beans	33
	Maize	10
	Sweet potatoes	10
Venezuela	Rice	16
	Maize	33
	Beans	20
	Cassava	20
	Cotton	50

[1]Modified after Francis (1986).

Many of these agroecosystems are small-scale, geographically discontinuous, and located on a multitude of slopes, aspects, microclimates, elevational zones, and soil types. They also are surrounded by many different vegetation associations. The combinations of diverse physical factors therefore are numerous and are reflected in the diverse cropping patterns chosen by farmers to exploit site-specific characteristics. Many of the systems are surrounded by physical barriers (e.g., forest, river, mountain) and therefore are relatively isolated from other areas where the same crops are grown in large scale. Descriptions of the species and structural diversity and management of these traditional systems are discussed elsewhere (Alcorn 1984; Altieri, Anderson, and Merrick 1987; Chang 1977; Clay 1988; Denevan et al. 1984; Francis 1986; Toledo, Carabias, and Toledo 1985).

Another important dimension of the local agroecosystem is related to the fact that many peasants utilize, maintain, and preserve natural areas within or adjacent to their properties. These contribute valuable food supplements, construction materials, medicines, organic fertilizers, fuels, and religious items among other items (Toledo 1980). In fact, the crop production units and adjacent ecosystems often are integrated into a single agroecosystem. The inclusion of "wildness" in and around agroecosystems not only promotes genetic diversity but also provides benefits to the cropping systems in the form of beneficial insects, organic matter, and nutrients.

Gathering is prominent among shifting cultivators whose cultivated fields are widely spaced throughout the forest. In traveling from one field to the next, many farmers collect wild plant food along the way to be added to the cooking pots of the family's compound (Lentz 1986). The Kayapo of central Brazil grow and/or sponsor useful plant species within "natural corridors" connecting swidden fields for harvest. In fact, the Kayapo gather a variety of forest plants, including several types of wild manioc, yams, and bush leaves, and they replant them near camps and major trails to reproduce artificial resource concentrations. These may be denoted as "forest fields" that allow the Kayapo to travel more frequently to more sites with greater ease (Posey 1985). In humid, tropical conditions the procurement of resources from primary and secondary forests is very impressive. For example, in the Uxpanapa region of Veracruz, Mexico, local peasants gather and exploit about 435 wild plant and animal species, of which 229 are used as food (Toledo et al. 1985).

Although gathering normally has been associated with conditions of poverty (Wilken 1969), recent evidence suggests that this activity is closely associated with the persistence of a strong cultural tradition. In addition, vegetation gathering has an economic and ecological basis, as collected wild plants afford a meaningful input to the peasant subsistence economy, especially during times of low agricultural production due to natural calamities or other circumstances (Altieri et al. 1987).

A number of plants within or around traditional cropping systems are wild relatives of crop plants. Several farmers in Mexico encourage weeds (*quelites*) in their fields, because many serve a useful purpose (Chacon and Gliessman 1982). The Tarahumara Indians practice a double crop system of maize and *quelites* (*Amaranthus, Chenopodium*, and *Brassica*) that provides edible green vegetables early in the season or when maize is lost to hailstorms (Bye 1981). In Tlaxcala, farmers sponsor wild Solanaceae in barley fields, harvesting up to 1.3 t/ha of *Solanum mozinianum* fruits with no apparent impact on barley yields (Williams 1985). Thus, through the practice of selective weeding, farmers inadvertently have increased the gene flow between crops and their relatives (Altieri and Merrick 1987). When Mexican farmers allow teosinte (*Zea diploperennis*) to remain within or near corn fields, some natural crosses occur between both wind-pollinated species (Wilkes 1977). The nature of these associations is important in determining the extent of gene flow and introgression between crop plants and wild races (de Wet and Harlan 1975).

Most traditional agroecosystems contain populations of variable landraces in addition to wild relatives of crops (Harlan 1976). Landrace populations consist of mixtures of genetic lines, all of which are reasonably well adapted to the region in which they evolved but that differ in their response to diseases and insect pests. Lines are differentially resistant or tolerant to certain races of pathogens (Harlan 1976). The resulting genetic diversity confers at least partial resistance to diseases that are specific to particular crop strains, and this allows farmers to exploit different microclimates and derive multiple nutritional (and other) uses from within-species genetic variation.

Ethnoecological Features

Ethnobiologists claim that the rich folk knowledge in traditional agriculture accounts for much of native crop diversity. This knowledge has many dimensions, including linguistics, botany, zoology, craft skills, and agriculture. It is derived from the direct interaction of humans and the environment. Information is extracted from the environment by special cognition and perception systems that select for the most adaptive or useful information. Finest discrimination can be observed in communities where environments have great physical and biological diversity, in communities living near the margins of survival, or both. Older members of the communities generally possess greater detailed knowledge than younger members.

Indigenous knowledge about the physical environment is often minutely detailed. For example, many farmers have developed traditional calendars to control the scheduling of agricultural activities and sow according to the phase of the moon, believing that there are lunar phases of rainfall. Many farmers also cope with climatic seasonality by utilizing weather indicators based on the phenologies of local vegetation. Soil types, degrees of soil fertility, and land-use categories are also discriminated in detail by farmers. Soil types are usually distinguished by color, texture, consistency, and even smell and taste.

Indigenous people have developed many complex systems to classify plants and animals (Berlin, Breedlove, and Raven 1973). In general, the traditional name of a plant or animal usually reveals that organism's taxonomic status. Ethnobotanies are the most commonly documented folk taxonomies (Alcorn 1984). The ethnobotanical knowledge of certain campesinos in Mexico is so elaborate that the Tzeltal, P'urepecha, and Yucatan Mayans can recognize more than 1200, 900, and 500 plant species, respectively (Toledo et al. 1985).

This rich ethnobotanical knowledge supports the vegetationally diverse farming systems. The development of such agroecosystems is not random, but rather is based on a deep understanding of the elements and interactions of the vegetation guided by complex ethnobotanical classification systems. Such nomenclature has allowed peasants to assign each landscape unit a specific productive practice, thus obtaining a diversity of plant products through a strategy of multiple use (Toledo et al. 1985). Although knowledge about the local adaptability of varieties is an important factor in the success of indigenous production systems, the intimate knowledge of microenvironments by indigenous cultivators is also important because it allows the placement of plants in precise sites delimited by soil, shade, drainage, and associated crops (Clay 1988).

In Mexico, for example, Huastec Indians manage a number of agricultural and fallow fields, complex home gardens, and forest plots totaling about 300 plant species including vegetables, medicinals, spices, fuelwood species, and cash crops. Small areas around the houses commonly average 80 to 125 useful plants, mostly native medicinal plants (Alcorn 1984). Similar systems also have been developed by the Kayapo and Bora Indians in the Brazilian (Posey 1985) and Peruvian Amazon (Denevan et al. 1984), respectively, and by the Lacandones in the forests of Chiapas, Mexico (Nations and Nigh 1980). The diverse home gardens concentrated in areas around the households in most farming systems throughout the Latin American tropics and subtropics are perhaps the maximum expression of agricultural management, resembling a natural forest structure (Gliessman 1988; Niñez 1985).

In the Andes, farmers cultivate many potato varieties in their fields (Brush 1982), and for classifying potatoes have a four-tiered taxonomic system that is important in the selection of different potato varieties (Brush 1982). In the Montaro Valley of Peru, Carney (1980) found a mean of fourteen potato varieties per field, but peasants could name up to seventy-seven varieties, including modern hybrids for the market (*papa mejorada*), native varieties (*papa de regalo* for home consumption), and bitter potatoes (*papa shiri* to make *chuno*).

A number of cultural factors influence the diversity and distribution of potato cultivars within and among fields. Peasants generally plant fields with a selected combination of varieties, mostly guided in their selection by culinary quality. By rotating potato and the tuber crop *oca*, the introduction of new potato genotypes is favored. During the *oca* harvest, potato tubers derived from sexually propagated true seed are collected along with *oca* and subsequently used. In addition, seed networks among farmers serve to diffuse preferred clones and thereby amplify the gene pools of individual fields (Carney 1980).

Clearly the maintenance of crop genetic diversity in peasant agroecosystems is dependent on the interaction of human and biological factors. Environmental heterogeneity, cropping system diversity, farmers' seed selection criteria, and socioeconomic factors are all factors that determine opportunities for seed exchange, hybridization, and isolating mechanisms that may amplify genetic diversity. Figure 3.1 depicts the cultural, agroecological and socioeconomic factors affecting the selection, maintenance,

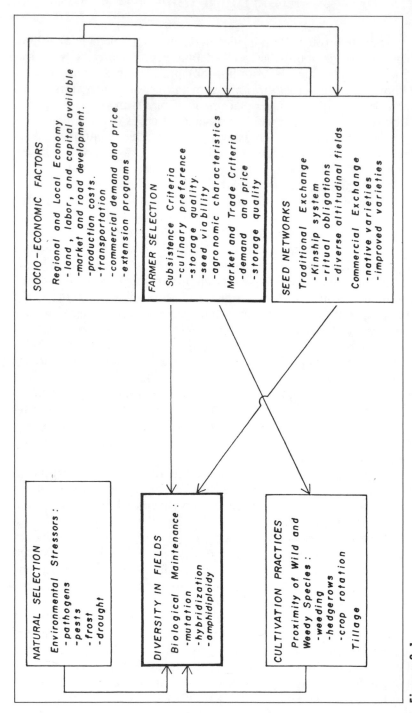

Figure 3.1
Summary of the Selection and Maintenance of Potatoes by Peruvian Highland Peasants (from Brush, Carney, and Huaman 1980).

and distribution of potato germplasm in central Peru, suggesting that the maintenance of native potato diversity is far more complex than the random planting of numerous varieties (Brush et al. 1980). Clearly there is a strong cultural and ecological basis for the maintenance of potato diversity.

Agricultural Development And In Situ Crop Genetic Conservation

With populations rising, land holdings becoming smaller, environments degrading, and per capita food production remaining static or declining, the peasantry in Latin America is undergoing a process of systematic impoverishment. To prevent the collapse of the peasant economy in the region, a strategy is needed that enables peasant agriculture to become a more sustainable and productive one. Developing ecologically and culturally based agricultural practices that raise land productivity while conserving and regenerating the resource base is the only sensible strategy to improve the quality of life of resource-poor farmers.

Previously we have argued that the design of sustainable farming systems aimed at upgrading peasant food production should incorporate locally adapted native crops, wild/weedy relatives within and around agroecosystems, and relatively undisturbed ecosystems to complement the various production processes (Altieri and Merrick 1987; Altieri et al. 1987). In this way crop genetic resource conservation is linked to rural development efforts within the context of sustainable development. In Latin America several agricultural improvement programs initiated by non-government organizations (NGOs) have been successful because they have combined both approaches. Specifically, the improvement programs have (a) utilized and promoted indigenous knowledge and resource-efficient technologies; (b) emphasized the use of local and indigenous resources, including valuable crop germplasm, firewood resources, and medicinal plants; and (c) promoted a self-contained, village-based effort with active participation of peasants (Altieri and Anderson 1986). NGOs learned that subsidizing peasant agricultural systems with external resources (e.g., pesticides, fertilizers, irrigation water) can bring high levels of productivity through dominance of the production system, but these systems are sustainable only at high external costs and are dependent on the uninterrupted availability of commercial inputs that are usually not available to resource-poor farmers. Conversely, an agricultural strategy based on a diversity of plants and cropping systems can bring moderate to high levels of agricultural productivity through manipulation and exploitation of the resources internal to the farm, and it can be sustainable at a much lower cost for a longer time.

Any attempt at *in situ* crop genetic conservation must struggle to preserve the agroecosystem in which those resources occur (Nabhan 1979, 1985). In the same vein, preservation of traditional agroecosystems cannot be achieved isolated from maintenance of the sociocultural organization of the local people (Altieri 1983). Given these facts, agroecologists, anthropologists, and ethnoecologists (especially ethnobotanists) have an important yet unrealized role in agricultural development and genetic resource conservation (Alcorn 1981). Through interdisciplinary efforts they can assess traditional "know-how" to guide the use of modern agricultural science in the improvement of small-farm productivity. Ethnobotanists and ecologists can provide critical information for policymakers about resources needing protection and about the ecological and management factors that determine the persistence of elements of natural vegetation in the traditional agroecosystems (Alcorn 1984).

We suggest that multidisciplinary teams that work under the paradigms of ethnoecology and agroecology can prove effective in integrating farmers' knowledge with Western scientific approaches (Figure 3.2). The application of agroecology can ensure that *in situ* crop genetic conservation is achieved by preserving the agroecosystems in which these resources occur. Ethnoecological research and surveys should remind us that preservation of traditional agroecosystems cannot be achieved isolated from maintenance of the ethnoscience and sociocultural organization of the local people (Altieri 1989).

CONCLUSIONS

For decades it has been the responsibility of governments, genetic resource organizations, and plant breeders, both public and private, to salvage germplasm before it is lost and to assure its introduction into germplasm banks (Brown 1983). It is time to recognize the active role of peasants in genetic resource conservation (Alcorn 1984). Sociocultural issues make it impossible to view the resources merely as a set of genes that one can conserve by placing them into a gene bank. If isolated from the folk science and traditional uses of the cultures that have nurtured them, they lose part of their value or cultural-historical meaning (Nabhan 1979, 1985; Sarukhán 1985).

A main research concern for agroecologists and ethnoecologists will be to determine the factors that influence farmers' maintenance of genetic diversity or, conversely, rates of genetic erosion, as well as the impacts of such losses on peasant welfare. Given the cultural, ecological, and socioeconomic heterogeneity that characterizes peasant agriculture in Latin America, it is expected that the levels of genetic diversity as well

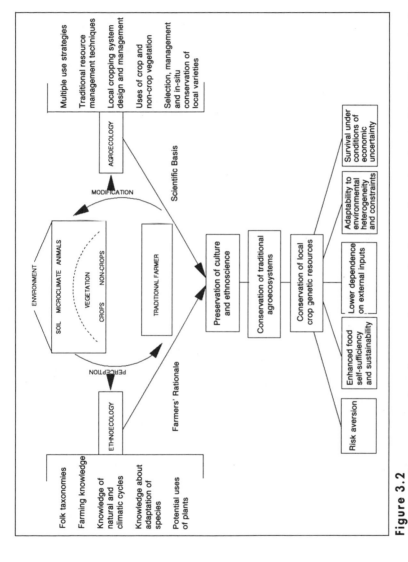

Figure 3.2
The complementary roles of agroecology and ethnoecology in the making of a method-ological framework for the conservation of traditional agroecosystems and their local crop genetic resources (from Altieri 1989).

as key factors explaining such diversity will vary in the different farming systems. Understanding the site-specific dynamics of genetic diversity maintenance in peasant cropping systems is vital to the design of meaningful *in situ* crop genetic conservation strategies adapted to the local conditions and needs of farmers inserted in different ecological and socioeconomic realities.

The core of such strategies should be the incorporation of indigenous crops and other native plant germplasm in the design of self-sustained agroecosystems, as this is the only sensible avenue to assure maintenance of local genetic diversity available to resource-poor farmers. This approach sharply contrasts with current efforts by international centers that tend to concentrate on fewer varieties, potentially eroding genetic diversity and making farmers increasingly dependent on seed companies for their seasonal seed supply. A major concern is that when impoverished peasants become dependent on distant institutions for inputs, rural communities tend to lose control over their production systems.

LITERATURE CITED

Alcorn, J.B. 1981. Huastec noncrop resource management. *Human Ecol.* 9:395–407.

———. 1984. Development policy, forest and peasant farming: Reflections on Huastec-managed forests' contributions to commercial production and resource conservation. *Econ. Bot.* 38:384–406.

Altieri, M.A. 1983. *Agroecology: The scientific basis of alternative agriculture.* Berkeley: University of California, Division of Biol. Control.

———. 1989. Rethinking crop genetic resource conservation: A view from the South. *Cons. Biol.* 3:1–3.

Altieri, M.A., and M.K. Anderson. 1986. An ecological basis for the development of alternative agricultural systems for small farmers in the Third World. *Amer. J. Alternative Agriculture* 1:30–38.

Altieri, M.A., M.K. Anderson, and L.C. Merrick. 1987. Peasant agriculture and the conservation of crop and wild plant resources. *J. Cons. Biol.* 1:49–58.

Altieri, M.A., and L.C. Merrick. 1987. *In situ* conservation of crop genetic resources through maintenance of traditional farming systems. *Econ. Bot.* 4:86–96.

Beadle, G.W. 1980. The ancestry of corn. *Scientific American* 242:112–19.

Berlin, B., D.E. Breedlove, and P.H. Raven. 1973. General principles of classification and nomenclature in folk biology. *Amer. Anthro.* 75:214–42.

Brown, W.L. 1983. Genetic diversity and vulnerability—an appraisal. *Econ. Bot.* 37:2–12.

Brush, S.B. 1982. The natural and human environment of the central Andes. *Mountain Research and Development* 2:14–38.

Brush, S.B., H.J. Carney, and Z. Huaman. 1980. *The dynamics of Andean potato agriculture.* Working Paper Series no. 1980–5. Lima: International Potato Center.

Bye, R.A. 1981. *Quelites*—ethnoecology of edible greens—past, present and future. *J. Ethnobiol.* 1:109–23.

Carney, H.J. 1980. *Diversity, distribution and peasant selection of indigenous potato varieties in the Mantaro Valley, Peru: A biocultural evolutionary process.* Working Paper Series no. 1980–2. Lima: International Potato Center.

Chacon, J.C., and S.R. Gliessman. 1982. Use of the "non-weed" concept in traditional tropical agroecosystems of south-eastern Mexico. *Agroecosystems* 8:1–11.

Chambers, R. and B.P. Ghildyal. 1985. Agricultural research for resource poor farmers: the farmers first and best model. *Agric. Admin.* 20:1–30.

Chang, J.H. 1977. Tropical agriculture: Crop diversity and crop yields. *Econ. Geogr.* 53:241–54.

Clay, J.W. 1988. *Indigenous peoples and tropical forests.* Cambridge, Mass.: Cultural Survival.

de Janvry, A. 1981. *The agrarian question and reformism in Latin America.* Baltimore: The John Hopkins University Press.

Denevan, W.M., J.M. Treace, J.B. Alcorn, C. Padoch, J. Denslow, and S.F. Paitan. 1984. Indigenous agroforestry in the Peruvian Amazon: Bora Indian management of swidden fallows. *Interciencia* 9:346–57.

de Wet, J.M., and J. Harlan. 1975. Weeds and domesticates: Evolution in man-made habitat. *Econ. Bot.* 29:94–107.

Ford-Lloyd, B., and M. Jackson. 1986. *Plant genetic resources: an introduction to their conservation and use.* London: Edward Arnold.

Francis, C.A. 1986. *Multiple cropping systems.* New York: Macmillan.

Gliessman, S.R. 1988. *The home garden agroecosystem: A model for developing sustainable tropical agricultural systems.* Proceedings of the 6th International Scientific Conference of the International Federation of Organic Agriculture Movements. ed. P. Allen and D. van Dusen, 445–50. Santa Cruz: University of California.

Harlan, J.R. 1976. Genetic resources in wild relatives of crops. *Crop Sci.* 16:329–33.

Hawkes, J.G. 1983. *The diversity of crop plants.* Cambridge: Harvard University Press.

Jackson, M.T., J.G. Hawkes, and P.R. Rowe. 1980. An ethnobotanical field study of primitive potato varieties in Peru. *Euphytica* 24:107–13.

Jennings, D.L. 1976. Cassava. In *Evolution of crop plants.* ed. N.W. Simmonds. London: Longman.

Lentz, D.I. 1986. Ethnobotany of the Jicaque of Honduras. *Econ. Bot.* 40:210–19.

Mooney, P.R. 1983. The law of the seed. *Development Dialogue* 1:1–172.

Nabhan, G.P. 1979. Cultivation and culture. *Ecologist* 9:259–63.

Nabhan, G.P. 1985. Native crop diversity in Aridoamerica: Conservation of regional gene pools. *Econ. Bot.* 39:387–99.

Nations, J.D., and Nigh, R.B. 1980. The evolutionary potential of Lacandon Maya sustained-yield tropical forest agriculture. *J. Anthropol. Res.* 36:1–30.

Niñez, V. 1985. Household gardens and small scale food production. *Food and Nutrition Bull.* 7:1–5.

Ortega, E. 1986. *Peasant agriculture in Latin America.* Santiago, Chile: Joint ECLAC/FAO Agriculture Division.

Posey, D.A. 1985. Indigenous management of tropical forest ecosystems: The case of the Kayapo Indians of the Brazilian Amazon. *Agroforestry Systems* 3:139–58.

Posner, J.L., and M.F. McPherson. 1982. Agriculture on the steep slopes of tropical America: Current situation and prospects for the year 2000. *World Development* 10:341–53.

Sarukhân, J. 1985. Ecological and social overviews of ethnobotanical research. *Econ. Bot.* 39:431–35.

Toledo, V.M. 1980. La ecologia del modo campesino de producion. [The ecology of the peasant's way of farming.] *Anthropologia y Marxismo* 3:35–55.

Toledo, V.M., J. Carabias, and C. Toledo. 1985. *Ecologia y autosuficiencia alimentaria. [Ecology and food self-sufficiency.]* Mexico: Siglo Veintiuno Editora.

Wilken, G.C. 1969. The ecology of gathering in a Mexican farming region. *Econ. Bot.* 24:286–95.

Wilkes, H.G. 1977. Hybridization of maize and teosinte in Mexico and Guatemala and the improvement of maize. *Econ. Bot.* 31:254–293.

———. 1979. Mexico and Central America as a center for the origin of agriculture and the evolution of maize. *Crop Improvement* 6:1–18.

Williams, D.E. 1985. *Tres Arvenses solanâces comestibles y su proceso de domesticaciòn en Tlaxcala, Mexico. [The edible wild Solanaceae and their domestication process in Tlaxcala, Mexico.]* Master's thesis, Colegio de Postgraduados, Chapingo, Mexico.

CHAPTER
4

The New Paradigm in Ecology: Implications for Conservation Biology Above the Species Level

Essay by STEWARD T.A. PICKETT

V. THOMAS PARKER

and PEGGY L. FIEDLER

INTRODUCTION

Conservation biology has traditionally focused on the fine scale and the species level of biological organization (Soulé and Wilcox 1980), and biotic conservation is only one of the various goals that has directed the preservation and management of natural areas and resources. Resource management goals have largely been utilitarian or commercial, such as the maintenance of large areas for watersheds; preservation of marshes, bogs, and seasonal wetlands for flood water storage, nursery areas, and flow continuity in river systems, and the sustained yield of forests, fisheries, and wildlife (Temple et al. 1988; Aplet, Laven, and Fiedler in press). There have, however, been both practically and theoretically motivated calls for widening the focus to include scales beyond that of the individual site and levels of organization above that of the species (Soulé 1989; Western 1989) to approach new goals. This essay will explore the relationship of this new frontier for conservation biology to advances in ecology. Much of the new ecological focus of conservation biology is driven by the shift in the overarching paradigm of ecology itself. We begin by defining the concept of paradigm and characterizing the classical paradigm of ecology. The classical paradigm had definite implications for conservation, and these are shown to be problematical. Therefore, we outline the contemporary paradigm in ecology and show how the science of ecology has been affected by the shift in paradigm. Our central concern in the essay is to show what the contemporary paradigm of ecology suggests for conservation biology. Because the application of a new scientific paradigm that does not necessarily match popular assumptions of nonspecialists about how nature works, we make some suggestions for how the new paradigm might best be presented to the public and to policymakers.

THE CLASSICAL PARADIGM
OF ECOLOGY

"Paradigm" is a term that is associated with great excitement and controversy in science. Since the publication of Thomas Kuhn's (1962) *The Structure of Scientific Revolutions*, the term has been invoked in a hopeful way by those who seek to promote some scientific advance they propose as a truly revolutionary one. Rarely, however, is the term used very carefully in the informal scientific discourses in which it usually appears.

Indeed, Kuhn (1972), in clarifying the term in his second enlarged edition, admitted to using the term rather loosely in his first edition. The refined definition considers a paradigm to be (1) the world view shared by a scientific discipline or community, and (2) the exemplars of problem solutions used in that discipline. To expand on these basic ideas, a world view consists of the beliefs, values, and techniques used by a discipline. More appropriately, perhaps, the world view would better be thought of as the family of theories that undergird a discipline. However, the term "theory" in this context should be thought of much more broadly than the specific quantitative models or sets of equations that most ecologists think of when they invoke the term (cf. Pickett and Kolasa 1989). Taking the viewpoint component of paradigm as a family of broad theories is, we believe, much more objective and scientifically valuable a concept than Kuhn's dependence on beliefs or values alone. Of course, we admit that such things affect the theories that scientists develop and use, but it is those explicit, communicable, and assessable theories that can actually be tested, refined, applied, discarded, or replaced in a science. The concept of exemplars also can be expanded. Exemplars include not only the patterns of how to attack problems, but also considerations of what problems are thought to be important in a discipline.

The classical paradigm in ecology (Simberloff 1982) can be called for convenience here the "equilibrium paradigm." We characterize the earlier manifestations of the classical paradigm rather than its evolving later manifestations in order to draw the clearest contrast with the contemporary paradigm that will be presented later.

The equilibrium paradigm emphasized the stable point equilibrium of ecological systems. Thus, the focus and driving concern was with the end points of ecological processes and interactions. An example appears in the first coherent theory to be elaborated in ecology, that of community succession (Jackson 1981). The major engine of succession was considered to be the attainment of the climax state (Clements 1916). The processes involved necessarily led to that state, and deviations were of little fundamental interest.

A second emphasis of the equilibrium paradigm was on the closure of ecological systems. Because systems were considered to be functionally and structurally complete in and of themselves, they could also be thought to be self-regulating. Hence, early models of communities or of populations emphasized internal competition or facilitation, internally driven density dependence, and so forth.

The classical equilibrium paradigm was consonant with the cultural metaphor of the "balance of nature." This idea has deep roots in Western culture and may well have influenced both scientists and the public (Botkin 1990). Indeed, the public still uses terms and ideas of "balance" as central aspects of its arguments and discourse about the natural world in general, and conservation and urban-wildland interactions in particular (Letters to the Editor, "Deer on your Doorstep," *New York Times Magazine*, 19 May 1991).

THE CLASSICAL PARADIGM AND CONSERVATION

Implications of the classical paradigm and its lay referent, the cultural metaphor of the balance of nature, have direct implications for conservation biology (Ehrenfeld 1972). For clarity, we present the implications in their most extreme form. The classical paradigm can be considered to suggest that any unit of nature is, in and of itself, conservable. Thus, any unit of a landscape can be an adequate nature reserve. Second, systems will maintain themselves in balance. Systems conserved and isolated from direct human perturbation will maintain themselves in the desirable state for which they were originally conserved. Furthermore, systems disturbed from equilibrium will return to that same equilibrium. Any excursion from the desired state for which an area was set aside will be corrected by natural processes generated in and acting at that site. All that is necessary to do is to protect the site from people. This previous point leads to the final implication of the classical paradigm for conservation—people and their activities are not a part of natural systems, and conservation should strive to exclude them.

We can exemplify the application of the classical ecological paradigm to conservation. One such case is Mettler's Woods, the last remaining uncut upland forest in central New Jersey. The value of this site was recognized early by plant ecologists at Rutgers University. The forest had enjoyed the admiration and protection of a single family since European establishment. When farming became untenable with spreading urbanization, many people joined together to preserve Mettler's Woods, leading to the establishment of the Hutcheson Memorial Forest Center (HMFC) at Rutgers University. The forest was considered to be "climax" and thus self-perpetuating. No manipulation or human disturbance was to be permitted in the old-growth Mettler's Woods. Access by both the public and researchers was closely monitored and regulated. Public access currently is restricted to a single trail and then only with a qualified guide.

In many ways, the HMFC has been a great success. Much has been learned about the structure and functioning of old-growth, mixed-oak forests, and many generations of graduate and undergraduate students, primary and secondary school students, and educators at all levels have been influenced by the lessons Mettler's Woods offers.

However, in one spectacular way, the conservation goals of the project have not been met. This is not because of bad management or abuse of the site by visitors or researchers. Rather, the equilibrium paradigm led the people who established the HMFC to assume that protecting the forest but otherwise leaving it alone would be an adequate and appropriate conservation strategy. In fact, the reality has been quite different. The old oaks (*Quercus* spp.), the average age of which in the early 1980s was on the order of 235 yr, are senescing and becoming sensitive to the combined stresses and disturbances that periodically occur. The overstory is now quite open, the mid-layer dogwoods (*Cornus florida*) have succumbed to a regional epidemic of anthracnose, and many species of wildflowers are quite rare (Sulser 1971). After the severe defoliation by gypsy moth (*Porthetria dispar*) caterpillars in the early 1980s, several sun-loving shrubs and herbaceous species invaded the forest understory and have maintained high populations. Spraying for gypsy moths was considered inappropriate in a nature reserve in general, and in particular because the native forest moth and butterfly species (as well as many other arthropods that are a part of the forest fauna and nutrient processing) would have been adversely affected by the spraying of either pesticides or the available biotic control agents.

Perhaps most surprising from the perspective of the classical paradigm is the absence of successful oak regeneration. Although the oaks do periodically produce large seed crops, as is expected, there is an almost complete failure of seedling oaks to graduate to sapling or larger sizes. As foresters and ecologists begin to unlock the secrets of oak forest dynamics, it is becoming clear that oak species in many parts of the eastern United States require some sort of ground fire for successful regeneration. Indeed, the ecologists who were so instrumental in the preservation of the forest learned that ground fires had been a regular part of the history of the forest before it fell into European hands in 1711. Fire scars preserved within the trunks of white oak (*Quercus alba*) trees document light fires on an interval of roughly every 10 yr before settlement (Buell 1957). Perhaps it was just such fires that permitted the oaks that are dominant in the forest today to establish themselves initially. The original conservation strategy and the mandate forbidding manipulation or disturbance of the forest that are enshrined in the deed reflect the classical paradigm, with its assumption that any natural system is self-maintaining and closed. Thus, the paradigm and the strategy derived from it have failed to preserve the forest that the people who in 1955 dedicated the preserve (Buell 1957) loved so well.

The forest today has considerable conservation and educational value in the densely urbanized Eastern Seaboard. And the visitors and custodians today still love the forest, but they must necessarily do so for different reasons because of the natural changes and the unexpected external disturbances and influences that have come to affect it.

THE CONTEMPORARY
PARADIGM IN ECOLOGY

The contemporary paradigm arose for a number of reasons (Simberloff 1982). There were, for example, empirical problems with the classical paradigm. The dynamics of natural communities were found to have multiple persistent states, so that a local climax was not often to be found (Botkin and Sobel 1975). Likewise, there were multiple pathways of vegetation change, violating the dominant role of the climax state in guiding system changes (Pickett 1989). Natural systems thus were found to have many states or "ways to be" and many ways to arrive at those states.

Another powerful challenge to the classical paradigm came with the recognition that natural systems are subject to physical disruption from a wide range of natural forces and events (White 1979). Natural disturbances such as fires, windstorms of various intensities, earth movements, floods, volcanism, outbreaks of native herbivores, etc., all can influence the structure and function of ecological systems (Pickett and Thompson 1978). Episodes of drought are associated with successful establishment in swamps, while episodic fires are associated with regeneration in forest and shrubland. Episodic events that alter the composition and performance of natural systems also were found on much longer time scales. In high latitudes and altitudes, glaciation obliterated vegetation and caused species to migrate across the continental scale, often in unique, noncongruent patterns (Davis 1983). This wide array and variety of events and processes that can have a dominant or trigger effect in determining the species composition and the horizontal and vertical structure of communities and landscapes as well as the size, age, and genetic structure of populations all suggest that natural systems are contingent. They reflect, to a very large degree, their initial conditions, the prevailing and often changing boundary conditions, the order of events within them, and the influence of adjacent or distant systems.

A second important aspect of the shift from the classical to the contemporary paradigm is a shifting scale of focus in ecology. The classical paradigm arose from a focus at a strictly coarse scale. Many of the contingencies and disturbances that contemporary ecology has come to include in its toolbox were simply invisible at the coarse scale on which early theoreticians like Clements (1916) focused. Gleason (1917), Clements's primary dialectical partner, did in fact focus on the fine scale and so emphasized the processes and results of contingency. He was, for all his insight and clarity, essentially driven out of ecology because of his iconoclasm, becoming relatively early in his career a plant systematist and phytogeographer.

The shift from an equilibrium paradigm in ecology is mirrored by major changes in perspective in other sciences over the same time period. Quantum mechanics arose in physics and accepts probability and indeterminacy. In geomorphology, the classical erosion cycle has given way to process geomorphology that looks at actual mechanisms and interactions in landform evolution rather than recurring, idealized landforms. In evolutionary biology, strict gradualism has opened to accommodate "punctuated equilibrium" and nonselective agents of evolutionary change. Plate tectonics has revolutionized geology, and in climatology the Malankivitch climate cycles are invoked to explain long-term patterns in climate, including glacial periodicity.

The new paradigm in ecology can be labeled, for convenience, the "nonequilibrium paradigm." Note, however, that the new paradigm can accept equilibrium or a point stable state as a special case, so it is inclusive rather than exclusive of important components of the older view. The nonequilibrium paradigm suggests several important ideas about natural systems. First, it accepts natural systems as open; that is, they must be put into the context of their surroundings, from which fluxes of organisms and materials may come (Whittaker and Levin 1977). Unexpected periodic or unique arrivals are especially important to consider because of their potential influence on the structure and functioning of the systems. Regulation may arise in part or completely outside a focal system.

The nonequilibrium paradigm emphasizes process rather than end point (Vitousek and White 1981). For instance, in the pure form of Clementsian succession, the climax was taken as essentially an Aristotelian final cause. Whether or not this should have been accepted, it was a powerful expectation in research into system dynamics throughout much of the first six decades of this century. Now, however, the concern is with how systems actually behave, how their structure and trajectories are actually determined. Of course, some systems may exist for greater or lesser periods of time in a compositional or functional steady state (not, however, a single invariant stable state). And in keeping with the capacity of the contemporary paradigm to focus on small or large scales, it is proper and plausible to seek an equilibrium distribution of states or patches in a system (Figure 4.1). For instance, a landscape may be in compositional equilibrium even though individual patches may be in a variety of states, and individual patches change state through time. The forested landscape of mesic portions of the eastern United States, with their disturbance primarily by wind that generates small gaps, may well have been in an equilibrium distribution (Bormann and Likens 1979).

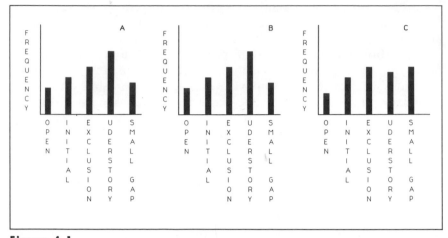

Figure 4.1
Distributions of patch types contrasting equilibrium and non-equilibrium patch distributions in a landscape. **A.** *Frequencies of patches representing different successional states.* Open *refers to recently disturbed,* initial *refers to early invasion,* exclusion *refers to the stem exclusion stage in forest,* understory *refers to patches in which the understory herbs and shrubs are establishing, and* small gap *refers to a closed canopy dominated by mature trees in which individual tree fall gaps appear (cf. Oliver and Larson 1990). The similar patch distributions at two different times (A and B) indicate the landscape supports an equilibrium distribution of patches, although specific plots of ground will likely be in different successional states at the two times represented by A and B. If the patch distribution at time 2 is different (C) from that at time 1 (A), the landscape is not at equilibrium.*

But some landscapes may be too small relative to the sizes of patches normally created by natural disturbances there to ever reach an equilibrium distribution of patches (Heinselman 1973). Such apparently is the case in the high-elevation Yellowstone Plateau coniferous forests (Romme 1982).

The aspects of the contemporary paradigm summarized previously can be encapsulated metaphorically. The scientific metaphor is "patch dynamics" (Pickett and Thompson 1978) or the "shifting mosaic" (Bormann and Likens 1979). It is more difficult to suggest a popular metaphor, in part because of the power and historic weight of the balance of nature metaphor, with its attachment to deeply and widely held myths (Botkin 1990). We might suggest the "flux of nature" as a nontechnical substitute.

The contemporary paradigm has affected basic ecology in a number of significant ways. First, it has licensed new approaches to persistent problems, warranted new disciplines, and supported the inclusion of people in ecology. An example of a new approach within ecology is the modern use of succession. This pervasive and fundamental phenomenon in nature is now seen as having multiple causes (Connell, Noble, and Slatyer 1987). Likewise, now processes are viewed as being capable of advancing as well as retarding succession, depending on the environmental and historical context. In addition, the control of the process has been seen to frequently arise from outside the focal patch. And finally, the search for the climax has receded in importance, and, indeed, now the concept of climax is used as an idealization against which to compare and systematize the complexity of real successions and their causes (Pickett and McDonnell 1989).

Second, the new paradigm warrants new disciplines in ecology. One of the most prominent of these is landscape ecology. In the broadest sense, the subject of landscape ecology is heterogeneity in ecological systems (Milne 1991). This problem can be attacked at various spatial and temporal scales and various levels of organization, but landscape ecology is most often focused on a scale of kilometers. Thus, landscape ecology studies the nature, distribution, and abundance of patches; the dynamics of patches; and the local and distant interactions of patches (Forman and Godron 1986).

As a final example, the new paradigm permits the inclusion of humans in the scope of basic ecology. Once the openness of natural systems and their interaction with natural disturbances are recognized, it is a short logical step to include humans as agents of flux and disturbances in ecological systems. Ecologists and policymakers are now aware of the action of humans at a distance, as, for example, in acidic deposition, or the effect on migrant organisms. Furthermore, ecology is now beginning to unembarrassedly examine human populated areas in the same way that it might approach any ecological system—as an arena to generate new patterns, to examine the structure and function of ecological systems, and to test general theories (McDonnell and Pickett 1990). Anthropogenic disturbances are being compared with natural ones, and the combined effects of the two agents are being integrated in surprising new ways. Finally, subtle (or unexpected) human effects are being examined by ecologists. These include causes of system organization that were established or triggered in the past, such as unrecorded selective removal of a species from a forest. Another kind of subtle effect can be the result of indirect interactions, as in the case of trophic cascades affecting primary productivity in aquatic systems triggered by fishing or stocking (Carpenter, Kitchell, and Hodgson 1987). Distant effects mediated by introduced organisms or by fluxes of chemicals in air or water are often initially subtle effects. The ozone hole is now well known but was previously unsuspected, and therefore a formerly subtle effect.

The implications for ecology of these new perspectives arising from the contemporary paradigm are great. We can reduce the implications and insight to a slogan: For ecology to understand natural systems, ecology must focus on process and context.

THE CONTEMPORARY PARADIGM AND CONSERVATION

The simplest translation of the contemporary paradigm into conservation biology is to translate its ecological slogan to this applied science. To conserve systems effectively, conservation biologists must focus on process and context. Following this advice requires conservation biology to look above the species level. This advice may best be illustrated by an example. To return to the case of Mettler's Woods, the old and new paradigms can be contrasted in two statements of conservation goals:

> *Old Paradigm*: We want to preserve the old-growth Mettler's Woods and will do so by establishing the Hutcheson Memorial Forest Center.
>
> *New Paradigm*: We want to maintain the integrity of the processes that have generated the old-growth Mettler's Woods and will establish and manage the Hutcheson Memorial Forest Center toward this end.

These two statements differ by focusing on a thing or place in the first case, and on the processes and their context in the second. In the remainder of this section, we will highlight some of the key processes and aspects of ecological context that are suggested by the second statement of conservation intent. Note that to determine what the relevant processes and context are, a clear conservation target must be in mind. Such a target could be any ecological system, including a population, a species, a community, an ecosystem, or a landscape.

The conservation of processes has, of course, been a major component of the relatively new science of conservation biology (White and Bratton 1980). However, most of the attention has been given to processes that reside at or below the population level of organization (e.g., Western and Pearl 1989). Genetics and breeding biology have been perhaps the most widely recognized processes relevant to conservation biology (Schoenwald-Cox et al. 1983). There are numerous excellent examples of studies clarifying and predicting the conservation needs in this realm of process.

Another process that is already widely recognized but needs development and still wider application is the role of herbivores on vegetation.

The exclusion of herbivores was at one time thought to be a requirement for the conservation of certain systems (Botkin 1990). There is no doubt that overgrazing or grazing by inappropriate organisms or at inappropriate times of the year can be extremely detrimental to a conservation effort. It took longer to recognize that conserving some landscapes or species actually required some degree of grazing (McNaughton 1989).

As we saw in the case of Mettler's Woods, and as is well documented in prairies and certain coniferous forests, a certain regime of fire is a critical ingredient in the recipe for successful conservation in many circumstances (Heinselman 1973). The Yellowstone ecosystem, the redwoods, chaparral, and various pine barrens are all well-known examples of natural systems that require episodic fires.

The two processes of fire and herbivory point to a generalization about successful conservation and land management. In any case where vegetation is involved as a part of the conservation project, the process of succession must be considered. All vegetation is capable of changing either compositionally or structurally through time as the result of interactions within the patch and between the patch and its surroundings. Luken (1990) suggests quite wisely that one does not manage vegetation, but rather manages the process of vegetation change, or succession. In the context of conservation, we could equally well say that one does not conserve vegetation, which is a thing, but rather one is attempting to conserve a dynamic. This successional or, more broadly in recognition of the contemporary paradigm, patch dynamic perspective gradient suggests that it is critical to know where on a successional gradient the target species or patch type is maximized, and where on that gradient it declines or is absent (Figure 4.2). In other words, the dynamic requirements of the entity targeted for conservation must be known.

Once the dynamic requirements of the target species, community, ecosystem, or landscape are known, management as well as preserve design must ensure that that dynamic state is present and available to the entity of interest. Management must provide either early, middle, or late successional habitat, as required (Foster 1980). For example, many mesic forest herbaceous perennials appear to require some intermediate-late phase of canopy closure but decline in very old forests (Whitford 1949; Collins, Dunne, and Pickett 1985). Therefore, exclusion of disturbance that opens the canopy would reduce the density and richness of such species.

The importance of managing processes rather than species is especially apparent when more than one target species exists in a habitat and their dynamic requirements are quite different. An example of this type comes from the California Ione soil formation, an unusual exhumed tropical oxisol that supports a number of restricted, endemic plant spe-

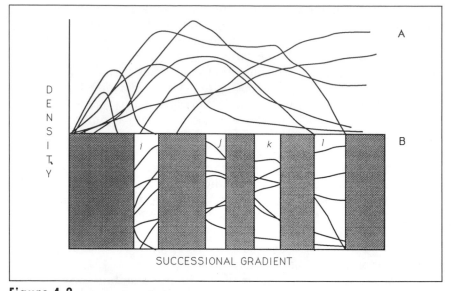

Figure 4.2
The successional niches of various species and their relationship to the successional patch structure of a landscape. **A.** *The niches of several species arrayed across a complex environmental gradient that appears through successional time.* **B.** *The distribution of species in a landscape having a specific distribution of successional patches* i *through* l. *A landscape having a different mix of successional patches would have a different mixture of species and abundances.* Successional *refers to the entire range of the process, including the idealized competitive equilibrium with which succession can end.*

cies, several of which are listed by the state of California as rare or endangered (Smith and Berg 1988; State of California 1990). One species is a chaparral shrub, the Ione manzanita (*Arctostaphylos myrtifolia*), and a second is a small perennial herb, the Ione buckwheat (*Eriogonum apricum*). These two species are sympatric, with the buckwheat occupying open spaces between stands of the larger shrubs or areas where the shrubs have died. When this chaparral community burns, surviving buckwheat plants quickly disperse seed into the newly opened habitat. In contrast, the manzanita subsequently reestablishes more slowly by seed from persistent seed banks. As the manzanitas increase in size over several decades and shade the smaller buckwheat plants, the latter populations decline. Management that maximizes the long-term persistence of one of these species necessarily results in the population decline of the other because they exist together dynamically, taking advantage of the same habitat at different times of the fire regime (Wood and Parker 1988). Therefore, management of just two rare plant species in this habitat requires a comprehensive approach that maintains key system processes like the precise natural fire regime, an approach that recognizes how the regeneration and long-term maintenance of populations are tied to short-term fluctuations in the system dynamics.

Another aspect of the successional approach to successful conservation is recognizing the need for certain episodic events as regeneration opportunities (Loucks 1970; Grubb 1977). In some swamps, periodic drawdown is required for successful regeneration of even the swamp forest dominants. We have already presented examples of how periodic fire of certain intensities can be associated with limited regeneration opportunities. Oddly, this whole lesson of designing and managing succession rather than a potentially static vegetation should have been learned much earlier, since succession is one of the oldest concepts in ecology. However, we suggest that the classical paradigm inhibited such a dynamic approach.

Conservation of context is the second strategy suggested by the contemporary paradigm. Resources, shelter, or sources of propagules for a target species may all reside in an adjoining patch to the one in which the organism is most obvious or usually found. An excellent example is that of the requirement of tropical passion flower (*Passiflora* spp.) vines for pollinators that garner a necessary portion of their resources from neighboring vegetation (Gilbert 1980). The classical example of wildlife species requiring one habitat for shelter and a second for forage is well known.

Water and wetland resources that flow between natural and developed areas provide perhaps the clearest example of context and certainly are one of the most controversial resources in land tenure worldwide. Moyle and Leidy (Chapter 6) provide a very clear example of context resource conflict in their discussion of the diversion of the Truckee River on the eastern slope of the Sierra Nevada and its devastating effects on the now-extinct Lahontan cutthroat trout (*Oncorhynchus clarki henshawi*) and the presently endangered cui-ui (*Chasmistes cujus*) of Pyramid Lake, Nevada.

An additional wetland example comes from the Florida Everglades. Everglades National Park is a 5670 km² reserve of freshwater and estuarine marsh and swamp habitat. It is bounded by undeveloped areas, such as Big Cypress National Preserve and three Everglades marsh reserves. Together these regions comprise over 10,000 km² (Kuslan 1979). Wading birds, including sixteen species of herons, bitterns, ibises, storks, and spoonbills, were abundant historically within this landscape. During the late nineteenth and early twentieth centuries, populations of these wading birds were drastically reduced, largely due to the hunting of feathers for ladies' fashions. More recently, in the 1960s, construction of a levee system that nearly encloses the Everglades National Park altered the flow of surface water into the park from upgradient areas. Specifically, hydrological alterations canalized water flows, moved discharge points out of the natural center of the Everglades ecosystem, released water on a regular (and not naturally variable) monthly delivery basis, and altered discharge–water level relationships (Kuslan 1979).

Kuslan (1979) reported severe reproductive declines in two birds from Everglades National Park, the white ibis (*Eudocimus albus*) and the wood stork (*Mycteria americana*), that correlated with these external hydrologic modifications. The white ibis, the most numerous species nesting in large colonies within this ecosystem, responded to the hydrologic modification by shifting colony sites from inside the park to outside areas where food was more available. The wood stork did not and behaviorally could not shift nesting locations within the park, and therefore its nesting population has declined significantly since the mid-1960s. In fact, the wood stork was reported to have nested successfully in Everglades National Park only twice between 1970 and 1979. In summary, Kuslan (1979) emphasized that design criteria, in addition to areal requirements, must incorporate environmental heterogeneity and maintenance of a reserve's functional characteristics in toto—that is, the ecological integrity of the reserve and its landscape context.

Less well appreciated as a conservation problem is the embedding of conservation areas in urban or suburban landscapes (McDonnell and Pickett 1990). Because so many of our important or best-loved conservation sites are near human population centers, the problem of urban sprawl, suburban spread, the conversion of agricultural to industrial or residential land, and the like can have serious implications for conservation. However, because ecologists have paid little attention to studying the effects of urbanization on natural areas, there is almost no empirical base for recommendations. Rather, we can only point to illustrative questions. For example, how will a nature reserve be affected by the conversion of neighboring agricultural land to housing, with its attendant increase in foot traffic, pets, subsidies of certain medium-sized predators, and introduction of horticultural plants and diseases? What management tactics can compensate for the appearance of a new land use adjacent to a nature reserve? How can local ordinances, traffic patterns, or changes in social attitudes in new suburbs or towns affect the management of processes necessary for conservation? Here, clearly, is a rich area for exploration by conservation biologists.

The policy of wildfire suppression in vegetation dependent on fire for its maintenance, for example, has its origin in various cultural and economic histories. In urban-wildland mosaics such as southern California urban and chaparral lands, processes that maintain the local native vegetation have been substantially altered by a century of fire suppression. Due in part to unnatural changes in the landscape's context, fire suppression, and habitat fragmentation, wildfires in southern California now have an extremely low probability of occurring naturally or, for human safety, are not permitted to burn. Interestingly, this policy may have the unique distinction of reducing a previously more variable fire regime to one with only large, catastrophic fires that endanger urban areas (Minnich 1988, 1989). Fire must be brought back into these systems, however, in order to maintain or restore the ecosystem's dynamics. But at urban boundaries, "controlled" burning in place of naturally occurring wildfires may not achieve the desired conservation objectives in fire-prone vegetation if it occurs outside of the historic fire regime (Parker 1987, 1989, 1990). Fire regime processes as a conservation issue can also occur in seemingly pristine areas like national parks, away from the urban areas, but still be strongly influenced by them (Elfring 1989; Christensen et al. 1989). The response of the general public, for example, can be dramatically negative when "objects" are placed in danger, as in the case of the 1988 fires in Yellowstone National Park (Elfring 1989).

The overall lesson that conservation biology must be responsive to process and context suggests that even well-established and successful reserves must be considered to be parts of conservation *systems* (Figure 4.3). One aspect of such a system is the geographic relationships between reserves (Harris and Eisenberg 1989). How to connect reserves or compensate for disrupted biotic communication among them in a changing, human-dominated landscape must be considered (Western 1989). Clustering reserves would, under such circumstances, have considerable value. Note that functional connections are the point, not mere spatial distances.

An additional important part of a conservation system cannot be seen on a map. As suggested previously, how adjoining land use, politics, and social attitudes affect the management and success of a reserve must be dealt with. Laws, zoning, education, and good neighborliness are examples of processes that must be considered just as important as the biological processes of succession, natural disturbance, and canopy regeneration.

Natural disturbance must be considered part of a conservation system. This is, of course, recognized in the concept of minimum dynamic area (Pickett and Thompson 1978) and the provision of refuges against catastrophic disturbances. Subtle or infrequent human disturbances should also be considered. Active management is not ruled out by the contemporary paradigm, as it was under the old paradigm.

Managing for processes rather than "objects" most often will demand a new concept of what is being preserved and managed (Christensen

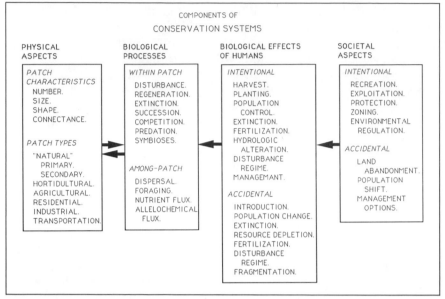

Figure 4.3
*Components of conservation systems indicating that successful conserva-
tion efforts will not only include the physical attributes and geographic
relationships of the nature reserves in an area but also will include settled
and production patches, account for a variety of biological processes that
occur within and among patches, and explicitly incorporate both the direct
and institutional impacts of humans. Note that the accidental impacts as
a result of direct human use or regulation and management of patches
not included in the reserve system are also a part of the overall conservation
system. Societal structures and regulations that are seemingly irrelevant
to conservation concerns can actually have a significant role in the larger
conservation system.*

1988). Not all the natural processes will come into play in nature preservation and management due to previous and ongoing human modification of the surrounding environments, especially near urban areas.
Managers will have to impose themselves on their systems at some point
in time or space.

A subset of considerations of the place of natural disturbance in a
conservation system is what to do when your treasured forest blows over,
or burns, or whatever. In the case of catastrophic wind, the immediate
temptation to salvage timber for either economic expediency or to tidy
up the site should be seriously resisted. There are several reasons for this.
First, salvage operations often scarify the soil surface in such a way as
to increase the success of invasive exotics or to extend the early succes-

sional phase. If in fact the desire of the conservation effort is for a mid-
or late-successional community, salvaging can be counterproductive. Second, the occurrence of natural disturbance provides a great opportunity
to educate the constituency of the reserve. Few people are able to appreciate the biological importance of episodic events firsthand because
they are infrequent at any particular spot. In addition, such catastrophic
events provide many people a rare opportunity to see *wilderness as a
process* in parts of the country where *wilderness as an entity* (e.g., a virgin
forest) is never to be seen again.

People and reserves are not anathema under the contemporary paradigm. Thus, the growing efforts to design reserves as rings of different
intensities of resource or recreational use of resources, extractive reserves,
reserves that incorporate agricultural patches, and reserves in metropolitan regions are all promising new approaches (Harris and Eisenberg
1989). The complexities of compromise and management required in
such situations are, of course, quite challenging.

PUBLIC KNOWLEDGE OF THE
NEW PARADIGM

What should the public and policymakers know about the new paradigm?
How can the complexities of the new views and insights be communicated
to nonspecialists? Indeed, not all specialists are fully aware of the profound changes in the ecological paradigm and its practical implications.

One major problem is that the old paradigm is still very much alive
and well in some quarters, and it resonates with the widely held cultural
bias of the balance of nature. Thus, in some cases, presentation of the
new paradigm can be difficult. The balance of nature concept and the
parallel classical paradigm take the pressure off humankind in its dealings
with nature. If nature will restore its balance, then all we have to do is
leave it alone, or tread lightly (Oelschlaeger 1991). A way to summarize
this view is that "nature has only one way to be." This idea, and its
associated false comfort, must be replaced. The classical paradigm has
led to demonstrably wrong conservation and management decisions (Botkin 1990).

The public and policymakers should know there is a new paradigm
in ecology. They should be given its metaphor, for instance, patch dynamics or the shifting mosaic. If one could wish for a replacement for
the all-too-comfortable and entrenched metaphor of the balance of nature,
it might be something like "the flux of nature." There is an immense
educational mission ahead.

The new paradigm in ecology can, like so much scientific knowledge, be misused. If nature is a shifting mosaic or in essentially continuous flux, then some people may wrongly conclude that whatever people or societies choose to do in or to the natural world is fine. The question can be stated as, "If the state of nature is flux, then is any human-generated change okay?"

The answer to this question is a resounding "No!" And the resonance is provided by the contemporary paradigm and the ecological knowledge that underwrites it. Human-generated changes must be constrained because nature has *functional, historical,* and *evolutionary* limits. Nature has a range of ways to be, but there is a limit to those ways, and therefore, human changes must be within those limits.

The limits to natural ways to be are set by functional constraints. There are physiological limits to what organisms, either alone or in some aggregation, can tolerate. Such limits are indicated by predictable declines in species richness or ecosystem productivity in certain environments, most often those with extreme levels of resources, environmental regulators such as temperature, or limited times for activity. These physiological limits are set by the chemistry and physics of the materials of which organisms are constructed. Such limits are fixed.

The second kind of limit is an evolutionary limit. Within the basic physiological limits to life, there may or may not be organisms that have arisen with the specific set of characteristics that permit them to tolerate a specified environment and deal with the biological interactions already present. This limit is affected by evolutionary processes such as the rate of mutation and the capacity to limit dilution of favorable new gene combinations. The particular combination of fortuitous events may or may not have occurred or been frequently repeated. In short, evolution may not have generated one or more organisms that can deal with any particular environment. The multiple probabilities involved take long times to accumulate.

The final kind of limit is a historic limit. Given that the environment of a situation is within the physiological tolerances, and given that evolution has produced a suite of reproductively isolated (or otherwise defined) species capable of dealing with that environment, the relevant questions are: Have the species reached the site, and have the competitively dominant ones excluded others that might tolerate the environment? The answer to these questions is determined by the specific history of the site.

These fundamental limits of organisms and the ecological systems containing them suggest that there are two limits humans must be aware of in their impacts on the flux of nature. The rates of natural change in natural systems can accommodate changes, sometimes catastrophic ones, in natural environments. Even though the surface of the Earth has been subject to repeated clusters of glacial advances and retreats, the continents skitter about on its surface, opening and closing seas and continental shelves, and apparent asteroid impacts may have caused global changes in environment and species complement, the planet is populated by a rich and diverse biota. This is a remarkable fact given that more than 90 percent of species that have ever lived are now extinct (Raup 1986)! The point is that evolution of new species and migration of species to tolerable sites has been able, over geologic time, to keep up with the immense and extensive changes in the Earth's environment.

Now, however, humans work their changes over continental and global scales at immensely faster rates. The mass extinctions of the geologic past are arguably phenomena of hundreds of thousands of years. Even the migrations of organisms after glacial and other climatic changes are on the order of thousands or hundreds of years. Anthropogenic changes are now numbered in decades or merely a few years.

In addition, the evolutionary and historical changes before the global spread of human technological effects were played out on an unobstructed stage. The relatively rapid postglacial migrations of trees in North America (Davis 1983) occurred in a landscape not cluttered with cities, highways, canals, suburbs, office parks, malls, and biocided agricultural land. And the refuges to which organisms might repair, in which the chance mutations and reproductive isolation might occur, are surely less large and less common than in the past.

Clearly, then, the new paradigm in ecology, and its lay translation into the flux of nature, does not license environmental abuse. The processes of nature that have generated the sorts of communities, ecosystems, and landscapes we wish to conserve have taken long times and large spaces. If our nature conservation is to be successful, indeed, if our tenure on this planet is to be successful, then we must respect the limits of physiology to deal with toxins and extreme climatic conditions, the limits of evolution to generate new gene combinations that are new ways of dealing with environment and that are recognized as new species, and the limits of history to assemble organisms into biological systems that deal with the environment at particular spots on the Earth's surface. Conservation biology is the basic science that must generate the knowledge for dealing with and, where possible, compensating for these limits. It must design and manage conservation systems by including both biology and politics and educating the public about these limits and the implications of their personal and political choices impinging on those limits. The new paradigm in ecology both frames the basic scientific tools and provides the justification for this task.

CONCLUSIONS

The classical paradigm in ecology, with its emphasis on the stable state, its suggestion of natural systems as closed and self-regulating, and its resonance with the nonscientific idea of the balance of nature, can no longer serve as an adequate foundation for conservation. The new paradigm, with its recognition of episodic events, openness of ecological systems, and multiplicity of locus and kind of regulation, is in fact a more realistic basis for conservation planning and management. The new paradigm suggests scientific metaphors that are properly embedded in explicit theories not presented here, such as patch dynamics and the shifting mosaic. A nontechnical metaphor of "the flux of nature" might ultimately replace the outmoded metaphor of "the balance of nature," a concept loosely associated with the discarded classical paradigm in ecology.

The lesson of the new paradigm for conservation biology is that scientists, policymakers, and managers must know and understand much more about the target entities they desire to conserve. Thus, for the effective practice of conservation biology, much more needs to be known about *process* and *context*. Indeed, it is the processes that have generated or that maintain the species, community, ecosystem, or landscape, and the spatial context and the functional connections within that context that must be maintained. Conservation of isolated target species or other ecological entities is not likely to be successful. Rather, conservation above and beyond the species level must become increasingly employed.

To recapitulate and amplify, conservation biologists must know and apply the following things to their problems:

1. The processes governing the system
2. The context in which it is embedded
3. The historical range of flux in the system
4. The evolutionary and physiological limits of the organismal components of the system
5. The nature and impact of episodic and long-term phenomena, including the nonobvious or subtle roles of people in the past and present

ACKNOWLEDGMENTS

We thank the students at San Francisco State University for stimulating discussions in the formative period of this paper. John A. Reed kindly read an early version. This essay represents a contribution to the program of the Institute of Ecosystem Studies with partial financial support from the Mary Flagler Cary Charitable Trust.

LITERATURE CITED

Aplet, G., R.D. Laven, and P.L. Fiedler. In press. The relevance of resouce management to conservation biology. *Cons. Biol.*

Bormann, F.H., and G.E. Likens. 1979. *Pattern and process in a forested ecosystem.* New York: Springer-Verlag.

Botkin, D.B. 1990. *Discordant harmonies: A new ecology for the twenty-first century.* New York: Oxford University Press.

Botkin, D.B., and M.J. Sobel. 1975. Stability in time-varying ecosystems. *Am. Nat.* 109:625–46.

Buell, M.F. 1957. The mature oak forest of Mettler's Woods. *Bull. Wm. L. Hutcheson Mem. For.* 1:16–19.

Carpenter, S.R., J.F. Kitchell, and J.R. Hodgson. 1987. Cascading trophic interactions and lake productivity. *BioScience* 35:634–39.

Christensen, N.L. 1988. Succession and natural disturbance: Paradigms, problems, and preservation of natural ecosystems. In *Ecosystem management for parks and wilderness.* ed. J.K. Agee and D.R. Johnson, 62–86. Seattle: University of Washington Press.

Christensen, N.L., J.K. Agee, P.F. Brussard, J. Hughes, D.H. Knight, G.W. Minshall, J.M. Peek, S.J. Pyne, F.J. Swanson, J.W. Thomas, S. Wells, S.E. Williams, and H.A. Wright. 1989. Intepreting the Yellowstone fires of 1988. *BioScience* 39:678–85.

Clements, F.E. 1916. *Succession.* Washington, D.C.: Carnegie Institution of Washington.

Collins, B.S., K.P. Dunne, and S.T.A. Pickett. 1985. Responses of forest herbs to canopy gaps. In *The ecology of natural disturbance and patch dynamics,* ed. S.T.A. Pickett and P.S. White, 218–34. Orlando: Academic Press.

Connell, J.H., I.R. Noble, and R.O. Slatyer. 1987. On the mechanisms producing successional change. *Oikos* 50:136–37.

Davis, M.B. 1983. Holocene vegetational history of the eastern United States. In *Late Quarternary environments of the United States: Vol. 2, The Holocene,* ed. H.E. Wright, 116–81. Minneapolis: University of Minnesota Press.

Ehrenfeld, D.W. 1972. *Conserving life on Earth.* New York: Oxford University Press.

Elfring, C. 1989. Yellowstone: Fire storm over fire management. *BioScience* 39:667–72.

Forman, R.T.T., and M. Godron. 1986. *Landscape ecology.* New York: John Wiley & Sons.

Foster, R.B. 1980. Heterogeneity and disturbance in tropical vegetation. In *Conservation biology: An evolutionary-ecological perspective,* ed. M.E. Soulé and B.A. Wilcox, 75–92. Sunderland, Mass.: Sinauer.

Gilbert, L.E. 1980. Food web organization and the conservation of neotropical diversity. In *Conservation biology: An evolutionary-ecological perspective,* ed. M.E. Soulé, and B.A. Wilcox, 11–33. Sunderland, Mass.: Sinauer.

Gleason, H.A. 1917. The structure and development of the plant association. *Bull. Torrey Bot. Club* 43:463–81.

Grubb, P.J. 1977. The maintenance of species-richness in plant communities: The importance of the regeneration niche. *Biol. Rev. Cambridge Phil. Soc.* 52:107–45.

Harris, L.D., and J.F. Eisenberg. 1989. Enhanced linkages: Necessary steps for success in conservation of faunal diversity. In *Conservation for the twenty-first century*, ed. D. Western and M.C. Pearl, 166–81. New York: Oxford University Press.

Heinselman, M.L. 1973. Fire in the virgin forests of the Boundary Waters Canoe Area, Minnesota. *Quart. Res. (N.Y.)* 3:329–82.

Jackson, J.B.C. 1981. Interspecific competition and species distribution: The ghosts of theories and data past. *Am. Zool.* 21:889–901.

Kuhn, T.S. 1962. *The structure of scientific revolutions.* Chicago: University of Chicago Press.

——. 1972. *The structure of scientific revolutions*, 2nd. enlarged ed. Chicago: University of Chicago Press.

Kuslan, J.A. 1979. Design and management of continental wildlife reserves: Lessons from the Everglades. *Biol. Cons.* 15:281–90.

Loucks, O.L. 1970. Evolution of diversity, efficiency and community stability. *Am. Zool.* 10:17–25.

Luken, J.O. 1990. *Directing ecological succession.* New York: Chapman and Hall.

McDonnell, M.J., and S.T.A. Pickett. 1990. The study of ecosystem structure and function along gradients of urbanization: An unexploited opportunity for ecology. *Ecology 71*:1232–37.

McNaughton, S.J. 1989. Ecosystems and conservation in the twenty-first century. In *Conservation for the twenty-first century*, ed. D. Western and M.C. Pearl, 109–20. New York: Oxford University Press.

Milne, B.T. 1991. Heterogeneity as a multiscale characteristic of landscapes. In *Ecological heterogeneity.* ed. J. Kolasa and S.T.A. Pickett, 69–84. New York: Springer-Verlag.

Minnich, R.A. 1988. The biogeography of fire in the San Bernadino Mountains of California. *Univ. Calif. Publ. Geogr.* 28:1–120.

——. 1989. Chaparral fire history in San Diego County and adjacent northern Baja California: An evaluation of natural fire regimes and the effects of suppression management. In *The California chaparral: Paradigms reexamined*, ed. S.C. Keeley, 37–47. Publication no. 34, Los Angeles: Natural History Museum Los Angeles County.

Oelschlaeger, M. 1991. *The idea of wilderness: From prehistory to the age of ecology.* New Haven: Yale University Press.

Oliver, C.D., and B.C. Larson. 1990. *Forest stand dynamics.* New York: McGraw-Hill.

Parker, V.T. 1987. Effect of wet-season management burns on chaparral regeneration: Implications for rare species. In *Proceedings of a conference on the conservation and management of rare and endangered plants*, ed. T.E. Elias, 233–37. Sacramento: California Native Plant Society.

——. 1989. Maximizing vegetation response on management burns by identifying fire regimes. In *Proceedings of a conference on fire and watershed management,* N.H. Berg, (technical coordinator, 87–91.) U.S.D.A. Forest Service General Technical Report PSW-109.

————. 1990. Problems encountered while mimicking nature in vegetation management: An example from a fire-prone vegetation. In *Ecosystem management: Rare species and significant habitats,* ed. R.S. Mitchell, C.J. Sheviak, and D.J. Leopold, 231–34. New York State Mus. Bull. no. 471. Albany, New York.

Pickett, S.T.A. 1989. Space-for-time substitution as an alternative to long-term studies. In *Long-term studies in ecology: Approaches and alternatives,* ed. G.E. Likens, 110–35. New York: Springer-Verlag.

Pickett, S.T.A. and J. Kolasa. 1989. Structure of theory in vegetation science. *Vegetatio* 83:7–15.

Pickett, S.T.A., and M.J. McDonnell. 1989. Changing perspectives in community dynamics: A theory of successional forces. *Trends Ecol. Evol.* 4:241–45.

Pickett, S.T.A., and J.N. Thompson. 1978. Patch dynamics and the design of nature reserves. *Biol. Cons.* 13:27–37.

Raup, D.M. 1986. Biological extinction in Earth history. *Science* 231:1528–33.

Romme, W.H. 1982. Fire and landscape diversity in subalpine forests of Yellowstone National Park. *Ecol. Monogr.* 52:199–221.

Schoenwald-Cox, C.M., S.M. Chambers, B. MacBryde, and L. Thomas, eds. 1983. *Genetics and conservation: A reference for managing wild animal and plant populations.* Menlo Park, Calif.: Benjamin/Cummings.

Simberloff, D. 1982. A succession of paradigms in ecology: Essentialism to materialism and probabilism. In *Conceptual issues in ecology,* ed. Esa Saarinen, 63–99. Boston: Reidel (Kluwer).

Smith, J.P. Jr., and K. Berg. 1988. *Inventory of rare and endangered vascular plants of California.* California Native Plant Society Special Publication no. 1, 4th ed. Sacramento, Calif: California Native Plant Society.

Soulé, M.E. 1989. Conservation biology in the twenty-first century: Summary and outlook. In *Conservation for the twenty-first century,* ed. D. Western and M.C. Pearl, 297–313. New York: Oxford University Press.

Soulé, M.E., and B.A. Wilcox, eds. 1980. *Conservation biology: An evolutionary-ecological perspective.* Sunderland, Mass.: Sinauer.

Sacramento, Calif.: State of California, Department of Fish and Game, Natural Heritage Division, Endangered Plant Program. 1990. *List of designated endangered or rare plants, March 1990.*

Sulser, J.S. 1971. Twenty years of change in the Hutcheson Memorial Forest. *Bull. Wm. L. Hutcheson Mem. For.* 2:15–24.

Temple, S.A., E.G. Bolen, M.E. Soulé, P.F. Brussard, H. Salwasser, and J.G. Teer. 1988. What's so new about conservation biology? In *Transactions of the 53rd North American Wildlife and Natural Resources Conference,* 609–12. Louisville, Kentucky: Wildlife Management Institute.

Vitousek, P.M., and P.S. White. 1981. Process studies in succession. In *Forest succession: Concepts and application,* ed. D.C. West, H.H. Shugart, and D.B. Botkin, 267–75. New York: Springer-Verlag.

Western, D. 1989. Conservation without parks: Wildlife in the rural landscape. In *Conservation for the twenty-first century,* ed. D. Western and M.C. Pearl, 158–65. New York: Oxford University Press.

Western, D., and M.C. Pearl, eds. 1989. *Conservation for the twenty-first century.* New York: Oxford University Press.

White, P.S. 1979. Pattern, process, and natural disturbance in vegetation. *Bot. Rev.* 45:229–99.

White, P.S., and S.P. Bratton. 1980. After preservation: Philosophical and practical problems of change. *Biol. Cons.* 18:241–55.

Whitford, P.W. 1949. Distribution of woodland plants in relation to succession and clonal growth. *Ecology* 30:199–208.

Whittaker, R.H., and S.A. Levin. 1977. The role of mosaic phenomena in natural communities. *Theor. Pop. Biol.* 12:117–39.

Wood, M.K., and V.T. Parker. 1988. Management of *Arctostaphylos myrtifolia* at the Apricum Hill Ecological Reserve. Unpublished report to the California Department of Fish and Game, Region 2, and Endangered Plant Project, Sacramento, California.

PART

II

. . .

PROCESSES AND PATTERNS OF CHANGE

● ● ●

CHAPTER
5

Reptilian Extinctions:
The Last Ten Thousand Years

TED J. CASE

DOUGLAS T. BOLGER

and ADAM D. RICHMAN

ABSTRACT

We summarize what is known of reptilian extinctions worldwide over the last 10,000 years. Two patterns are evident: the great majority of these extinctions have occurred on islands, and they are usually due to human-related disturbance. The effects of human-related disturbance are calibrated by measuring the rate of extinction for two sets of Holocene landbridge islands where human impact has been minimal. Extinction rates for islands with a history of human habitation are also determined for comparison. The results of this investigation indicate that human-related disturbance increases extinction rates by roughly an order of magnitude for smaller islands. Interestingly, very large islands and continental areas exhibit lower rates of extinction than predicted from the landbridge island data. More detailed consideration of the cause of species extinction in particular cases strongly implicates introduced predators, chiefly mongoose, rats, cats, and dogs as the agents of many human-related extinctions, whereas competition with introduced reptiles appears to have had little impact on native species.

INTRODUCTION

The fossil record of the earth shows that faunal and floral extinctions increased dramatically during certain periods. These "paleo" upheavals like those at the end of the Permian and Cretaceous have long provided the punctuations that geologists and paleontologists use to divide the geological periods. A challenging question in conservation science is whether the processes affecting extinction rates today are helpful in interpreting extinction in the past, and conversely, whether prehistoric extinctions are useful for understanding recent extinctions.

One overriding pattern affecting historical extinctions that may not be true for prehistoric extinctions is that they are concentrated on islands. Diamond (1984) has recently summarized the modern extinctions of birds and mammals from compilations in IUCN Redbooks. For birds, 171 species and subspecies have gone extinct since about 1600, and over 90 percent of these extinctions have occurred on islands. For mammals, out of 115 documented historical extinctions, 36 percent of these have occurred on islands. The smaller proportion of island extinctions for mammals is in part simply a reflection of their poor representation on islands relative to birds. Many islands (e.g., New Zealand, Hawaii, Fiji, the Mascarenes, and the Seychelles) with large numbers of bird species and many avian extinctions simply have no native mammals except for bats.

Unfortunately, the IUCN Redbook is not yet complete for reptiles (Honegger 1975). Here we attempt to tally many of the historical and Holocene (Recent) prehistorical extinctions and compare the emerging pattern with that for birds and mammals. We find that, as with birds, the proportion of island extinctions compared to continental extinctions is very high. This pattern is in part an unsurprising consequence of island populations. They are small and isolated; thus they cannot recover from local extirpation following environmental perturbations or long-term climatic changes by immigration from other areas (MacArthur and Wilson 1967; Leigh 1981).

A growing body of evidence for birds and mammals suggests that over the last few thousand years, the most important agent of directed change in the environment is not climatic change but human disturbance and alteration of habitats (see Diamond (1984) for recent review). Most extinctions of entire species in recorded history are attributable to some aspect of human intervention. For example, paleontological investigations in the West Indies and Pacific link the extinction of numerous species of vertebrates with human colonization of these islands in recent prehistory (Steadman et al. 1984; Olson and James 1982; Steadman and Olson 1985). For birds and mammals, the major mechanisms are habitat destruction; human hunting; effects of introduced taxa, particularly predators, and trophic cascades (i.e., secondary extinctions caused by previous extinctions; Diamond and Case 1986). Here we look for the generality of these findings by evaluating the evidence for the human impact on Holocene reptilian extinctions.

ISLAND REPTILES AND THE PREHISTORICAL LEGACY

Evidence for extinctions of reptiles in historical time is more fragmentary than for other taxa. For example, while we have a specimen of the dodo from Mauritius residing in a museum, the contemporaneous giant skinks also from Mauritius are known only from subfossils. Careful taxonomy and biogeographic documentation of reptiles lagged somewhat behind that for birds and mammals; consequently, early extinctions of reptiles may have gone without detection. Because reptiles are not as generally conspicuous or noisy as birds, they often pass unnoticed even when they are relatively plentiful. Thus, we must rely more on subfossil evidence for extinctions rather than accurate taxonomic descriptions of extant species. It is not often easy to pin an exact date on a species' extinction, and therefore we are forced to rely on an accumulation of evidence rather than a single survey. For these reasons, choosing the year 1600 as a starting point for historical extinctions, as was done with birds and mammals, is rather arbitrary, and we will review all extinctions dating over the Holocene (or Recent), about the last 10,000 years. In what follows we use the term *prehistoric* to refer to extinctions that occurred prior to the arrival of Europeans to the locality, and so the exact dates delimiting this period vary from place to place.

In the rest of this chapter we will focus more closely on the big questions raised in this introduction. What are the geographic patterns in reptilian extinctions? Are extinctions less common on continents and on large islands than on small islands, as predicted by theory and demonstrated for bird extinctions? We will also explicitly examine the effect of the presence of man on island extinction rates and shed some light on the mechanisms by which humans impact reptile populations. Specifically, we will consider the effect of human-introduced predators and competitors. How much of a role do they play relative to habitat destruction, and is the evolutionary "predator naïveté" of island species important?

A WORLD TOUR OF HOLOCENE REPTILIAN EXTINCTIONS

Continents

The most striking observation about Holocene reptilian extinctions in continental North America is that few occurred. Although the mammalian megafauna was severely depleted, only three reptilian extinctions are known out of perhaps 130 fossil species known for the continental United States since the Pleistocene: a large tortoise (*Geochelone wilsonii*), a horned lizard (*Phrynosoma josecitensis*), and a largish rattlesnake (*Crotalus potterensis*) (Moodie and Van Devender 1979; Gehlbach 1965; Estes 1983). As climates changed and plant communities shifted, reptiles underwent local extirpations, range contractions, or range expansions. These sometimes led to drastic changes in species associations of reptiles (Van Devender 1977, 1987; Van Devender and Mead 1978), but surprisingly only these three extinctions.

The situation is similar for mainland Australia. The largest varanid lizard in the world, *Megalania*, which dwarfed the extant Komodo dragons, went extinct in the Pleistocene, probably sometime after the entry of the Aborigines in Australia 30,000 to 50,000 years ago; exactly how recently is uncertain, but a date of 10,000 B.P. would not be unreasonable (Hecht 1975). The largest boids in Holocene times, the Australian *Wonambi*, became extinct sometime during the same period as did the Australian meiolanid horned tortoises (Molnar 1984a, 1984b). These three extinctions are the only late Pleistocene fossil forms that cannot be assigned unambiguously to living reptile species (Molnar 1984b).

Unfortunately, it is nearly impossible to even begin to make Holocene tallies for South America, Africa, and Eurasia. The extant fauna is not completely known, let alone those species that have failed to survive.

Islands

Islands can be grouped into three categories with respect to human settlement histories and thus to the possible influence of man on extinction.

1. Islands first colonized in prehistory by aboriginal people and then later colonized by Europeans. Many birds and mammals became extinct on these islands during the aboriginal period and are known only as subfossils (Martin and Klein 1984). This pattern also holds for reptiles.

In New Zealand, thirty-eight species of native reptiles are now known from the Holocene period. Three species of lizards are extinct—a species of *Cylodina* larger than any extant form and known only from subfossil deposits in Northland, and *Leiolopisma gracilocorpus* and *Hoplodactylus delcourti* known only from unique museum specimens (Bauer and Russell 1986; Worthy 1987a; Hardy 1977). *Hoplodactylus delcourti* is the largest known gecko, with a snout-vent length of 370 mm. Nine reptiles today are found only on the off-lying islands (five skinks, three geckos, and the tuatara). These include all the relatively large extant species (Hardy and Whitaker 1979). Evidence for a mainland distribution for the tuatara as recently as 1,000 years ago, and for some of the other species as well (Cassels 1984; Crook 1973), suggests that the present distributions are relictual. Only one species, *Leiolopisma fallai* of the Three Kings Islands (which are not landbridge islands but are much older), is regarded as a nonrelictual island endemic (Robb 1986). Since the other islands were connected to New Zealand at the end of the Pleistocene, these unique offshore species probably indicate mainland extinctions. In the case of *Cyclodina macgregori*, *C. alani*, and *Hoplodactulus duvaucelii*, subfossils indeed establish them as formerly occurring on the North Island as recently as 1000 A.D. (Worthy 1987).

The time of disappearance of the tuatara and most of the lizards coincides well with the date for human arrival on New Zealand and the subsequent introduction of the Polynesian rat. On islands where the rat is present, the tuatara is either absent or not breeding (Crook 1973). The three largest of six species of frog (*Leiopelma*) have gone extinct in New Zealand in the Holocene (Worthy 1987b), and the largest surviving frogs (*L. hamiltoni*) occur only on two rat-free islands.

Shifting to the Caribbean region, the pattern is similar. The large herbivorous iguanine *Cyclura* has gone extinct on a number of Caribbean islands in the recent past (Pregill 1981, 1986). The giant *Cyclura pinguis*, which probably became extinct on Puerto Rico in Holocene times, survives on the off-lying small island of Anegada. Two species of iguanids in the genus *Leiocephalus* (the curly-tailed lizards), *L. eremitus* and *L. herminieri*, became extinct in the last 100 years (Pregill 1992). Both occupied small islands in the Caribbean; *L. eremitus* is known only from the type specimen, a female 63 mm snout-vent (SV), which is moderate to large for the genus. *Leiocephalus herminieri* was very large (up to 140 mm SV). Six other relatively large species for the genus (reaching 200 mm SV) are known only from fossil material and probably became extinct during aboriginal occupation on Hispaniola, Jamaica, Puerto Rico, and the Barbuda bank, but other smaller species survive in the Bahamas, on Hispaniola, and Cuba (Pregill 1990).

The extinction of larger endemic forms occurs in other reptile groups in the Caribbean. The giant gecko (*Aristelliger titan*) disappeared from Jamaica sometime before European settlement (Hecht 1951). More recently, the very large legless lizard *Celestus* (*Diploglossus*) *occiduus* vanished from Jamaica. The last specimens were collected around fifty years ago. Greg Pregill (pers. com.) found ample fossil material dating no more than 800 years B.P. when presumably it was much more common. The giant anole, *Anolis roosevelti*, is known only from a few specimens from tiny Culebra Island off the east coast of Puerto Rico and has not been seen since 1932 (Pregill 1981). Finally in the Caribbean, we have Holocene fossils of giant tortoises (*Geochelone*) from the Bahamas, Mono Island, and Curaçao.

The Canary Islands in the East Atlantic are home to the largest lacertid lizards in the world and in the recent past were occupied by even larger species. In 1974, the large *Gallotia* (*Lacerta*) *simonyi*, long thought to be extinct, was rediscovered on Hierro (Böhme and Bings 1975). Hierro is the smallest major island of the Canaries and the most distant from the African mainland. Before the Spanish arrived in the fourteenth century, all the islands were occupied by an aboriginal people, the Guanches, whose ancestors probably arrived around 2,000 to 4,000 years ago (Schwidetzky 1976; Mercer 1980).

In Figure 5.1, we compare the extant lacertids to the fauna that probably existed before the arrival of humans. We restrict our attention to the five relatively mesic western islands where fossil forms have been collected. Four of these islands originally were inhabited by two or three *Gallotia* species in the late Pleistocene and Holocene. There was a small species (*G. galloti*) sympatric with a larger species (*G. simonyi* or *G. stehlini*), and/or a still larger *G. goliath* (Mertens 1942; Bravo 1953; Arnold 1973; Marrero Rodriguez and Garcia Cruz 1978; Hutterer 1985; López-Jurado 1985). The exception is Gran Canaria where no small *G. galloti* exists and none is evident in existing fossil deposits. At least on Tenerife, an even larger species, *G. maxima*, existed from probably the Pliocene to the early Pleistocene, although it is still unclear whether *G. maxima* evolved into *G. goliath* or went extinct. In any event, this very large form seems to have disappeared before the islands were colonized by humans (Bravo 1953).

The other large *Gallotia* species became extinct at the end of the Pleistocene or even more recently and many, if not all, of these extinctions were contemporaneous with human colonization. Fossils of the now extirpated *G. simonyi* on Gomera have been found at one 500 year old, pre-hispanic site (Hutterer 1985). Elsewhere fossil lizards are found in association with abundant human artifacts (Böhme et al. 1981; Bings 1985). There are also a few historical references to the presence of gigantic

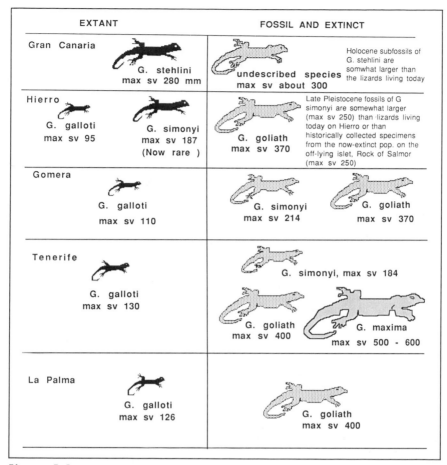

EXTANT	FOSSIL AND EXTINCT

Gran Canaria — G. stehlini, max sv 280 mm | undescribed species, max sv about 300 — Holocene subfossils of G. stehlini are somwhat larger than the lizards living today

Hierro — G. galloti, max sv 95; G. simonyi, max sv 187 (Now rare) | G. goliath, max sv 370 — Late Pleistocene fossils of G simonyi are somewhat larger (max sv 250) than lizards living today on Hierro or than historically collected specimens from the now-extinct pop. on the off-lying islet, Rock of Salmor (max sv 250)

Gomera — G. galloti, max sv 110 | G. simonyi, max sv 214; G. goliath, max sv 370

Tenerife — G. galloti, max sv 130 | G. simonyi, max sv 184; G. goliath, max sv 400; G. maxima, max sv 500 - 600

La Palma — G. galloti, max sv 126 | G. goliath, max sv 400

Figure 5.1

Body sizes (maximum snout-vent length, SVL) of extant and extinct lizards of the genus Gallotia (Lacerta) on the Canary Islands. The size data was gathered for extant species during field work by TJC in the Canaries during 1980 and were supplanted by measurements of preserved specimens at the British Museum and records in Machado (1985a) and López-Jurado (1989) for Hierro, and Thorpe (1985) for Lanzarote and Fuerteventura. Extinct species' maximum sizes are based on estimates in Mertens (1942), Bravo (1953), Marrero Rodriguez and Garcia Cruz (1978), Böhme et al. (1981), Hutterer (1985), Izquierdo et al. (1989), and López-Jurado (1985). The length of the lizards is drawn roughly to scale.

This figure illustrates the preferential extinction of larger forms, probably due to man, and the downward size shifts of surviving large species on Hierro and Gran Canaria. The extinct and extant species were all contemporaneous in the late Pleistocene and Holocene, with the possible exception of G. maxima.

lizards on Hierro, Gomera, and Gran Canaria (see review in Machado 1985b). We know that the early Canary Island aborigines hunted and ate lizards (Hooton 1925; Schwidetsky 1976; Bings 1985), but lizards were not a major portion of their diet. Indirect anthropogenic influences such as the introduction of rats, goats, pigs, and especially dogs by the aborigines may have been more important in the large lizards' eventual demise. Machado (1985b) attributes the exceptional survival of the large *G. stehlini* on Gran Canaria to the absence of any smaller-sized lacertid competitors in the face of introduced predators; he speculates that the tenuous survival of *G simonyi* on Hierro might be due to the apparent absence of Guanche dogs.

Madagascar is the largest nonpolar island in the world and, as with continents, reptilian extinctions are rare. The sole incidents are the giant tortoises (probably two species) that are known only from subfossils and probably became extinct around the same time as the giant elephant birds, after the arrival of humans in about 500 A.D. (Dewar 1984). Some authorities, however, believe the giant tortoises actually went extinct before human contact (Paulian 1984). Three smaller species still survive on Madagascar, and other giant tortoises survived until European settlement in the Seychelles and still survive today on Aldabra (Arnold 1976). The absence of fresh water on Aldabra restricted permanent human settlement, a gratuitous benefit to the tortoises. In the nearby Mascarenes (Mauritius, Reunion, and Rodrigues), at least six species of large tortoises also became extinct shortly after human contact (Cheke 1987; Arnold 1980).

Giant tortoises occurred on Sicily, Malta, and the Balearic islands in the Mediterranean, but their time of extinction is unclear. It may have occurred earlier than the arrival of humans (Reese 1989). The giant lizard *Lacerta siculimelitensis* (220 mm SV) occurred on Malta and Sicily but became extinct sometime toward the end of the Pleistocene (Böhme and Zammit-Maempel 1982). It is not known if it survived to the Holocene or if it was contemporaneous with humans. Recent excavations on nearby Cyprus, however, suggest an earlier date for human influence and a later date for the extinction of that island's megafauna than was previously thought (Reese 1989); here the megafauna and humans are known to have been contemporaneous.

On Tonga in the South Pacific, Pregill and Dye (1988) recently found subfossils of an extinct large iguanine in the genus *Brachylophus* on the island of Lifuka. The length of these lizards is about twice that of the extant *Brachylophus* on nearby Fiji. The fossils are about 2,000 years old. They are directly associated with human artifacts and bear distinctive marks that testify to their use as human food. In New Caledonia, fossils of now-extinct large varanid lizards, giant meiolaniid turtles, and crocodiles occur near or with sediments containing human artifacts (Gaffney, Balouet, and DeBroin 1984; Rich 1982; Gifford and Shutler 1956).

2. Islands with a colonial period but no aboriginal history. Here, unique island species survived to be described as living species, only too often, to meet their demise shortly thereafter. For example, Rodrigues Island in the Indian Ocean experienced a period of intensive European settlement in the late seventeenth century. At that time large numbers of some spectacular endemic geckos were found. *Phelsuma edwardnewtonii* was a large diurnal species, bright green with blue spots. It was described as being so tame that it inhabited houses and would eat fruits from the owners' hands (Leguat 1708). However, the species was devastated by rats and cats on the main island around the mid-nineteenth century. It survived for a short time on small outlying islets but finally disappeared from these too as they became infested with rats. An even larger species, *P. gigas*, reaching nearly a half meter in total length, vanished from the main island prior to the disappearance of *edwardnewtonii*. It too survived on uninhabited off-lying cays only to disappear later when rats were introduced (Vinson and Vinson 1969).

A huge skink, *Leiolopisma mauritiana* (300 mm SV), inhabited nearby Mauritius. This species is known today only as subfossils, and the cause and chronology of its extinction is not known. A likely guess is that they went the way of their contemporary, the dodo, for much the same reasons.

Many of the other endemic reptile species on Mauritius became extinct after conversion of habitat to agriculture and the introduction of rats, cats, and other predators in the seventeenth century. The presence of these "missing species" was only recently confirmed from fossil deposits of quite recent age (Arnold 1980). Some species survived on satellite islands that lie on the same island bank and were connected in times of lowered sea level. Most important in this respect is rat-free Round Island, where four species survive that have gone extinct on Mauritius. This includes the three largest lizards known from Mauritius (*Phelsuma guentheri*, *Leilopisma telfairii*, and *Nactus sepensinsula*) and one snake (*Casarea dussumieri*) in a distinct group of primitive boas, the Bolyerinae. The only other species in the Bolyerinae (*Casarea dussumieri*) also occurs only on Round Island. Yet, since Round was connected to Mauritius less than 12,000 years ago, an extinction is implicated although confirming fossil evidence is so far lacking. The small skink, *Scelotes bojerii*, occurs on Round Island and surrounding islets but was known from Mauritius in the last century and was once thought to be extinct there. It was rediscovered, however, in the Macabé forest (Vinson 1973; Arnold 1980).

The Cape Verde Islands off the west coast of Africa are the home of the second-largest living skink, *Macroscinus coctei* (320 mm SV), the "end product" of a small adaptive radiation of *Mabuya* skinks dating back to the Cretaceous, when these islands were probably formed. The skinks have been described by residents as tame and easy to catch (Greer 1976). Perhaps this is why this giant species is now restricted to two tiny (total area 10 km²) uninhabited islands in the archipelago, Branco and Razo (Mertens 1956). These islands were severed from the other larger islands on the bank about 10,000 years ago with rising sea levels. Thus, the skink's absence from the larger adjacent islands suggests a recent extirpation. The islands were first colonized in the late fifteenth century by the Portuguese, who left no records of a broader range for this species. Interestingly, the gecko *Tarentola delalandei* is divided into two subspecies on the Cape Verdes. The large form *T. d. gigas* (max 125 mm SV) inhabits the same two islands as *Macroscinus*, whereas the substantially smaller form *T. d. rudis* (max 70 mm SV) occurs on most of the remaining islands in the Cape Verdes (Mertens 1956; Greer 1976).

None of the endemic reptile species of the Galapagos have become extinct, but population densities have declined and local extirpations have occurred in association with introduced cats, rats, dogs, and pigs (Honegger 1975). The land iguana (*Conolophus subcristatus*) is extinct on Baltra and James islands. The cause of extinction on James Island is not known; the species was abundant in Darwin's time but known only as fossils seventy years later in 1905–1906 during the California Academy of Science expedition. Although feral dogs are not now present, they were in the nineteenth century and perhaps drove the local extinction. Land iguanas on Santa Cruz were thought to have been exterminated by feral dogs before 1906, but small populations remained at Conway Bay, Cerro Colorado, and East Tortuga Bay. These populations persisted until they were heavily attacked by feral dogs in the 1970s. Today, only captive individuals remain. Giant tortoises (*Geochelone elephantopus*) have become extinct on Barrington and Floreana and are rare on all the other major islands except Isabela, Duncan, and Santa Cruz (Steadman 1986; Kramer 1984; Thorton 1971).

3. Islands with no permanent human settlement to date. Usually these islands are too small or too bleak and isolated to support human settlement (e.g., tiny islets in the Caribbean and Pacific, Malpelo Island off South America, most of the desert islands in the Sea of Cortez or off arid Australia, and many polar islands). Many of these islands are not well studied for obvious reasons, and the high-latitude islands are too cold to support any reptiles. Moreover, such islands usually support few endemics, so local extirpations do not result in the extinction of a species.

Yet these islands are extremely important for calibrating the magnitude of natural extinctions apart from the effects of human disturbance.

Richman, Case, and Schwaner (1988) used some relatively undisturbed arid landbridge islands to estimate the local extinction rate in the absence of human disturbance. Landbridge islands were formed as a consequence of rising sea levels at the end of the Pleistocene. They are convenient in this regard because one may estimate the rate of extinction for a particular taxon using (a) information on the number of species in the taxon of interest on the island today; (b) an estimate of the number of species on the island at its time of isolation, as determined by counting the average number of species on the mainland today in a similar-sized area; and (c) an estimate of the time elapsed since island isolation. These data are then fit to an a priori model that describes the dynamics or "relaxation" of species loss over time (Diamond 1972).

Richman et al. (1988) estimated the relaxation rate for reptilian faunas of two landbridge island groups, one off Baja California and the other near South Australia. If relaxation rates on landbridge islands are to provide a valid estimate of a natural background rate of extinction, it is essential to evaluate the importance of human-related effects on these islands. The islands of Baja California are arid, extreme environments, and the establishment of human settlements or introduced animals have been severely limited as a result (Bahr 1983). Most of the islands of South Australia are similarly uninhabited, though a few of the largest islands have been settled for some time. However, in these instances initial surveys of the resident faunas began quite early. Thus, species lists for the islands used in this analysis have not been impoverished by anthropogenic extinctions.

Compared to other vertebrate taxa, reptiles present particular advantages for partialling out the contribution of extinction to observed relaxation rates. They are poor overwater dispersers and thus rarely recolonize these islands subsequent to their isolation from the mainland. In addition, they are relatively resistant to extinction compared to warm-blooded vertebrates (Wilcox 1980; Case and Cody 1987), presumably because of their lower metabolic requirements and often higher densities. Thus, the observed disparity between current island censuses and estimates of species number at the time of island isolation may be attributed largely to extinctions occurring in the absence of confounding immigration events.

Conclusions from these relaxation studies are as follows:

1. Even in the absence of much habitat disturbance or climatic change, a substantial number of extinctions may occur, and the rate of extinction declines with increasing area. Figure 5.2 shows a significant

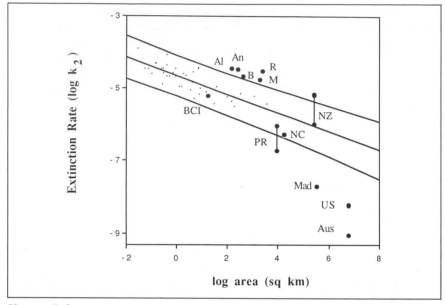

Figure 5.2
The extinction rate for recent reptile faunas (log of the relaxation parameter k^2) is plotted against the log of area for selected areas discussed in the text: Al (Aldabra), An (Antigua), B (Barbuda), M (Mauritius), R (Reunion), PR (Puerto Rico), US (continental United States). The small, unlabeled points represent landbridge islands off Baja California and Australia. The regression line and its 95 percent prediction intervals are calculated for all landbridge islands. Data are from major reference sources in Richman et al. (1988) with additions and changes as follows:

Place	Reference	Spp now	Spp extinct	Time (yr)
New Zealand (NZ)	Robb 1986; Bauer and Russel	30	13	10,000
(two main islands)	1986		7	1,000
Madagascar (Mad)	Blanc 1972, 1984	247	2	2,000
New Caledonia (NC)	Gaffney et al. 1984; Sadlier 1986; 1988; A. Bauer, pers. comm.	42	3	300
Mauritius (M)	Cheke 1987	4	9	10,000
Reunion (R)	Cheke 1987	2	3	10,000
Aldabra (Al)	Gardner 1986; Arnold 1976	2	8	10,000
Australia (Aus)	Molnar 1984b	628	3	10,000
Barro Colorado (BCI)	Myers and Rand 1969	66	2	77

The two extinction rates for New Zealand are both based solely on the main two islands (not including their satellites). The first is based on

(continued)

Figure 5.2 *(continued)*
the roughly thousand-year period since the islands were first occupied by Maoris and includes known historical extinctions plus identified and dated subfossil extinctions. The second calculation makes the assumption that the endemic satellite landbridge island species were also present on the two main islands 10,000 years ago when these islands would have been connected to the main islands. In all cases, species whose incidence on an island is probably man-aided are excluded from the species count. This excludes the following species for all these sites: Hemidactylus frenatus, Hemidactylus mercatorius, Lepidodactylus lugubris, Gehyra mutilata, Cryptoblepharus boutonii, *and* Typhlops braminus *(see Case and Bolger 1991 and Darlington 1957). The point for the USA is based on only the contiguous 48 states.*

negative correlation between extinction risk as measured by the extinction rate parameter k_2 and increasing island area. (The rate of species loss is empirically unlike radioactive decay, where there is a constant half-life independent of initial abundance. Instead, species loss is better described by nonlinear models of faunal relaxation. The constant k_2 is the rate parameter from an equation describing faunal relaxation, $dS/dt = -k_2S^2$. Greater extinction rates correspond to higher levels of k_2.) This pattern is a general one. Case and Cody (1987) calculated extinction rates based on Baja island mammals and found the regression line of k_2 with an island area to have the same slope (but higher overall magnitude) as that for reptiles.

In addition, experimental support for declining rates with increasing island area comes from the study of Schoener and Schoener (1983) who introduced *Anolis sagrei* or *Leiocephalus* spp. onto 30 very small islands in the Bahamas having no lizards naturally. Small island populations quickly became extinct while larger island populations still survive today (Schoener, pers. com.).

2. In spite of widely different faunas with little taxonomic overlap even at the family level, the two island groups (Australia and Baja California) display the same pattern and magnitude of extinction rate as a function of island area; the two regressions are not significantly different (test for coincidence of regression lines, $p > 0.5$). Similarly, Figure 5.2 plots the known reptilian extinctions for Barro Colorado Island (BCI). Since its inception about 80 years ago as a reserve in the Panama Canal, BCI has lost two reptile species (Myers and Rand 1969). This translates to about 3 percent of the initial fauna and compares to 23 percent for bird species (Willis 1974). When plotted in Figure 5.2, the point for BCI falls roughly on the regression line for the undisturbed arid islands. Since these islands all have minimal

human impact and little climatic change over the interval measured, we believe that these extinctions primarily represent "background" extinction rates in the absence of significant human intervention.

Worldwide, a few islands have both a reasonable historical record documenting environmental changes and abundant fossil records documenting past extinctions. Here we are not inferring extinctions but actually have the "smoking gun" in subfossil form. Richman et al. (1988) calculated the extinction rates for these islands assuming that the original species number is the present species number (minus all species introduced by humans) plus the number of extinct species (or forms) as determined by subfossil evidence. If anything, this will give a conservative estimate of extinction because undiscovered fossils may include new extinct forms. Of course, there are many islands with no known extinctions simply because geological conditions are not favorable for their deposition or discovery or researchers have not yet looked.

The results are superimposed as points on Figure 5.2. It is apparent that the effects of disturbance are most telling on the smallest islands; the per-species extinction rate for Antigua, Barbuda, Mauritius, and Reunion is approximately ten times that on islands of similar size in Australia or Baja California. It is impossible from these data alone to entangle the causative role of introduced predators, competitors, and the like from simple habitat destruction.

Significantly, the now-familiar trend of decreasing extinction risk with increasing area is preserved even for these disturbed areas where extinction rates are calculated on the basis of known fossils. The observed elevation of the extinction parameter decreases with area, with no elevation in risk for the very large "islands" of Australia, the continental United States, and Madagascar. Indeed, these points lie far below the predicted extinction rate based on relatively undisturbed landbridge islands. The low number of extinct reptiles recorded in the continental U.S. contrasts with the large numbers of Holocene extinctions of reptiles in the nearby West Indies (Etheridge 1964; Pregill 1981). Although data for Madagascar are probably much less complete because of its larger area, fewer fossil digs, and a reptilian fauna that is still incompletely documented, the calculated extinction rate (two species in 2,000 years or about 0.4% of the fauna) yields an extinction rate roughly similar to that of the continental U.S. and substantially lower than that for the next smaller islands of New Zealand and New Caledonia. This low extinction rate for reptiles is all the more surprising given the tremendous amount of habitat destruction on Madagascar; about 80 to 90 percent of the original vegetation has been cleared (Jolly, Oberlé, and Albignac 1984) and along with other human impacts has resulted in an extinction of at least 13 of the 75 native mammal species (approximately 17%; Jolly et al. 1984).

For completeness, it would be nice to compare extinctions on the continents of Africa, Eurasia, and South America with those from North America and Australia, but unfortunately the fossil record for these areas is not well known for the Holocene and late Pleistocene. Based on present knowledge, the record for South America is like that of North America in that practically all fossils known for the past 10,000 years are referable to extant taxa (Baez and Gasparini 1979).

It probably is the case that many undocumented large-island and continental extinctions simply await discovery, but we see no reason that there should be any particular bias against fossil discovery on mainlands compared to islands. Moreover, abundant evidence for mammalian extinctions on continents occurs worldwide over this same time frame and from the same deposits. The greater number of mammalian species is probably at least partly due to overhunting by Pleistocene humans (Martin and Klein 1984). Certainly it is often difficult to distinguish taxa at the level of species from fossil material alone but this problem befalls islands as well as mainlands.

Conventional explanations for island extinctions emphasize the extreme vulnerability of native island species to introduced predators, and this effect cannot be denied (see Predation section). Mainlands and large islands have endemic predators with which the fauna has presumably coevolved. The prey have probably evolved better defenses and the predators' populations are in turn kept in check by higher-order predators and parasites. Perhaps equally important, low extinction rates are expected given large area and thus increased opportunity for immigration after local extirpation.

3. Reptiles have lower rates of extinction than birds or mammals. Case and Cody (1987) showed that on the same islands in the Sea of Cortez, mammals have extinction rates about an order of magnitude higher than those for reptiles. Additionally, Schoener (1983), based on a wide review of the literature, finds that species turnover rates for reptiles generally fall below those of birds and mammals and most arthropod systems. Lizard populations should be expected to be more resistent to extinction because their lower metabolic rate should allow higher densities than either birds or mammals and thus larger population sizes.

We have made a tally of reptile extinctions over the last 10,000 years, both historic and prehistoric. This count is only an approximation because of the previously mentioned lack of fossils from Asia and South America and also because of the difficulty in deciding whether fossil finds in certain taxa such as turtles and tortoises represent one or several extinct species. This minimum estimate is likely to be revised upward in the future but presently there is evidence for 60 Holocene species extinctions, eight (or 13 percent) of which involve continental species. When we compare this to 115 mammalian (64% continental) and 171 bird extinctions (10% continental) just since 1600 (Diamond 1984), we must conclude (cautiously, given the caveats above) that Recent reptile extinctions are primarily on islands and that reptiles are less extinction-prone than the endothermic vertebrates.

Terrestrial mollusks are one of the few invertebrate taxa for which some compilation of extinctions has been attempted, although not comprehensively. In Hawaii alone, over twenty terrestrial snail species have gone extinct (Hadfield, Miller, and Carwile 1989). This vulnerability is probably caused by the limited geographical range of many of the endemic species and again the introduction of exotic species.

PREDATION

One of the most important factors influencing lizard abundance on islands is the variety and density of predators. On predator-free islands, lizards can achieve extremely high densities. For example, on small rat-free islands off New Zealand densities reach 1,390 lizards per acre, or nearly one lizard every 3 m^2 (Crook 1973; Whitaker 1968, 1973). In other parts of the world, one finds this pattern repeated. Up to 2,074 diurnal lizards per acre have been reported for rat-free Cousin Island in the Seychelles (Brooke and Houston 1983) and 1,214 per acre for San Pedro Martir in the Sea of Cortez, Mexico (Wilcox 1981; Case unpublished data).

That predation can have a large impact on lizard densities is tested by the introduction of lizard predators to some islands and not to others. Although strictly illegal in most places today, this "experiment" was conducted historically many times with rats, cats, dogs, and mongooses. The mongoose is one of the most potent predators on diurnal ground-foraging lizards. Mongooses have been introduced to various islands around the world with the hope of controlling rats and other vertebrate pests. Although their success in this regard has been mixed, their impact on native reptile (as well as bird populations), particularly ground-foraging forms, like skinks, teiids, lacertids, and snakes, has been devastating. In Puerto Rico, reptiles and insects, not rats, form the bulk of the mongoose diet (Pimentel 1955).

One of us (TJC) attempted to quantify the impact of the mongoose on diurnal lizard abundance on islands in the South Pacific by censusing lizards on islands with and without the mongoose (Case and Bolger 1991). Lizards were counted along 2–3 transects of about 1 km meter. There is nearly a 100–fold increase in diurnal lizard abundance on islands without mongooses compared to islands with mongooses (Figure 5.3).

The same qualitative pattern is evident in the West Indies. Nearly fifty years after the introduction of the mongoose to Jamaica, Barbour

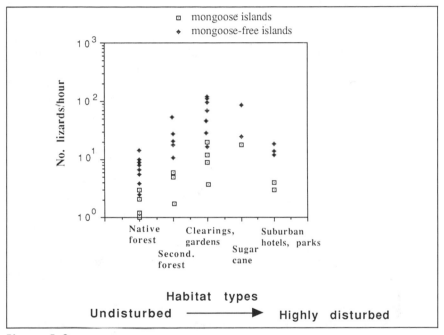

Figure 5.3
Crude lizard censuses (expressed as the average number of diurnal lizards seen per hour; also see Case 1975) in natural and human-modified habitats on mongoose-inhabited and mongoose-free islands in the tropical Pacific. All censuses were conducted during sunny days from 1984 to 1988 by TJC. No attempt was made to capture any of the lizards so that a constant search speed could be maintained. Nearly all lizards seen were skinks and included both native and introduced species. Each point represents the average of two to four censuses. The islands and the number of habitats censused are mongoose-inhabited islands in Hawaii (2), Oahu (2), Molokai (2), and Maui (1); in Fiji, Vite Levu (5), Rabi (2), and Vanua Levu (1). Mongoose-free islands are New Caledonia (3), Kauai (3), Efaté (2), Espiritu Santo (2), Tahiti (3), Moorea (1), Roratonga (1), and Atiu (2); in Fiji, Kadavu (4), Taveuni (3), and Ovalau (1).

(1910, 273) noticed the "almost complete extinction of many species which were once abundant ... true ground inhabiting forms have, of course, suffered most ... snakes have perhaps suffered more than lizards." This effect on lizard abundance is also seen today on small cays off Jamaica and elsewhere in the West Indies. Where the mongoose is absent, terrestrial lizards and snakes are much more common (Barbour 1930; Schmidt 1928; Pregill 1986; Mittermaier 1972).

On St. Lucia island in the Lesser Antilles, for example, three reptile species have been extirpated in historical time coincident with the introduction of mongoose: one skink (*Mabuya mabuya*) and two colubrid snakes (*Clelia clelia* and *Liophis ornatus*; Corke 1987). *Liophis ornatus* survives only on the tiny offshore island of Maria Major along with the ground-foraging lizard, *Cnemidophorus vanzoi*, which is curiously absent from St. Lucia itself. It seems unlikely that this lizard was not once present on St. Lucia in that the islands are so close inshore, yet no specimens were ever deposited in museum collections. Similarly, the colubrid snake, *Alsophis antillensis*, once occupied Barbuda and Antigua but today can be found only on mongoose-free offshore cays (Pregill et al. 1988).

The last major experiment with rat control by mongoose introduction took place in Mauritius but not until about 1900, after most of the large endemic species had already become extinct (see earlier). Today, the only surviving ground-dwelling species on Mauritius, the skink *Scelotes bojerii*, is extremely rare and until recently, was thought to be extinct. The other surviving species are relatively common but are arboreal (the endemic *Phelsuma* day geckos) or are widely-distributed non-endemic species (the skink *Cryptoblepharus boutonii* and the house gecko *Hemidactylus frenatus*) of continental origin whose introductions here and elsewhere have been man-aided (Cheke 1984, 1987).

Domestic cats and dogs have also had devastating effects on island species. We have already mentioned the role of dogs in the local extinction of land iguanas in the Galapagos. Dogs have also reduced populations of marine iguanas as well, but to date no extirpations are known. Because they are more arboreal than dogs or mongooses, cats and tree rats (*Rattus rattus*) affect prey species that the mongooses and dogs are less likely to capture. Gibbons and Watkins (1982) suggest that cats may have been even more damaging than mongooses to highly arboreal Fijian lizards and in particular to the now rare endemic Fijian iguanas. Today, substantial populations can be found only on small islands lacking both mongooses and cats. The combination of cats and mongooses on the two largest islands of Fiji, Viti Levu and Vanua Levu, has resulted in the local extinction of the ground-foraging skinks *Emoia nigra* and *E. trossula*. These are the two largest skinks in Fiji, and they have not been seen in over 100 years (Gibbons pers. com.; Zug 1992), although they survive quite well on mongoose-free islands in the archipelago (e.g., Ovalau, Rotuma, and Taveuni). Interestingly, nearly all these islands have rats (*Rattus rattus* and *R. exulans*). The ground-nesting banded rails (*Rallus phillippensis*) and ground doves are also absent from the mongoose-inhabited islands and are presumed to be extinct (Gibbons 1984).

The large herbivorous iguanine *Cyclura carinata*, was nearly extirpated on Pine Cay in the Caicos islands (West Indies) during the three years following construction of a hotel and tourist facility (Iverson 1978). Predation by cats and dogs introduced during the hotel construction resulted in the decline from about 5,500 adults to only around five. Iverson also presents evidence suggesting that population declines of *Cyclura* elsewhere in the Turks and Caicos Banks stem directly from cat and dog predation.

Thomson (1922) noted that New Zealand lizards became much less common after the mid-nineteenth century, and he attributed this decline to loss of cover and predation by cats. Today reptile density and species numbers are almost invariably higher on predator-free islands than on mainland New Zealand or islands with introduced mammalian predators (Whitaker 1982). Off-lying islands with *Rattus exulans*, the Polynesian rat, support small populations of lizards and tuataras than do islands without rats (Crook 1973; Whitaker 1973). The only exceptions occur in predator-proof habitat at some mainland sites, such as deep boulder banks where local lizard densities may exceed one per m^2 (Whitaker 1982; Towns 1972).

If introduced predators reduce reptile densities low enough, extinction follows, particularly on smaller islands. The role of predators in causing many of the extinctions documented here is circumstantial but voluminous. The tuatara is the last remaining representative of a widespread Mesozoic order of reptiles known as the Ryncocephalia. Today it is found on uninhabited landbridge islands off New Zealand, but subfossils, less than a thousand years old, are found on both of the main islands (Cassels 1984). In all, ten species of lizards (about 1/3 of the New Zealand lizard fauna), in addition to the tuatara, are restricted to small off-lying islands formerly connected to the main islands (Robb 1986; Newman 1982; Cassels 1984). Predation by introduced animals, predominantly rats, is thought to be responsible for this pattern of extinctions. Whitaker (1973) found that small islands off New Zealand with the introduced Polynesian rat have fewer lizard species (all natives) for their size than islands without rats. McCallum (1986) documents the changes to the herpetofauna following the colonization of Lizard Island by the Polynesian rat in 1977. Two lizard species appeared to go locally extinct, and overall lizard densities dropped by at least one order of magnitude. Norway rats colonized Whenuakura Island in 1983–84, and by 1985 the previously thriving tuatara population had disappeared as had nearly all the lizards (Newman 1986).

This pattern of endemic lizards being restricted or at least much more common on smaller rat-free islands off-lying larger rat-infested islands is repeated in the Mascarenes (Vinson and Vinson 1969), the Seychelles (Gardner 1986), the Canary Islands (Klemmer 1976), the Cape Verdes (Greer 1976), Norfolk (Cogger, Sadlier, and Cameron 1983) and Lord Howe Islands (Cogger 1971). On most of these islands, rats arrived so early historically that we do not have adequate pre-rat reptile records or census data. In the case of Lord Howe Island, however, the numerical decline of the only two native lizards, *Phyllodactylus guentheri* and *Leiolopisma lichenigerum*, on the main island seems to have occurred after the arrival of rats in 1918 (Cogger 1971).

In the Seychelles, all the populations of the largest extant skink, *Mabuya wrightii,* are on rat-free islands that usually also have nesting seabirds (Cheke 1984; Gardner 1986). When Lanz visited Marianne in 1877 both seabird colonies and *M. wrightii* were present and rats were not found (Cheke 1984). Subsequently rats were introduced and today neither *M. wrightii* nor breeding seabirds are present.

Mammals are not the only taxa implicated in causing reptile extinctions or extirpations. The introduced brown tree snake, *Boiga irregularis,* which has become infamous for decimating populations of endemic birds on Guam (Savidge 1987), has also severely impacted the lizard fauna there (Engbring and Fritts 1988). Juvenile snakes prey predominantly on lizards and are suspected of being a major factor in the possible extirpation of three species of skinks and two geckos (T.H. Fritts, pers. com.). They have also apparently reduced the numbers of forest populations of some other geckos (*Gehyra oceanica, G. mutilata,* and *Lepidodactylus lugubris*), species that are usually abundant in these habitats in the absence of the tree snake (pers. obs.). This snake originates from New Guinea, the Solomons, northern Australia, and Indonesia, where it is not particularly common (e.g., McCoy 1980 describes it as uncommon in the Solomons) and where it coexists with a rich landbird and reptilian fauna. The havoc that it is causing on Guam could stem from its high densities due to release from its own predators or prey naïveté or both.

Most human-introduced predators are brought to islands shortly after they are colonized, either by aborigines or Europeans. Pregill (1986) has correlated the settlement time of islands with the extinction times of a number of insular reptiles. The overall picture is quite convincing; the arrival of humans to an island is closely associated with increased reptile extinction rates, especially of large endemic species. The inference is that habitat destruction and predation by humans and/or their entourage of introduced animals is responsible for these extinctions.

Another feature of these extinctions and extirpations is that they seem to occur most often within the endemic component of the fauna and on islands with high levels of endemism. The fossil species going extinct on the islands in Figure 5.2 are typically forms that are endemic at the species or genus level on the island or archipelago. This trend is explored further in a subset of islands in Table 5.1 that have both endemic and nonendemic species as well as good fossil evidence for extinction. For each island the proportion of species becoming locally extinct on the island is broken down into three categories: (1) species endemic to the island (or its immediate satellites); (2) species that are nonendemic; and, (3) species that are endemic to the island group (e.g., a species endemic to Aldabra and the Comoros would be counted in category 3 but not category 1 or 2). Endemic species have significantly higher extinction rates than nonendemics ($p < 0.034$; Mann-Whitney U test).

Table 5.1

Terrestrial Reptile Extinctions on Selected Islands for Endemic and Nonendemic Species over Roughly the Last 10,000 Years
The table's entries give the number of species in each category becoming locally extinct on the island divided by the number of species in that category that were initially present on the island. Human-introduced species are excluded. Species are grouped into three categories: endemic to the island or its satellites, endemic to the region, or nonendemic to the island or region. Species that are "endemic to the island" occur only on the island in question. Species that are "endemic to the island or region" include species confined to the island plus species endemic to the island and nearby islands or archipelagoes. The total number of species in each category on the island appears in parentheses. For example, Aldabra had a total of five species that were not endemic to the island or nearby islands. Forty percent, or two, of these species have become extinct. Overall, extinction rates are significantly higher in the endemic species component of the fauna compared to the nonendemic.

| | Extinction Probabilities | | |
Island	Nonendemic	Endemic	Endemic to Island or Region
Aldabra	0.40 (5)	1.00 (4)	0.83 (6)
Tenerife, Canary Is.	0.00 (1)	1.00 (1)*	0.33 (6)*
Mauritius	0.00 (4)	0.89 (9)	0.75 (12)
Reunion	0.00 (3)	1.00 (1)	0.50 (4)
Rodriques	— (0)	1.00 (4)	1.00 (4)
Puerto Rico	0.00 (4)	0.13 (30)	0.13 (30)
Means	0.08	0.84	0.59

*This includes *Gallotia maxima*, so far only known from Tenerife, although it is not clear whether it survived into the Holocene. We follow the taxonomy of Machado, López-Jurado, and Martin 1985.

Endemic species have been isolated on islands lacking mammalian predators for long periods and have presumably become relatively defenseless to introduced predators. Few attempts have been made, however, to quantify this supposition, although Shallenberger (1970) measured the flushing distance of insular and mainland iguanid lizards and found that a human can get up to ten times closer to the insular varieties.

A similar predator naïveté, this time in birds, is apparent when one compares the effects of introduced predators on islands that previously had no similar predators to those that did. On Hawaii, Midway, Lord Howe, New Zealand, and others, introduced rats have led to the extinction of many native bird species. Yet on others, like Fiji, Tonga, Samoa, Marquesas, Rennell, the Solomons, Aldabra, Christmas (Indian Ocean), and the Galapagos, the introduction of rats was not accompanied by a wave of avian extinctions. Atkinson (1985) points out that the extinction-resistant islands all have native rats or land crabs. Atkinson argues that land crabs fill an ecological niche very similar to that of the rat. Birds on predator-free islands are easy prey for the rats, whereas on the other islands the birds have presumably acquired a more effective predator-avoidance behavior.

Rats are similarly implicated in reptile extinctions or extirpations on many of the same islands as for birds (except, of course, those where no native reptiles occur). Rats have had the greatest effect (in terms of lizard densities and numbers of extinctions) in the Mascarenes, Seychelles, Lord Howe, Norfolk, and New Zealand but interestingly seem to have had little effect in most of the Central Pacific, for example, Fiji, New Caledonia, Tonga, Samoa, the Solomons, the Galapagos, and Australia.

Why should the reptile fauna in these different places exhibit such varying susceptibilities to rat introductions? This question needs further study. Invading rats freed from their continental predators and parasites can reach high densities on islands and often invade forest habitats; whereas in more continental faunas they are nearly restricted to man-modified habitats. This factor however does not readily explain the apparent differences between islands lacking rat predators. One likely factor is the high frequency of introduced reptile species on islands where rats have not had a big impact. Introduced reptiles generally come from mainland areas where they have had a long coevolutionary history with predators. In Fiji, for example, 46 percent (11/24) of its reptile fauna is introduced, and rats have had little apparent effect. Contrast this with New Zealand, with no introduced reptiles, and a reptile fauna severely impacted by rat introduction. Introduced mongoose has played a large role in the Central Pacific in affecting overall lizard densities but it has not greatly affected the number of *introduced* reptile species on islands. After partialling out differences between islands in their area and maximum elevation, islands in the mid-Pacific with mongooses do not have significantly fewer species of introduced reptiles than do mongoose-free islands (Case and Bolger 1991). (A similar analysis cannot be done for the native species because about half the islands have none at all.)

In the Seychelles and the Comoros, rat-free islands have approximately equal numbers of introduced lizard species as do rat-infested islands, although densities may be very different (Cheke 1984; Evans and Evans 1980; Brooke and Houston 1983). Because nearly all introduced species come originally from predator-rich continental areas, they may be less susceptible to introduced predators than the sympatric endemic predator-naive species.

Another contributing factor to the apparent vulnerability of island endemics could be the lack of recolonization sources. When a population of nonendemics or regional endemics becomes locally extinct on an island, the island can potentially be recolonized from individuals still surviving on other nearby islands. For a single-island endemic, however, extirpation and extinction are synonymous.

COMPETITION FROM AND AMONG INTRODUCED SPECIES

Unlike birds, lizards have not been able to colonize on their own the remotest islands of the world, such as those in the mid-Pacific (e.g., Hawaii and the Marquesas). For the most part, reptiles reached these islands when the Polynesians and Melanesians inadvertently began spreading a set of geckos and skinks throughout much of the Pacific about 4,000 years ago. Additions to this set of aboriginal introductions have occurred more recently during European settlement. The reptilian faunas of somewhat less-isolated islands (e.g., Guam, Fiji, Vanuatu) today are a mixture of native and introduced species. These introductions, although unconscious, poorly documented, and not as well controlled as a manipulative experiment, can be used to sort out competitive relationships among species because of the huge sample sizes involved (i.e., literally hundreds of island and mainland locations).

Case and Bolger (1991) reviewed this literature for reptiles and found no documented case in which a native reptile species was reduced to extinction by the introduction of a reptilian competitor. We are aware of only one example where an introduced species seemed to numerically supplant a native species. South Florida has only two native anoles (*A. carolinensis* and *A. distichus*). In recent years it has been a beachhead for at least six introduced anoles from the more anole-rich Greater Antilles. Most of these introductions are still highly localized in urban areas, but

Anolis sagrei is successfully displacing the native *A. carolinensis* as the most common anole in urban areas, penetrating agricultural and even native habitats (Wilson and Porras 1983; Salzburg 1984).

Case and Bolger (1991) also found evidence that (1) native species-rich faunas seem to resist invasion by exotics, and (2) the densities of resident introduced species may decline after the introduction of new competing species, but the mechanisms behind both these effects are not well understood.

A striking example apparent competitive displacement occurs among introduced species in Hawaii. Until about 1940 one of the most common skink was *Emoia cyanura*, a Polynesian introduction, that is still the most common skink in Fiji, Samoa, the Marquesas, and nearly everywhere else in the eastern Pacific where it occurs (Oliver and Shaw 1953; McKeown 1978; Jones 1979). It is also common in subfossil deposits on Hawaii from the Polynesian period (G.K. Pregill pers. com.). This pattern changed when *Lampropholis delicata* was accidentally introduced to Hawaii from southeast Australia (McKeown 1978; Case pers. obs.). Today *E. cyanura* is rare, whereas *Lampropholis* is the most frequently seen ground-dwelling skink on the islands. Since *L. delicata* was not introduced elsewhere in the Pacific, we have no replicates, but on the many "control" islands in the Pacific *E. cyanura* is still very common. Moreover, in lizard-rich southeast Australia and Tazmania where *L. delicata* originated, it is neither particularly common nor widespread (Case pers. obs.; Cogger 1983).

Another example of apparent competitive exclusion has also occurred in Hawaii. After World War II, a new gecko appeared in Hawaii: the common house gecko, *Hemidactylus frenatus*, native to Asia and the Indo-Pacific. It subsequently increased in numbers in urban/suburban habitats, while three other Polynesian-introduced geckos, the fox or Polynesian gecko (*Hemidactylus garnotii*), the mourning gecko (*Lepidodactylus lugubris*), and the stump-toed gecko (*Gehyra mutilata*), formerly occupying this niche, became scarce in these habitats (Oliver and Shaw 1953). Today, the most common association on lighted building walls is the house gecko, alone or sometimes in association with the smaller and typically less abundant mourning gecko (Table 5.2). The pattern is complicated by two additional factors. The house gecko has spread beyond Oahu to other Hawaiian islands but this spread has been recent and the situation is not at equilibrium. Secondly, based on our studies in progress in Hawaii and Fiji, it is apparent that climatic factors also impinge on the competitive interaction between house geckos and mourning geckos. The competitive displacement goes slower in the more mesic habitats on the windward sides of islands. Today one can still find good numbers of mourning geckos on building walls in Hilo (on Hawaii) for example,

Table 5.2

Gecko introductions and faunal affects in the guild of gecko species occupying human structures. (a), (b)

Archipelago	Island	Historical house geckos	Recent invaders (a)	House geckos today (b)
Fiji	w. Viti Levu	(GO) (LL) (c)	HF in 1960 (d)	HF:LL 30:1 (n=90)
	e. Viti Levu	GO, LL (c)	HF in 1982 (e)	HF:LL:GO 30:20:1 (n= several hundred)
	Other islands	GO, LL (c)	none	LL:GO 7:3 (n=several hundred)
Western Samoa	Upolu	GO, LL, (GM) (f)	HF to Upolu in 1960's (g)	Upolu—HF dominant. Other islands—LL and GO (g)
	Savai'i	GO, GM, LL (f)	None (except local invasion of HF at Saleloga Wharf)	LL:GO:GM 10:10:1 (n= 41)
American Samoa		GO, LL (f)	HF to Tuitilla in mid-1960's (h)	Tuitilla—HF dominant. Other islands—LL, GO, GM (h)
Vanuatu	Espiritu Santo	GO, LL (i)	HF post 1971	HF:LL 17:3 (n=99)
	Efaté (Port Vila)	GO, LL (i)	HF post 1971	HF:LL:GO 85:15:1 (n= 99)
	north. Efaté	GO, LL	none	LL:GO 2:1 (n=32)
	Emao	GO, LL	none	LL:GO 1:1 (n=33)
Hawaii	Oahu	GM, LL, (HG) (j)	HF 1951 (k)	HF:LL:GM:HG 30:1:0:0 (n= 31)
	e. Hawaii	GM, LL, HG (j)	HF post 1965 (l)	HF:LL:GM:HG 10:1:0:0 (n= 37)
	w. Hawaii	GM, LL, HG (j)	HF post 1965 (l)	HF:LL:GM:HG 2:1:0:0 (n= 72)

(continued)

Table 5.2 *(continued)*

Archipelago	Island	Historical house geckos	Recent invaders (a)	House geckos today (b)
Society Islands	Tahiti, Papeete port area	GO, GM, HG, LL (m)	mid 1980's	HF:LL:GO 30:1:2 (n= 36)
	Tahiti, elsewhere	GO, GM, HG, LL (m)		LL:GO:GM 27:12:1 (n= 40)
	Moorea	GO, GM, HG, LL (m)		LL:GO:GM 12:9:1 (n= 22)
Mainland Mexico (San Blas and Mazatlan)		GM (n)	HF post 1963 (n)	HF dominant (n)

(a) In most cases the exact date of the invasion is unknown, the date given is the date of the last survey that did not find the invader.

(b) Except where noted these are personal observations by the authors. GO = *Gehyra mutilata;* GO = *Gehyra oceanica;* LL = *Lepidodactylus lugubris;* HF = *Hemidactylus frenatus;* HG = *Hemidactylus garnotii*

References: (c) Pernetta and Watling 1978; (d) Watling pers. com.; (e) Gibbons pers. com.; (f) Burt and Burt 1932; (g) Zug pers. comm.; (h) Amerson et al. 1982; (i) Medway and Marshall 1975; (j) Stejneger 1899; (k) Hunsaker and Breese 1967; (l) Jones 1979; (m) I. Ineich pers. com.; (n) N. Scott pers. com.

although they are typically far outnumbered by house geckos. This same leeward/windward difference is also evident on islands in Fiji.

Frogner (1967) found that the house gecko could displace the mourning gecko from favored shelter sites in laboratory experiments and that it would eat juvenile *Lepidodactylus*. The reverse is not true, however, in that hatchling house geckos are larger than the largest prey taken by *Lepidodactylus* in the field. Laboratory experiments have shown that *H. frenatus* is behaviorally dominant to both the smaller *L. lugubris* and the equivalently sized *H. garnotii* (Bolger and Case in press).

Elsewhere in the Pacific where the house gecko has yet to invade, for example, most of the Societies, Tuamotus, and Marquesas, most of the Cooks, and most of Fiji, *G. mutilata* or *G. oceanica* with *Lepidodactylus lugubris*, and/or *Hemidactylus garnotii* have remained dominant in the "human building" niche (Table 5.2). This appears to be changing, however, on the main Fijian island, Viti Levu. Although unrecorded until recently, the house gecko has been in the Nadi area on the west for at least twenty years (Pernetta and Watling 1979; D. Watling pers. com.) and now is the only gecko common in towns along the west. It appeared in the major port city of Suva on the southeast windward side in about

1983 and already has become the most frequent gecko on walls at the University of the Pacific in Suva with the concomitant decline of the previous resident geckos on the same walls (Bolger 1991; J. Gibbons pers. com.; D. Watling pers. com.). Today the area around Suva is a mosaic, with *H. frenatus* already dominant in some areas but absent in others, and instead the other geckos are found in high numbers. In areas where *H. frenatus* is present but not common, its numbers have been increasing over the last two years (Bolger 1991). Transplantation experiments are under way to determine whether these enclaves have simply not yet been reached by *H. frenatus* and to uncover the mechanism behind the competitive interaction.

The house gecko fauna also changed rapidly in Vanuatu (New Hebrides). In 1971, the Royal Society did not find a single *H. frenatus* in Vanuatu (Medway and Marshall 1975). Today it is virtually the only urban gecko seen in the major city of Port Vila on Efate (although it is still restricted to the Port Vila area on Efate) and is by far the most common gecko in the town of Santo on Espiritu Santo (Table 5.2). Despite much recent work on the geckos of the Society Islands, Ineich (1987) and Ineich and Blanc (1988) did not find any *Hemidactylus frenatus*. In 1989 we recorded the presence of this species on Tahiti for the first time; it presently is restricted to the wharf area of Papeete where it is already the most common gecko on buildings. A second census of Papeete was performed in 1991; *H. frenatus* had now spread about 10 km beyond where we found it localized in 1989 (Bolger and Case, pers. obs.).

Bermuda has no native lizards other than a single endemic skink. Wingate (1965) documents the introduction of *Anolis grahami* from Jamaica in 1905. After about 35 years the lizard had spread throughout the main island. Sometime around the early 1940s a second anole (*Anolis leachi*) was introduced from the Barbuda Bank in the Lesser Antilles; the exact circumstances are unknown. The rate of spread of this second species was considerably slower than that of *A. grahami* and today the range of *A. leachi* is encompassed within that of *A. grahami*. *Anolis leachii* is much larger and is behaviorally and numerically dominant, consequently the two species are allotopic on a fine scale. Finally, a third anole, *A. roquet*, from Barbados, was introduced sometime prior to 1945. *Anolis roquet* has not spread yet into the range of *A. leachii* but is sympatric with *A. grahami*, which it resembles in body size and habits. In spite of this ecological similarity, Wingate (1965) found no obvious displacement as for the previous size-dissimilar species pair, suggesting that competition among these anoles might have more to do with overt interference interactions between size-dissimilar lizards than competition for limited food resources.

CONCLUSIONS AND SUMMARY

A worldwide survey of Holocene (Recent) reptile extinctions yielded several conclusions: (1) Humans are implicated either directly or indirectly in many extinctions, extirpations, and population declines. (2) In the absence of much environmental change, either climatically or due to humans, a background extinction rate still exists, and the magnitude of local extinction decreases with increasing island area. (3) In the presence of humans, this background rate is exaggerated—sometimes by an order of magnitude for small islands, but the effect of island area is, if anything, accentuated because very large islands and mainlands have lower extinction rates than predicted from landbridge island extrapolations. (4) Island extinctions are more common than mainland extinctions. This resilience of continental faunas may in part be an artifact of the difficulties in finding fossils spread throughout larger areas. Yet, native island species seem more vulnerable to introduced predators. Perhaps equally important, low extinction rates are expected given large area and thus increased opportunity for immigration after local extirpation. (5) Species becoming extinct are usually those with relatively large body size and a long history of island isolation resulting in endemic status. (6) Predators, chiefly mongooses, rats, cats, and dogs, are often implicated in the extinction of reptiles. (7) Introduced reptiles do not usually competitively affect native reptiles, although they have sometimes had dramatic impacts on the densities of other introduced species. (8) Reptile extinction rates are often lower than those calculated for mammals and birds.

ACKNOWLEDGMENTS

At various stages during the development of this paper, many people contributed ideas and suggestions and useful unpublished information: Nick Arnold, Aaron Bauer, Harold Cogger, Ronald Crombie, Robert Fisher, John Gibbons, Mike Gilpin, Greg Pregill, Tom Schoener, Dick Watling, Ernest Williams, Tony Whitaker, Richard Zweifel, and George Zug. We thank them and we thank Allison Alberts, Jared Diamond, Peggy Fiedler, Mike Gilpin, Greg Pregill, and Tom Schoener for improving the presentation.

LITERATURE CITED

Amerson, A.B., W.A. Whistler, and T.D. Schwaner. 1982. *Wildlife and wildlife habitat of American Samoa. II. Accounts of flora and fauna.* Washington, D.C.: U.S. Fish and Wildlife Service.

Arnold, E.N. 1973. Relationships of the palaearctic lizards assigned to the genera *Lacerta, Algyroides,* and *Psammodromus* (Reptilia: Lacertidae). *Bull. Brit. Mus. Nat. Hist.* 25:291–366.

————. 1976. Fossil reptiles from Aldabra Atoll, Indian Ocean. *Bull. Brit. Mus. Nat. Hist.* 29:83–116.

————. 1980. Recently extinct populations from Mauritius and Reunion, Indian Ocean. *Jour. Zool. Lond.* 191:33–47.

Atkinson, I. 1985. Effects of rodents on islands. In *Conservation of island birds,* ed. P.J. Moors, 35–81. Cambridge: International Council for Bird Preservation.

Baez, A.M., and Z.B. Gasparini. 1979. An evaluation of the fossil record. In *The South American herpetofauna: Its origin, evolution, and dispersal,* W.E. Duellman, 29–54. Univ. Kansas Mus. Nat. Hist. Monograph 7.

Bahre, C.J. 1983. Human impact: The Midriff Islands. In *Island biogeography in the Sea of Cortez,* ed. T.J. Case and M.L. Cody, Chap. 10. Berkeley: University of California Press.

Barbour, T. 1910. Notes on the herpetology of Jamaica. *Bull. Mus. Comp. Zool.* 52 (no. 15).

————. 1930. Some faunistic changes in the Lesser Antilles. *Proc. New England Zool. Club* 11:73–85.

Bauer, A.M., and A.P. Russell. 1986. *Hoplodactylus delcourti* n. sp. (Reptilia: Gekkonidae), the largest known gecko. *N.Z. J. Zool.* 13:141–48.

Bings, W. 1985. Zur früheren Verbreitung von *Gallotia simonyi* auf Hierro, mit Vorschlägen zur Wiederansiedlung. *Bonn. Zool. Beitr.* 36:417–27.

Blanc, C.P. 1972. Les reptiles de Madagascar et des iles voisines. In *Monographiae Biological,* vol. 21, ed. J.I. Schlitz, 501–614. The Hague: Junk.

————. 1984. The reptiles. In *Madagascar (key environments),* ed. A. Jolly, P. Oberlé, and R. Albignac, Chap. 6. Oxford: Pergamon Press.

Böhme, W., and B. Bings. 1977. Nachträge zur Kenntnis der kanarischen Rieseneidechsen (*Lacerta simonyi*-Gruppe) *Salamandra* 13:105–11.

Böhme, W., W. Bischoff, H. Nettmann, S. Rykena, and J. Freundlich. 1981. Nachweis von *Gallotia simonyi* (Steindachner, 1889) (Reptilia: Lacertidae) aus einer frühmittelalterlichen Fundschicht auf Hierro, Kanarische Inseln. *Bonn. Zool. Beitr.* 32:157–66.

Böhme, W., and G. Zammit-Maempel. 1982. *Lacerta siculimelitensis* sp. n. (Sauria: Lacertidae), a giant lizard from the Late Pleistocene of Malta. *Amphibia-Reptilia* 3:257–68.

Bolger, D.T. 1991. Community perturbation: Introduced species and habitat fragmentation. Ph.D. thesis, University of California at San Diego.

Bolger, D.T., and T.J. Case. N.d. Intra-specific and inter-specific interference behavior among sexual and asexual geckos. *Anim. Behav.* In press.

Bravo, T. 1953. *Lacerta maxima* n. sp. de la fauna continental extinguida en el Pleistoceno de las Islas Canarias geol. *Inst. Invest. Geol. Lucas Mallada* 9:7–34.

Brooke, M.L., and D.C. Houston. 1983. The biology and biomass of the skinks *Mabuya sechellensis* and *Mabuya wrightii* on Cousin Island, Seychelles (Reptilia: Scincidae). *J. Zool. Lond.* 200:179–95.

Bull, P.C., and A.H. Whitaker. 1975. The amphibians, reptiles, birds, and mammals. In *Biogeography and ecology in New Zealand,* ed. G. Kushel. The Hague: Junk.

Burt, C.E., and M.D. Burt. 1932. Herpetological results of the Whitney south sea expedition. VI. Pacific island amphibians and reptiles in the collection of the American Museum of Natural History. *Bull. Amer. Mus. Nat. Hist.* 63:461–597.

Case, T.J. 1975. Species numbers, density compensation, and colonizing ability of lizards on islands in the Gulf of California. *Ecology* 56:3–18.

———. 1983. The reptiles: Ecology. In *Island biogeography in the Sea of Cortez,* ed. T.J. Case and M.L. Cody, Chap. 7. Berkeley: University of California Press.

Case, T.J., and D.T. Bolger. 1991. The role of introduced species in shaping the distribution and abundance of island reptiles. *Evol. Ecol.* 5:272–90.

Case, T.J., and M.L. Cody. 1987. Island biogeographic theories: Test on islands in the Sea of Cortez. *Am. Sci.* 75:402–11.

Cassels, R. 1984. The role of prehistoric man in the faunal extinctions of New Zealand and other Pacific islands. In *Quaternary extinctions,* ed. P.S. Martin and R.G. Klein, Chap. 34. Tucson: University of Arizona Press.

Cheke, A.S. 1984. Lizards of the Seychelles. In *Biogeography and ecology of the Seychelles Islands,* ed. D.R. Stoddart, Chap. 19. The Hague: Junk.

———. 1987. An ecological history of the Mascarene Islands with particular reference to extinctions and introductions of land vertebrates. In *Studies of Mascarene Island birds,* ed. A.W. Diamond and A.S. Cheke, 5–89. Cambridge: Cambridge University Press.

Cogger, H.G. 1971. The reptiles of Lord Howe Island. *Proc. Linn. Soc. New South Wales* 96:23–38.

———. 1983. *Reptiles and amphibians of Australia,* 3rd ed. Wellington, New Zealand: Reed.

Cogger, H.G., R. Sadlier, and E. Cameron. 1983. *The terrestrial reptiles of Australia's Island Territories.* Australia National Parks and Wildlife Service Special Publ. no. 11.

Corke, D. 1987. Reptile conservation on the Maria Islands (St. Lucia, West Indies). *Biol. Conserv.* 40:263–79.

Crook, I.G. 1973. The tuatara, *Sphenodon punctatus* on islands with and without populations of the Polynesian rat, *Rattus exulans. Proc. N.Z. Ecol. Soc.* 20:115–20.

Darlington, P.J. 1957. *Zoogeography, the geographical distribution of animals.* New York: John Wiley & Sons.

Dewar, R.E. 1984. Extinction in Madagascar: The loss of the subfossil fauna. In *Quaternary extinctions,* ed. P.S. Martin and R.G. Klein, Chap. 26. Tucson: University of Arizona Press.

Diamond, J.M. 1972. Biogeographic kinetics: Estimation of relaxation times for avifaunas of Southwest Pacific Islands. *Proc. Nat. Acad. Sci.* 69:3199–203.

———. 1984. Historic extinctions: A rosetta stone for understanding prehistoric extinctions. In *Quaternary extinctions,* ed. P.S. Martin and R.G. Klein, Chap. 38. Tucson: University of Arizona Press.

Diamond, J.M., and T.J. Case. 1986. Overview: Introduction, extinctions, exterminations, and invasions. In *Community ecology*, ed. J.M. Diamond and T.J. Case, Chap. 4. New York: Harper & Row.

Engbring, J., and T.H. Fritts. 1988. Demise of an insular avifauna: The brown tree snake on Guam. *Trans. W. Sect. Wldlf. Soc.* 24:31–37.

Estes, R. 1983. *Handbuch der Paläoherpetologie, Teil 10A. Sauria terrestria Amphisbaenia.* Stuttgart, W. Germany: Gustav Fischer Verlag.

Etheridge, R. 1964. Late Pleistocene lizards from Barbuda British West Indies. *Bull. Florida State Mus.* 9:43–75.

Evans, P.G., H., and J.B. Evans. 1980. The ecology of lizards on Praslin Island, Seychelles. *J. Zool. Lond.* 191:171–92.

Frogner, K.J. 1967. Some aspects of the interaction between the gecko species *Hemidactylus frenatus* and *Lepidodactylus lugubris* in Hawaii. M.S. thesis, University of Hawaii at Honolulu.

Gaffney, E.S., J.C. Balouet, and F. DeBroin. 1984. New occurrences of extinct meiolaniid turtles in New Caledonia. *Amer. Mus. Nov.* 2800:1–6.

Gardner, A.S. 1986. The biogeography of the lizards of the Seychelles Islands. *J. Biogeography.* 13:237–53.

Gehlbach, F.R. 1965. Amphibians and reptiles from the Pliocene and Pleistocene of North America: A chronological summary and selected bibliography. *Texas J. Science* 17:56–70.

Gibbons, J.R.H. 1984. Iguanas of the South Pacific. *Oryx* 18:82–92.

Gibbons, J.R.H., and I.F. Watkins. 1982. Behavior, ecology, and conservation of South Pacific banded iguanas, *Brachylophus*, including a newly discovered species. In *Iguanas of the world, their behavior, ecology, and conservation*, ed. G.M. Burghardt and A.S. Rand, Chap. 23. Park Ridge, N.J.: Noyes.

Gifford, E.W., and D. Shuter, Jr. 1956. Archeological excavations in New Caledonia. *Anthropol. Records* 18:1–148.

Gilpin, M.E., and M.E. Soulé. 1986. Minimum viable populations: Processes of species extinction. In *Conservation biology*, ed. M.E. Soulé, Sunderland, Mass.: Sinauer.

Greer, A.E. 1976. On the evolution of the giant Cape Verde scincid lizard *Macroscincus coctei. J. Nat. Hist.* 10:691–712.

Hadfield, M.G., S.E. Miller, and A.H. Carwile. 1989. *Recovery plan for the Oahu tree snails of the genus* Achatinella. Washington, D.C.: U.S. Fish and Wildlife Service.

Hardy, G.S. 1977. The New Zealand Scincidae (Reptilia: Lacertilia): A taxonomic and zoogeographic study. *N.Z. J. Zool.* 4:221–325.

Hardy, G.S., and A.H. Whitaker. 1979. The status of New Zealand's endemic reptiles and their conservation. *Forest and Bird* 13:34–39.

Hecht, M.K. 1951. Fossil lizards of the West Indian genus *Aristelliger* (Gekkonidae). *Amer. Mus. Nov.* 1538:1–33.

———. 1975. The morphology and relationships of the largest known terrestrial lizard, *Megalania prisca* Owen, from the Pleistocene of Australia. *Proc. Royal Soc. Victoria* 87:239–49.

Honegger, R.E. 1975. *Red data book. Vol. 3. Amphibia and Reptilia.* Morges, Switzerland: IUCN.

Hooton, E.A. 1925. Ancient inhabitants of the Canaries. *Harvard Afr. Studies* 7:1–401.

Hunsaker, D., and P. Breese. 1967. Herpetofauna of the Hawaiian Islands. *Pacific Science* 21:168–72.

Hutterer, R. 1985. Neue Funde von Rieseneidechsen (Lacertidae) auf der Insel Gomera. *Bonn. Zool. Beitr.* 36:365–64.

Ineich, I. 1987. Recherches sur le peuplement et l'evolution des reptiles terrestres de polynesie francaise. Ph.D. thesis, Université des Sciences et techniques du Languedoc, Academie de Montpellier, France.

Ineich, I., and C.P. Blanc. 1988. Distribution des reptiles terrestres en Polynesie Orientale. *Atoll Res. Bull.* 318:1–75.

Iverson, J.B. 1978. The impact of feral cats and dogs on populations of the West Indian iguana, *Cyclura carinata. Biol. Conserv.* 14:63–73.

Izquierdo, I., A.L. Medina, and J.J. Hernández. 1989. Bones of giant lacertids from a new site on El Hierro (Canary Islands). *Amphib.-Reptil.* 10:63–69.

Jolly, A., P. Oberlé, and R. Albignac, eds. 1984. *Madagascar (key environments).* Oxford: Pergamon Press.

Jones, R.E. 1979. Hawaiian lizards—their past, present and future. *Bull. Maryland Herpet. Soc.* 15:37–45.

Klemmer, K. 1976. The amphibia and reptilia of the Canary Islands. In *Biogeography and ecology in the Canary Islands,* ed. G. Kunkel, Chap. 15. The Hague: Junk.

Kramer, P. 1984. Man and other introduced organisms. In *Evolution in the Galapagos,* ed. R.J. Berry, 253–58. London: Academic Press.

Leguat, F. 1708. Voyages et aventures de Francois Leguat & de ses compagnons en deux iles désertes des Indes Orientales. Vol. 1 and 2. London: David Mortier.

Leigh, E.G. 1981. Average lifetime of a population in a varying environment. *J. Theor. Biol.* 90:213–39.

López-Jurado, L.F. 1985. Los reptiles fósiles de la Isla de Gran Canaria (Islas Canarias). *Bonn. Zool. Beitr.* 36:355–65.

———. 1989. A new Canarian lizard subspecies from Hierro Island (Canarian archipelago). *Bonn. Zool. Beitr.* 40:265–72.

MacArthur, R.H,. and E.O. Wilson. 1967. *The theory of island biogeography.* Princeton, N.J.: Princeton University Press.

Machado, A. 1985a. New data concerning the Hierro giant lizard and the lizard of Salmor (Canary Islands). *Bonn. Zool. Beitr.* 36:429–70.

———. 1985b. Hypothesis on the reasons for the decline of the large lizards in the Canary Islands. *Bonn. Zool. Beitr.* 36:563–75.

Machado, A., L.F. López-Jurado, and A. Martin. 1985. Conservation status of reptiles in the Canary Islands. *Bonn. Zool. Beitr.* 36:585–606.

Marrero Rodriguez, A., and C.M. Garcia Cruz. 1978. Nuevo yacimiento de restos subfósiles de dos vertebrados extintos de la Isla de Tenerife (Canarias), *Lacerta maxima* Bravo 1953 y *Canariomys bravoi* Crus. et Pet., 1964. *Vieraea* 7:165–74.

Martin, P.S., and R.G. Klein, eds. 1984. *Quaternary extinctions.* Tucson: University of Arizona Press.

McCallum, J. 1986. Evidence of predation by kiore upon lizards from the Mo-kohinau Islands. *N.Z. J. Ecol.* 9:83–87.

McCoy, M. 1980. Reptiles of the Solomon Islands. Wau, Papua New Guinea: *Wau Ecology Inst. Handbook,* no. 7.

McKeown, S. 1978. *Hawaiian reptiles and amphibians.* Honolulu: Oriental.

Medway, L., and A.G. Marshall. 1975. Terrestrial vertebrates of the New He-brides: Origin and distribution. *Phil. Trans. Roy. Soc. Lond. B.* 272:423–65.

Mercer, J. 1980. *The Canary Islands, their prehistory, conquest, and survival.* London: Rex Collins.

Mertens, R. 1942. *Lacerta goliath* n. sp., cine ausgestorbene Rieseneidechse von den Kanaren. *Senckenbergiana* 25:330–39.

———. 1956. Dei Eidechsen der Kapverden. *Commentat. Biol.* 15:1–16.

Mittermeier, R.A. 1972. Jamaica's endangered species. *Oryx* 11:258–62.

Molnar, R. 1984a. Cainozoic reptiles from Australia (and some amphibians). In *Vertebrate zoology and evolution in Australia,* ed. M. Archer and G. Clayton, 337–41. Perth: Hesperian Press.

———. 1984b. A checklist of Australian fossil reptiles. In *Vertebrate zoology and evolution in Australia,* ed. M. Archer and G. Clayton, 405–6. Perth: Hesperian Press.

Moodie, K.B., and T.R. Van Devender. 1979. Extinction and extirpation in the herpetofauna of the Southern High Plains with emphasis on *Geochelone wil-sonii* (Testitudinae). *Herpetologica* 35:198–206.

Myers, C.W., and A.S. Rand. 1969. Checklist of amphibians and reptiles of Barro Colorado Island, Panama, with comments on faunal change and sampling. *Smithsonian Contr. Zool.* no. 10.

Newman, D.G., ed. 1982. *New Zealand herpetology: Proceedings of a symposium held at Victoria University of Wellington, January 1980.* New Zealand Wild-life Service Occasional Publ. no. 2.

———. 1986. Can tuatara and mice coexist? The status of the tuatara, *Sphenodon punctatus* (Reptilia: Ryncocephalia), on the Whangamata Islands. In *The offshore islands of northern New Zealand,* ed. A.E. Wright and R.E. Beever, 179–85. New Zealand Dept. of Lands and Survey Info. Series no. 16.

Oliver, J.A., and C.E. Shaw. 1953. The amphibians and reptiles of the Hawaiian Islands. *Zoologica* 38: 65–95.

Olson, S.L., and H.G. James. 1982. Prodromus of the fossil avifauna of the Hawaiian Islands. *Smithsonian Contr. Zool.* 365:1–59.

Paulian, R. 1984. Madagascar: A micro-continent between Africa and Asia. In *Madagascar (key environments),* ed. A. Jolly, P. Oberlé, and R. Albignac, Chap. 1. Oxford: Pergamon Press.

Pernetta, J.C., and Watling, D. 1979. The introduced and native terrestrial ver-tebrates of Fiji. *Pacific Science* 32:223–44.

Pimentel, D. 1955. Biology of the Indian mongoose in Puerto Rico. *J. Mamm.* 36:62–68.

Pregill, G.K. 1981. *Late Pleistocene herpetofaunas from Puerto Rico.* University of Kansas Museum of Natural History Miscellaneous Publ. no. 71:1–72.

———. 1986. Body size of insular lizards: A pattern of Holocene dwarfism. *Ev-olution* 40:997–1008.

————. 1992. Systematics of the West Indian lizard genus *Leiocephalus* (Squamata: Iguania: Tropiduridae). *Univ. Kansas Mus. Nat. Hist. Misc. Publ.* no. 84.

Pregill, G.K., D.W. Steadman, S.L. Olson, and F.V. Grady. 1988. Late Holocene fossil vertebrates from Burman Quarry, Antiqua, Lesser Antilles. *Smithsonian Contr. Zool.* no. 463.

Pregill, G.K., and T. Dye. 1989. Prehistoric extinction of giant iguanas in Tonga. *Copeia* 1989:505–8.

Reese, D.S. 1989. Tracking the extinct pygmy hippopotamus of Cyprus. *Field Mus. of Nat. Hist. Bull.* 60:22–29.

Rich, P.V. 1982. Kukwiede's revenge: A look into New Caledonia's distant past. *Hemisphere* 26:166–71.

Richman, A., T.J. Case, and T.D. Schwaner. 1988. Natural and unnatural extinction rates of reptiles on islands. *Am. Nat.* 131:611–30.

Robb, J. 1986. *New Zealand amphibians and reptiles,* 2nd ed. Auckland: William Collins.

Sadlier, R.A. 1986. A review of the Scincid lizards of New Caledonia. *Records of the Austr. Mus.* 39:1–66.

————. 1988. *Bavayia validiclavis* and *Bavayia septuiclavis,* two new species of gekkonid lizard from New Caledonia. *Records of the Austr. Mus.* 40:365–70.

Salzburg, M.A. 1984. *Anolis sagrei* and *Anolis cristatellus* in southern Florida: A case study in interspecific competition. *Ecology.* 65:14–19.

Savidge, J.A. 1987. Extinction of an island forest avifauna by an introduced snake. *Ecology* 68:660–68.

Schmidt, K.P. 1928. Amphibians and land reptiles of Porto Rico, with a list of those reported from the Virgin Islands. *New York Acad. Sci.* 10:1–160.

Schoener, T.W. 1983. Rates of species turnover decreases from lower to higher organisms: A review of the data. *Oikos* 41:372–77.

Schoener, T.W., and A. Schoener. 1983. The time to extinction of a colonizing propagule of lizards increases with island area. *Nature* 302:332–34.

Schwidetzky, I. 1976. The prehispanic population of the Canary Islands. In *Biogeography and ecology in the Canary Islands*, ed. G. Kunkel, Chap. 2. The Hague: Junk.

Shallenberger, E.W. 1970. Tameness in insular animals: A comparison of approach distances of insular and mainland iguanid lizards. Ph.D. thesis, University of California, Los Angeles.

Steadman, D.W. 1986. Holocene vertebrate fossils from Isla Floreana, Galapagos. *Smithson. Contr. Zool.* no. 413.

Steadman, D.W., and S.L. Olson. 1985. Bird remains from an archaeological site on Henderson Island, South Pacific: Man-caused extinctions on an "uninhabited" island. *Proc. Natl. Acad. Sci.* 82:6191–95.

Steadman, D.W., G.K. Pregill, and S.L. Olson. 1984. Fossil vertebrates from Antigua Lesser Antilles: Evidence for late Holocene human-caused extinctions in the West Indies. *Proc. Natl. Acad. Sci.* 81:4448–51.

Stejneger, L. 1899. *The land reptiles of the Hawaiian Islands.* Proc. U.S.N.M., vol. 21, no. 1174.

Thomson, G.M. 1922. *The naturalization of animals and plants in New Zealand.* Cambridge: Cambridge University Press.

Thorpe, R.S. 1985. Body size, island size and variability in the Canary Island lizards of the genus *Gallotia. Bonn. Zool. Beitr.* 36:481–87.

Thorton, I. 1971. *Darwin's islands: A natural history of the Galapagos.* New York: American Museum of Natural History Press.

Towns, D.R. 1972. Ecology of the black shore skink, *Leiolopisma suteri* (Lacertilia: Scincidae), in boulder beach habitats. *N.Z. J. Zool.* 2:389–407.

Van Devender, T.R. 1977. Holocene woodlands in the Southwestern Deserts. *Science* 198:189–92.

———. 1987. Holocene vegetation and climate in the Puerto Blanco Mountains, southwestern Arizona. *Quaternary Res.* 27:51–72.

Van Devender, T.R., and J.M. Mead. 1978. Early and Late Pleistocene amphibians and reptiles in Sonoran Desert packrat middens. *Copeia* 1978:464–75.

Vinson, J. 1973. A new skink of the genus *Gongylomorphus* from Macabé forest (Mauritius). *Revue Agric. Sucr. Ile Maurice* 52:39–40.

Vinson, J., and J. Vinson. 1969. The saurian fauna of the Mascarene Islands. *Mauritius Inst. Bull.* 6:203–320.

Whitaker, A.H. 1968. The lizards of the Poor Knights Islands, New Zealand. *N.Z. J. Sci.* 11:623–51.

———. 1973. Lizard populations on islands with and without Polynesian rats, *Rattus exulans. Proc. N.Z. Ecol. Soc.* 20:121–30.

———. 1982. Interim results from a study of *Hoplodactylus maculatus* (Boulenger) at Turakirae Head, Wellington. In *New Zealand herpetology*, ed. D.G. Newman, 363–74. Auckland: New Zealand Wildlife Service Occasional Publ. no. 2.

Wilcox, B.A. 1980. Insular ecology and conservation. In *Conservation biology: An evolutionary-ecological perspective*, ed. M.E. Soulé and B.A. Wilcox, Chap. 6. Sunderland, Mass.: Sinauer.

———. 1981. Aspects of the biogeography and evolutionary ecology of some island vertebrates. Ph.D. diss., University of California, San Diego.

Willis, E.O. 1974. Populations and local extinction of birds in Barro Colorado Island, Panama. *Ecol. Monog.* 44:153–69.

Wilson, L.D., and L. Porras. 1983. The ecological impact of man on the South Florida herpetofauna. University of Kansas Museum of Natural History Spec. Publ. no. 9.

Wingate, D.W. 1965. Terrestrial herpetofauna of Bermuda. *Herpetologica* 21:202–18.

Worthy, T.H. 1987a. Osteological observations on the larger species of the skink *Cyclodina* and the subfossil occurrence of these and the gecko *Hoplodactylus duvaucelii* in the North Island, New Zealand. *N.Z. J. Zool.* 14:219–29.

———. 1987b. Paleoecological information concerning members of the frog genus *Leiopelma*: Leiopelmatidae in New Zealand. *J. Royal Soc. of N.Z.* 17:409–20.

Zug, G.R. 1992. *The lizards of Fiji: Natural history and systematics.* Honolulu: Bernice P. Bishop Museum Press.

CHAPTER
6

Loss of Biodiversity in Aquatic Ecosystems: Evidence from Fish Faunas

PETER B. MOYLE

and ROBERT A. LEIDY

ABSTRACT

Fishes are appropriate indicators of trends in aquatic biodiversity because their enormous variety reflects a wide range of environmental conditions. Fish also have a major impact on the distribution and abundance of other organisms in waters they inhabit. Examination of trends in freshwater fish faunas from different parts of the world indicate that most faunas are in serious decline and in need of immediate protection. Species most likely to be threatened with immediate extinction are either specialized for life in large rivers or are endemic species with very small distributions. We conservatively estimate that 20 percent of the freshwater fish species of the world (ca. 1800 species) are already extinct or in serious decline. Evidence for serious declines in marine fishes is limited largely to estuarine fishes, reflecting their dependence on freshwater inflows, or to fishes in inland seas. The proximate causes of fish species' decline can be divided into five broad categories: (1) competition for water, (2) habitat alteration, (3) pollution, (4) introduction of exotic species, and (5) commercial exploitation. Although one or two principal causes of decline can be identified for each species, the decline is typically the result of multiple, cumulative, long-term effects. Ways to protect aquatic biodiversity include the implementation of landscape-level management strategies, the creation of aquatic preserves, and the restoration of degraded aquatic habitats. Without rapid adoption of such measures we are likely to experience an accelerated rate of extinctions in aquatic environments as human populations continue to expand.

INTRODUCTION

Conservation biology has been focused mainly on the loss of biotic diversity in terrestrial environments, especially in tropical forests (Lovejoy 1980; Ehrlich and Ehrlich 1981; Myers 1980, 1983; Lanly 1982; Simberloff 1986; Wilson 1988). Loss of diversity in aquatic environments has received comparatively little attention, even though the physical, chemical, and biological degradation of aquatic environments is widely recognized as a major problem, usually in the context of spread of human disease; loss of fisheries; or loss of water quality for drinking, irrigation or recreation. Yet aquatic habitats support an extraordinary array of species, many of which are being lost as their habitats deteriorate. Unfortunately, the extent of species loss from aquatic systems is poorly known.

This lack of knowledge is caused by a number of factors. Most marine habitats are too remote or too deep for easy study; most freshwater habitats are too turbid, deep, or swift to monitor easily. In addition, most aquatic organisms are microscopic algae or small invertebrates that are difficult to identify. Fish are the organisms we know best in these habitats because of their size, abundance, economic importance, and comparative ease of capture and identification (Karr 1981). Presumably, the trends we see in fish diversity apply to other aquatic organisms as well. The purpose of this chapter, therefore, is to describe the decline of biodiversity in aquatic environments as reflected by fishes as well as to make suggestions for fish conservation. To do this, we will examine the following questions:

1. Why are aquatic environments so vulnerable to degradation?
2. How much diversity exists in aquatic environments?
3. How appropriate are fish as indicators of aquatic biodiversity?
4. How rapidly are species being lost in different fish faunas?
5. What are the primary causes of loss of aquatic biodiversity?
6. What can be done to protect aquatic biodiversity?

VULNERABILITY OF AQUATIC ENVIRONMENTS TO DEGRADATION

There are several reasons that aquatic environments are so vulnerable to degradation. These include the complex properties of water itself, the interactions between aquatic and terrestrial environments, and the proximity of human populations to aquatic systems.

The productivity of various aquatic environments is driven largely by the capacity of water as a solvent and its tendency to ionize dissolved substances. As a result, inland and near-shore marine environments are affected not only by internal biogeochemical processes, but also by processes in adjacent terrestrial environments. Such aquatic habitats are consequently extremely productive, and humans have exploited this productivity for food, especially fish. However, the great assimilative properties of water also have resulted in the use of inland and marine aquatic environments as seemingly endless "sinks" for cultural wastes produced in terrestrial biomes.

Some energy and material flow from aquatic to terrestrial environments, but it is much less than flow in the opposite direction. Thus, cutting rain forests adjacent to streams in the Pacific Northwest of North America or in the tropics causes major shifts in biotic communities of the streams through such processes as the filling in of pools by siltation, increased growth of algae in exposed reaches, and increased amounts of woody debris in the channel. In many instances, the effects of logging are felt far downstream as coastal lagoons and estuaries fill with sediment.

The effects of terrestrial changes on aquatic environments are particularly severe in areas where human populations are dense, as urbanization and agricultural development cause major alterations of aquatic communities. The vast majority of the human population is concentrated close to streams, lakes, estuaries, and coastal areas (Figure 6.1). In the United States, for example, roughly 50 percent of the population lives within 60 km of the coastline (U.S. Congress Office of Technology Assessment 1987), and most inland cities center on lakes and rivers. Coastal populations are expected to increase significantly by early next century (Culliton et al. 1990).

The fact that aquatic environments are recipients of virtually every form of human waste has resulted in their rapid and continuous degradation. It is an indication of their resilience that so many of them still contain rich native biotas, although the most altered environments are often dominated by introduced species that have displaced native species.

The concentration of people around freshwater systems has resulted in a much greater degree of degradation to these systems than to most open marine systems (International Institute for Environmental and Development and World Resources Institute [IIED and WRI] 1987). Increasing demand for freshwater resources generated by continued population growth, urbanization, industrialization, and irrigation will likely result in further declines of freshwater biotas. The loss of diversity we are seeing in freshwater systems is starting to spread seaward, especially

Figure 6.1
Most large estuaries have been modified and polluted by urbanization on their edges; Eureka, California, is on the estuary of the Eel River.

into shallow coastal areas that house a majority of marine species. The ability of freshwater environments to absorb so much abuse has buffered coastal marine environments from secondary effects of human civilization. This buffering capacity is rapidly being lost.

DIVERSITY IN AQUATIC ENVIRONMENTS

Because over two-thirds of the earth's surface is covered with water and because the average ocean depth is approximately 4,000 m, it might be expected that a large majority of aquatic species would be found in the open ocean. This is not the case, however; the surface and deep waters of the ocean are relatively uniform habitats of low productivity and contain few species. The greatest diversity of aquatic life is distributed overwhelmingly along continental shelves, in reefs associated with islands, or in fresh water. Thus, 41 percent of all fish species are exclusively freshwater, 1 percent move on a regular basis between the ocean and fresh water, 44 percent are shallow-water marine species, 12 percent are deep-sea fishes, and 1 percent are open-ocean fishes (Cohen 1970). Diversity of fishes in shallow-water marine environments and fresh water is caused by the same evolutionary processes that have created high diversity in terrestrial habitats: opportunities for speciation following events such as the rise of mountain ranges that isolate regions or sea level fluctuations that isolate bays. Considering that fresh water covers about 1 percent of the continents and that the continental shelves (down to 200 m) have an area of less than 10 percent of the continents, the diversity of aquatic species is surprisingly high.

Diversity can be viewed in terms of the number of species in an area (species richness) or in terms of the number of higher taxa, such as families, orders, or phyla. In general, aquatic environments have a greater number of higher taxa than do terrestrial environments. For example, roughly twice as many phyla live in oceans than in terrestrial environments (Ray 1988). This diversity reflects the long ancestry of life in the oceans. In terms of species richness, however, terrestrial habitats usually score much higher than aquatic ones, largely because of the diversity of insects, flowering plants, and their associated nematodes. Curiously, among vertebrates the breakdown is about even: 22,000 fishes versus 4,000 mammals, 8,700 birds, 6,000 reptiles and 3,000 amphibians, including the aquatic species within each group (21,700 total species).

FISH AS AN INDICATOR OF AQUATIC BIODIVERSITY

The main reason for using fish to monitor biodiversity is that we know more about them than about other aquatic organisms. They are also relatively easy to collect and identify. However, there are other good arguments for using fish. First, they are enormously diverse, with different species reflecting different environmental conditions (Moyle and Cech 1988). Second, fish often have major effects on the distribution and abundance of other organisms in the waters they inhabit. In lakes, for example, plankton-feeding fish can cause major changes in the abundance and species of zooplankton, which in turn causes cascading changes in the abundance of other organisms in the food web (Carpenter and Kitchell 1988). In streams, algae-feeding fish alter algal communities, except where their populations are kept low by predators (Power, Matthews, and Stewart 1985). Goulding (1980) presents evidence that some species of characins in the Amazon Basin may be important dispersers of tree seeds in seasonally flooded lowland forests of the Amazon. The presence of kelp forests off the coast of southern California is determined in part by the ability of fish to control invertebrates that graze on the kelp (Foster and

Schiel 1985). When kelp forests are large and widespread, many species of fish are more abundant than they would be without the forests (Figure 6.2). Fishes are thus often aquatic examples of, in the words of Terbough (1988), "the big things that run the world."

SPECIES LOSS IN FISH FAUNAS

In most of the world, the status of local fishes is poorly known. Clearly freshwater fishes and their habitats are being lost much more rapidly than marine fishes and habitats. This may be changing, however, as exploitation of oceanic resources, marine pollution, and coastal habitat alteration increase (Kaufman 1988). Marine fish communities also are monitored much less closely than freshwater communities. Therefore, this section will deal mainly with freshwater fishes and only briefly with marine fishes. For freshwater fishes, this section will deal only with fish faunas of selected countries or regions that have adequate information on the status of their fish faunas: (1) North America, with California and Arkansas as regional examples; (2) Europe; (3) Iran; (4) South Africa; (5) Sri Lanka; (6) Australia; and (7) Latin America.

The status of fishes can be divided into three overlapping categories: endangered, threatened, and of special concern (Table 6.1). *Endangered species* have populations that have declined to a point where extinction is imminent if action is not taken to protect the species. *Threatened species* are declining and are likely to become endangered in the near future if protective action is not taken. *Species of special concern* are those in decline or of very limited range but not faced with extinction in the immediate future (or not known to be so).

North America

The status of the freshwater fish fauna of North America is the best known for any zoogeographic region. Williams et al. (1989) list 397 fish taxa (species or subspecies) as either having or needing protection throughout or in some portion of their range. These 397 taxa represent 343 species, over one-third of the 950 species known from North America (Moyle and Cech 1988). Many of these fishes are listed by states on the periphery of their ranges but not by states more centrally located, an indication that increasingly, the ranges of fishes are shrinking to include only presumed "optimal" habitat. Since 1900, three genera, twenty-seven species, and thirteen subspecies of fish are known to have gone extinct in North American waters (Miller, Williams, and Williams 1989).

Figure 6.2
Fish have major effects on the distribution and abundance of kelp-grazing invertebrates and, in turn, the abundance of species of fish within the kelp forests. Kelp forest off southern California. Photography by P. Cotter.

Table 6.1
Status of Fish Faunas of Selected Countries or Regions.

Region	Endangered	Threatened	Special Concern	% of Total Fauna	Source
North America	103	114	147	31	Williams et al. (1989)
California	10	15	40	69*	Moyle and Williams (1990)
Arkansas	11	16	15	21	Robison and Buchanan (1988)
Europe	21	40	19	42	Lelek (1987)
Iran	1	15	19	22	Coad (1981)
South Africa	4	31	27	63	Skelton (1987) and pers. com.
Sri Lanka	8	3	7	28	Pathiyagoda (1991)
Australia	6	6	37	26	Harris (1988)
Latin America	1	1	12	9	Bussing (pers. com.)

*Includes eight species now extinct in California (updated to December 1991).

The U.S. Fish and Wildlife Service (1989) considers only seventy-three fish taxa to be endangered or threatened with extinction in the United States. This number is very conservative, reflecting the difficulty in listing species as federally threatened or endangered, especially in the unfriendly political climate for such listings that has existed since the 1980s. For example, of 214 native stocks of Pacific salmonids (*Oncorhynchus* spp.) identified from California, Oregon, Idaho, and Washington, Nehlsen, Williams, and Lichatowich (1991) list 101 stocks (47%) at a high risk of extinction, 58 stocks (27%) at a moderate risk of extinction, and 54 (25%) of special concern. However, only a single run, the winter run Chinook salmon (*O. tshawytschi*) of the Sacramento River, California, is listed as threatened under the Endangered Species Act of 1973. Additional difficulties that prevent the formal protection of fishes (and other organisms) are lack of supporting biological information and the uncertain taxonomic status of some taxa. Overall, it is likely that globally endangered or threatened species represent at least 10 percent of the fish fauna of North America.

California

California, because of its size, its aridity, and the high degree of endemism (50%) in its fishes, represents the problems facing freshwater fishes throughout the western United States. Of the 115 taxa native to the state, 7 percent are extinct, 12 percent are formally listed as threatened or endangered either by state or federal authorities, 28 percent either need listing or will be listed, and 22 percent have declining populations (Moyle and Williams 1990). Only 31 percent of the native fish fauna of California can be regarded as "secure." Even species now regarded as secure have to be monitored, however, as two species of California fishes have had their populations plummet from high abundance to endangered status in less than ten years (Moyle and Williams 1990).

Threatened or endangered fish taxa in California are most likely to be endemic forms with limited ranges, especially those confined to desert springs and small streams, or fishes that inhabit large rivers (Figure 6.3). Not surprisingly, these fishes occur most frequently in arid basins, where competition for water from humans is most severe. In the lower Colorado River, for example, all native fishes are either extinct or endangered. In the arid Los Angeles region, all native fishes need formal protection (Moyle and Williams 1990).

Arkansas

Arkansas is centrally located on the Mississippi River, and therefore the status of its fauna represents well the status of fishes of the eastern United States. According to Robison and Buchanan (1988), over 20 percent of the freshwater fishes of North America have been recorded in the state. Of the 198 native species, Robison and Buchanan (1988) list 48 species (24%) as either extinct (6 species), endangered (11 species), threatened (16 species), or of special concern (15 species). Nineteen of the 48 species (40%) have declined throughout their range either because they are highly localized endemics (14 species) or because they are inhabitants of the Mississippi River. The remaining 29 species are regarded as threatened or endangered mainly in Arkansas: five of these are big-river species; the remainder are widely distributed species for which Arkansas is the edge of their range. This analysis repeats the pattern found in California: Species most likely to be threatened with extinction either depend on big-river habitats or are localized endemics. Because of Arkansas' central location, another pattern is also apparent: Many species are becoming increasingly restricted in their distributions, as peripheral populations disappear.

Figure 6.3
Many of California's endangered fishes are endemic forms confined to small desert springs or large rivers. Top: *Big Spring, Nye County, Nevada, contains the endangered Ash Meadows Amargosa pupfish (*Cyprinodon nevadensis *ssp.* mionectes) *and speckled dace (*Rhinichthys osculus *ssp.* nevadensis). *A third species inhabiting the spring, the Ash Meadows pool-fish (*Empetrichthys merriami) *became extinct in the early 1950s, probably as a result of introduced predators and physical habitat alteration.* Bottom: *The construction of large storage reservoirs and channelization has contributed to the decline and/or extinction of all native fishes in the lower Colorado River. Davis Dam on the lower Colorado River, California/ Arizona.*

Europe

The status of the freshwater fish fauna of this heavily industrialized, temperate region is discussed by Lelek (1987). His classification of species' status is more inclusive than most other classifications of endangered species (e.g., a number of species he lists as endangered still support commercial fisheries). The overall numbers, however, provide a reasonable reflection of the status of the 193 species that comprise the European fish fauna: 21 (11%) are classified as endangered, 40 (21%) as threatened (vulnerable), and 19 (10%) are species of special concern (rare). An additional 21 species are classified as of "intermediate" or unknown status. Although probably only 4 or 5 species fit the definition of endangered used here, it is likely that all 80 species in Lelek's (1987) first three categories would at least fit the definitions of threatened or special concern species. In short, 42 percent of Europe's freshwater fish species need special protection to keep them from becoming extinct in natural habitats.

Iran

Iran is one of the older agricultural areas of the world. Humans have affected inland fish populations in this region through irrigation diversions, drainage of wetlands, and other activities for at least 5,000 years. These effects have increased greatly in the past fifty years because of rapidly increasing human population (Coad 1981). Coad (1981) lists 159 native inland fish species in Iran, of which 1 (0.6%) is endangered, 15 (9%) are threatened, and 19 (12%) are of special concern. In all, 22 percent of Iran's fishes have severely declining populations. Most of these species are associated with the Caspian Sea and its tributaries, where overfishing, pollution, habitat destruction, and water diversion have significantly reduced populations of five sturgeon (Acipenseridae) species, eight herring (Clupeidae) species, and six cyprinid (Cyprinidae) species (Coad 1981; Rozengurt and Hedgepeth 1989).

South Africa

The Republic of South Africa is as industrialized and as intensely farmed as its climatic counterparts in Europe and North America. Not surprisingly, its fish fauna is in equally poor condition. Skelton (1987 and pers. com.) considers 4 of the 94 freshwater fishes there to be endangered, 31 to be threatened, and 27 to be of special concern. In all, 63 percent of the fishes fall into one of the three categories. Many of these species occur in the Cape Province, where the fauna is depauperate (33 species) and largely endemic, and where the countryside is extensively developed. In this province, 9 percent of the species are endangered, 50 percent are threatened, and 39 percent are of special concern. Only one Cape Province species is regarded by Skelton (pers. com. 1989) to be secure!

Sri Lanka

Sri Lanka contains a freshwater fish fauna of 65 species. The southwestern third of this densely populated island is covered with rain forest and contains most of Sri Lanka's 29 endemic fishes (Senanayake and Moyle 1982). The remaining species are shared with India, although the differences between many Indian and Sri Lanka forms may be at the subspecies or species level. Eight of the endemic forms are endangered, 3 are threatened, and 2 are of special concern (Pathiyagoda 1991). This represents 28 percent of Sri Lanka's freshwater fishes. Fourteen of these fishes are confined to the rain forest streams.

Australia

The Australian fish fauna is unique because it is highly endemic and consists mostly of freshwater representatives of marine families. The fishes are found mainly in coastal drainages because the arid interior of the continent confines fishes to a few drainages with limited water. The status of many of its 188 freshwater species (Allen 1987) is poorly known, but human competition for limited water is causing species to decline in many areas (Arthington, Milton, and McKay 1983; Harris 1987). From Harris (1988) it can be determined that 3 percent of the fishes are endangered, 3 percent are threatened, and 20 percent are of special concern. An additional 9 percent of the fishes probably should be listed as well, but there is too little information on them to determine their status (Harris 1988).

Latin America

The fish fauna of Costa Rica is perhaps the best known of any region in Central and South America (Bussing 1987). Of the 127 freshwater fishes known from Costa Rica, 1 is endangered, 1 is threatened, and 10 are of special concern (W.A. Bussing, pers. com. 1989). Overall, 9 percent of the fauna need special management at present. This percentage is probably low compared to other countries in the region because in Costa Rica an exceptionally large proportion of the area consists of national parks. Chile, for example, with a relatively depauperate fauna, has 18 species of fish that are endangered, 23 that are threatened, and 3 that are of special concern (Glade 1987). Brazil, with between 2,000 and 5,000 fish species in the Amazon River drainage alone (M. Campos pers. com.), probably has many species that are threatened or endangered because of rapid development of rain forest lands. However, many of the species are not yet described. When stream degradation threatens indigenous fisheries, the governmental "solution" is to make fishing illegal to reduce protests (Petrere 1989).

Fresh Water Worldwide

There are few places in the world where the status of the fish fauna is as well documented as the examples just discussed. However, a reasonable, conservative extrapolation from this information is that 20 percent of the freshwater fish species of the world (ca. 1,800 species) are already extinct or in serious decline, with many of them in need of immediate protection to prevent extinction. Major ecological disasters quickly raise this estimate. For example, the introduction of predatory Nile perch (*Lates nilotica*) into Lake Victoria, Africa, has caused the probable extinction or endangerment of 100 to 200 species of cichlids (Hughes 1986).

Marine Fishes

In marine environments most species in serious decline are estuarine, reflecting their indirect dependence on freshwater inflows. For example, the totoaba (*Cynoscion mcdonaldi*) is one of the few marine fishes listed as endangered in the United States. It depends on the Colorado River estuary for spawning, a habitat severely degraded from freshwater diversions (Ono, Williams, and Wagner 1983). Similarly, populations of the delta smelt (*Hypomesus transpacificus*), endemic to California's Sacramento–San Joaquin estuary, have declined precipitously in recent years. Diversion of fresh water into the estuary has been implicated as a major cause of this decline (Moyle, Williams and Wikramanayake 1990). Skelton (1987) lists eighteen species of estuarine fishes in need of protection in South Africa but does not list any marine species, although South African marine fishes are comparatively well studied (Smith and Heemstra 1986).

The lack of listed endangered marine fishes in North America and South Africa may reflect the shortage of information on marine fishes but most likely means that few marine fishes are in danger of extinction worldwide. This is because most marine fishes are fairly widely distributed or occur in habitats that currently are not as disturbed as freshwater habitats. The exceptions are those species found in enclosed basins, such as the Black or Aral seas, that are more susceptible to habitat degradation and destruction. This pattern may change rapidly, however, as coastal development fragments marine habitats, pollution becomes more pervasive, introduced species become more abundant, and improved technology makes deep-water habitats more accessible to exploitation of mineral and biological resources. The invasions of introduced species of fish and invertebrates is a particularly thorny and poorly understood problem in that hundreds of species are transported long distances in ballast water of large ships (Carlton 1985). In some bays, a high percentage of the fauna today consists of nonnative species (Carlton 1989; Moyle 1991).

Exploitation of marine fishes is also a major factor altering marine fish communities. Although we have no direct evidence of marine fish species driven to extinction or near-extinction by overfishing, there are many examples of commercial fish populations becoming too low to support viable commercial fisheries (e.g., Pacific sardine, *Sardinops sagax*). Overharvesting of fish populations has repercussions throughout the entire local food web and may create problems for other species of which we have little knowledge. For example, there is growing concern over commercial and recreational overfishing of shark species and its potentially significant effects on marine food webs (Manire and Gruber 1990). In addition, some types of marine fishing physically alter the environment. Repeated bottom-trawling over an area literally plows the bottom like an agricultural field, altering its ability to support benthic life. Fishing on coral reefs using sodium cyanide and dynamite, which kills coral, is common in the Philippines and other tropical areas. This typically destroys much of the reef structure that enables it to support such a diversity of life. If such activities continue indefinitely, extinction of many marine fishes and invertebrates will follow.

CAUSES OF THE LOSS OF AQUATIC BIODIVERSITY

The ultimate cause of most of the loss of biodiversity is the exponential expansion of human populations. Until that expansion and the concomitant rapidly expanding use of natural resources cease, any efforts to protect species from extinction will be short-term "holding actions." If we are to invest wisely in such holding actions, we need to understand the proximate causes of species decline and loss. For fish and other aquatic organisms, we have divided these causes into five broad categories: (1) competition for water, (2) habitat alteration, (3) pollution, (4) introduction of exotic species, and (5) commercial exploitation. The first three factors, often acting in concert, are the principal causes of the loss of aquatic biodiversity in aquatic systems, but their effects typically are exacerbated by the introduction of exotic species and overexploitation. The effects of all these factors are also both additive and cumulative.

Competition for Water

Humans become successful competitors for water with fish and other organisms when they withdraw water from streams, lakes, springs or underground aquifers. Competition for water may take many forms. However, here we are concerned primarily with impacts on fish communities resulting from the total or partial desiccation of aquatic systems through various diversion, impoundment, and extraction practices. Water is withdrawn most often from aquatic environments for irrigation, flood control, and urban and industrial consumption. In arid western North America, such practices are a major direct cause of the decline of native fishes (Moyle and Williams 1990), and they are increasingly the cause of declines elsewhere, especially in semi-arid and arid regions of the world. This is illustrated by the following examples from Nevada, California, Kansas, Africa, and the Soviet Union.

Pyramid and Winnemucca Lakes, Nevada. As recently as the turn of the century, Pyramid and Winnemucca lakes covered approximately 200 and 100 km² in the northwestern Great Basin, respectively. These alkaline remnants of Pleistocene Lake Lahontan were interconnected periodically by a slough. Two endemic fishes, a Pyramid Lake strain of the Lahontan cutthroat trout (*Oncorhynchus clarki henshawi*) and cui-ui (*Chasmistes cujus*), a member of the sucker family (Catostomidae), were dependent on the inflowing Truckee River for spawning (Sigler et al. 1983; Scoppetone, Coleman, and Wedemeyer 1986). Diversions of the Truckee River to irrigate farmland in Nevada began in 1905 with the completion of Derby Dam, which also served as a barrier to fish migrating upstream to spawn. Continuing diversions reduced water levels in both lakes. and Winnemucca Lake became dry in 1938. Low flows of the Truckee River created a delta at its mouth that blocked spawning migrations from Pyramid Lake to the limited stream habitat still available (Figure 6.4; Scoppetone et al. 1986). As a result, cutthroat trout were extinct by 1943, and the cui-ui became listed as endangered (Ono et al. 1983). The cui-ui survives in Pyramid Lake only because it is long-lived (40+ years) and because extraordinary measures have taken made to insure its reproduction (e.g., hatchery production and construction of a fish elevator over a dam). A nonnative strain of cutthroat trout has been introduced into Pyramid Lake. The population is maintained primarily through fish hatcheries. However, these long-term measures may be of little avail; if present water diversions continue and the lake level continues to fall, then Pyramid Lake will become too saline to support any fish life. Unfortunately, plans to increase flows in the Truckee River to stabilize lake levels may result in increased water diversions from adjacent river drain-

Figure 6.4
Pyramid Lake, Nevada, showing the wide shoreline created by the drying up of the lake through water diversions and the canal constructed in an attempt to allow migrating cutthroat trout and cui-ui to pass human-made barriers in order to reach their spawning grounds in the Truckee River.

ages. This may have disastrous consequences for the biotas of these adjacent basins.

San Joaquin Valley, California. The San Joaquin Valley forms the southern portion of California's Central Valley and consists of two distinct drainages (San Joaquin and Tulare). Historically, the San Joaquin drainage supported spawning runs of about 300,000 chinook salmon (*Oncorhynchus tshawytscha*) (California Advisory Committee on Salmon and Steelhead Trout [CACSS] 1988). With the construction of Friant Dam on the mainstem San Joaquin River in 1946, spawning migrations were blocked to the upper headwaters and more than fifty miles of river downstream of the dam were dewatered (CACSS 1988). As a result, the entire spring run became extinct (Smith 1987a, 1987b). Other runs also declined; the overall decline of chinook salmon is estimated at 90 percent for the entire basin. The decline is tied not only to Friant Dam but to forty-one other major reservoirs with >4,000 acre-feet storage capacity on streams within the San Joaquin River basin (Smith 1987).

The Tulare Basin, consisting of Tulare, Buena Vista, and Kern lakes, was the largest contiguous wetland in California, totaling approximately 325,000 ha. These lakes supported a commercial fishery for native turtles and minnows in the nineteenth century (Moyle 1976). Conversion of these lakes to agricultural land was accomplished largely through diversion of inflowing streams and construction of drainage ditches. Several dams constructed on tributaries to the Tulare Basin to control flooding now supply irrigation water to farmlands located on the historic lakebed! Overall, of the twenty-four native fish taxa that originally lived in the San Joaquin drainage, 17 percent are extinct, 4 percent are endangered, 17 percent are threatened, and 21 percent are of special concern: 58 percent all together (Brown and Moyle 1989).

Kansas. In the Great Plains of central North America, a major cause of the decline of aquatic faunas is the pumping of groundwater for irrigation. This pumping has lowered groundwater levels so that springs and seeps that once kept streams flowing during dry periods have dried up (Cross and Moss 1987). The result has been a gradual loss of fish communities that inhabited small clear streams in addition to the loss of wetlands important to waterfowl. For example, of twenty-four species of fish collected from a Kansas stream in 1885, only seven of the most tolerant species were still there 100 years later (Cross and Moss 1987).

Africa. Dam construction in Africa has stabilized stream flows and lessened or eliminated annual flooding, with negative consequences to downstream riverine habitats, aquatic communities, and fisheries (Scudder 1989). For example, in northern Nigeria the construction of the Kainji, Bakolori, and Tiga dams greatly stabilized stream flows below the dams and eliminated much of the seasonal inundation of floodplains. As a result, fish landings declined over 50 percent in the region (Scudder 1989). There was also a dramatic change in species composition as fish species dependent on flooded areas or muddy backwaters became uncommon (Welcomme 1985). Negative effects on fisheries also have been documented below Kariba Dam in Zambia/Zimbabwe, Pongolapoort Dam in South Africa, Aswan High Dam in Egypt, and Akosombo Dam in Ghana (Scudder 1989).

Large water-diversion projects in Africa have not only affected freshwater fish and fisheries, but have also caused reductions in estuarine and marine fish populations. Reduction of freshwater inflows reduces the amount of nutrients flowing into downstream areas and increases salinities. For example, the Aswan High Dam impounds 50 to 80 percent of the flow of the Nile River. Its construction was associated with a drastic decline in pelagic fish populations in the eastern Mediterranean Sea (Ryder 1978). Annual landings of sardines (*Sardinella* spp.) and other fish in the eastern Mediterranean have declined roughly 96 percent and 36 percent, respectively, since the dam began operation in 1965 (Aleem 1972; White 1988). The drastic changes in the fisheries are indications that entire communities of organisms have changed.

Soviet Union. In the Soviet Union, construction of storage dams on large rivers for irrigation has reduced freshwater inflows to estuaries, inland seas, and shallow coastal zones and has caused the loss of fish species and near-elimination of major fisheries. Rozengurt, Hertz, and Feld (1987) concluded that the productivity and aquatic communities of the Azov, Black, Caspian, and Aral seas were drastically impacted by continuous water diversions exceeding 30 percent or more of the natural spring and annual runoff. Dam construction on the Volga River has been largely responsible for reduction in spring runoff into the Caspian Sea, contributing to a reduction in commercial fish harvests of up to 90 percent (Welcomme 1985; Rozengurt and Hedgepeth 1989). Similar patterns of declines in fish catches of up to 90 percent following construction of diversion projects on major rivers have been recorded for the shallow coastal zones of the Black Sea, the Sea of Azov, and the Aral Sea (Rozengurt et al. 1987; Micklin 1988).

Physical Habitat Alteration

Habitat alteration is the single biggest cause of loss of diversity of aquatic life because few aquatic habitats have not been affected either directly or indirectly by human activities. In this section we discuss several major types of habitat alteration and their impacts on aquatic habitats and fish.

Channelization. The physical, chemical, and biological impacts of channelization or bank stabilization on riverine habitats and their adjacent floodplains is well documented (Simpson et al. 1982). Channelization typically involves the realignment, clearing, widening, and lining of the stream channel, usually for flood control. Among other effects, channelization may reduce stream length, create uniform habitat conditions (reduce habitat heterogeneity), modify the hydrologic cycle, drain adjacent wetlands, eliminate instream cover and riparian vegetation, degrade water quality, and alter trophic relationships (Simpson et al. 1982; Figure 6.5).

In the United States, many small streams flowing through agricultural or urban lands have been channelized, resulting in major changes in the abundance and diversity of fishes and invertebrates. Channelization is often conducted in a piecemeal fashion. This results in fragmentation of

Figure 6.5
Top: *Channelized reach of the Sacramento River, California. Note the armouring of the banks and lack of riparian vegetation.* Bottom: *Many smaller streams, such as this tributary to San Francisco Bay, have been channelized for flood control.*

formerly continuous stream corridors and confines many less-tolerant species to short sections of undisturbed stream. Isolated populations within these habitat "islands" become more vulnerable to extinction by natural and human-caused events such as floods or pollution (Sheldon 1988). Recolonization of fragmented natural habitats often is not possible because interconnecting channelized sections will have high velocities during high flows and may dry up during low flow periods. For example, in small streams flowing into San Francisco Bay, California, channelized stream segments were found to be either devoid of fish altogether or dominated by introduced fishes (Leidy 1984; Leidy and Fiedler 1985). In general, habitat alteration encourages the invasion of "weedy" fish species that can tolerate degraded conditions, such as common carp (*Cyprinus carpio*), green sunfish (*Lepomis cyanellus*), and mosquitofish (*Gambusia affinis*). The high reproductive rates and aggressive behavior of such fishes may then allow them to invade adjacent, less disturbed habitats and displace native species.

Channelization also has dramatic effects on the fish faunas of larger rivers. For example, the Mississippi and Missouri rivers have been channelized over much of their lengths, eliminating shallow-water habitats within stream channels as well as adjacent riparian habitats that flooded annually. These habitats were vital to the reproduction and rearing of many riverine fishes. The channelization of the Missouri River between St. Louis, Missouri, and Sioux City, Iowa, resulted in the direct loss of over 40,000 ha of aquatic habitat and removal of 150,000 ha of wetland and terrestrial habitat from the active erosion zone. Channelization in combination with other flood-control projects (e.g., reservoirs) has led to agricultural, urban, and industrial encroachment on 95 percent of the Missouri River floodplain (Hesse, Chaffin, and Brabander 1989). Such activities have resulted in the shortening of the length of the Missouri River by at least 120 km and the impoundment and channelization of a minimum of ninety-five tributary streams (Hesse 1987). Constriction of the floodplain and reservoir construction have reduced the input of organic matter by 65 percent (Hesse 1987; Hesse et al. 1989). Channelized sections of river have fewer fish and lower species diversity and often lack characteristic big-river fishes such as paddlefish (*Polyodon spathula*), lake sturgeon (*Acipenser fulvescens*), and blue catfish (*Ictalurus furcatus*) (Welcomme 1985). Virtually all species of sand darter (Percidae, genus *Ammocrypta*) are now uncommon because they specialize in living in shifting sand microhabitats of swift, shallow-river reaches. This habitat is eliminated by dredging and levee construction and by reservoirs that often trap the sand necessary to maintain downstream darter habitats.

Channelization also has reduced inundation of floodplains of the lower Mississippi River Valley and other coastal drainages of the south-

Figure 6.6
Channelization of large rivers in the southeastern United States (top) and elsewhere has reduced the inundation of floodplain forests that serve as critical spawning and nursery areas for many fish species (bottom).

eastern United States (Figure 6.6). These floodplains are dominated by hardwood forests that serve as critical spawning and nursery areas for many species of fish and shellfish when they are flooded each spring. Twenty families represented by fifty-three species of fish have been recorded feeding and/or spawning on these floodplains (Wharton et al. 1982). These include menhaden (*Brevoortia* spp.), shad and alewife (*Alosa* spp.), pirate perch (*Apredoderus sayanus*), redfin pickerel (*Esox americanus*), and several species of catfish (*Ictalurus* spp.) and sunfishes (Centrarchidae) (Mitsch and Gosselink 1986). Despite their value to fisheries, bottomland hardwood forests have declined by 80 percent within the Mississippi River floodplain alone, and the land has been turned into farms (Tiner 1984; Gosselink and Lee 1987). As a consequence, only 20 percent of the Yazoo Basin of the Lower Mississippi River floodplain can sustain fish populations today, and standing biomass within this area declined from 170 to 340 kg/ha in historically undisturbed streams to 13 kg/ha by 1970 (McCabe, Hall, and Barkley 1981; Gosselink and Lee 1987). The cumulative effects of channelization and forest or wetland clearing is a watershed-level phenomenon not limited to the immediately adjacent floodplain. Rather, such activities often negatively effect hydrology, deposition of sediments, water quality, productivity, and biotic diversity of all downstream aquatic habitats. In the Mississippi River basin, these activities have increased catastrophic flooding, silted reservoirs, and eroded coastal wetlands, often creating the perceived need for further channelization and levee construction. Populations of many big-river fishes have reached a point where extinction in the next fifty years is a real possibility if present trends continue.

Dams and Reservoirs. As indicated in the section under competition for water, dams have been implicated in major changes in fish faunas worldwide. The magnitude of adverse effects of dams is indicated by at least 14,300 dams 15 m or more in height around the world (IIED and WRI 1987). In the United States alone there are more than 68,000 dams over 2 m high, with a water storage capacity of at least 50 acre feet (62,000 cubic m) and 2 to 3 million smaller dams (Figure 6.7; Interagency Task Force on Floodplain Management [ITFFM] 1989). Similarly, in China there are over 90,000 small hydroelectric dams. Such small dams can have significant effects on stream fishes. For example, small impoundments used for watering livestock in the western United States not only block the movements of stream fishes and degrade water quality, but also permit the survival of introduced predatory fishes that normally could not persist in the shallow stream pools of the dry season.

Even though the effects of dams on fish faunas in temperate areas are fairly well documented, their effects in tropical areas are poorly known

Figure 6.7
Two dams in the San Joaquin River drainage, California. Top: *Buchanan Dam on the Chowchilla River, which has dried up the river below it, except for stagnant, isolated pools.* Bottom: *Kings River, which releases water for irrigation downstream during the summer, maintaining a stream comparatively colder and larger than the original river in the summer months.*

but likely to be great (Balon and Coche 1974). In Brazil, for example, over forty major dams have been proposed, affecting every major river in the Amazon basin (Scott and Carbonell 1986). Completed reservoirs in the basin already range in surface area from 10,000 ha to 900,000 ha. Construction of additional dams can be expected to disrupt the extensive migrations of many riverine fishes and to reduce the annual flooding of floodplain forests on which many fishes depend for spawning and rearing.

Reservoirs behind dams do not provide habitats for most riverine fishes because their lakelike habitats favor different species of fish. Often these species are introduced. Although the number of species in a reservoir may be equivalent to the number inhabiting the original river at the reservoir site, native forms often disappear. Reservoirs on the Columbia River in the northwestern United States, for example, contain fish faunas bearing a greater resemblance to those of Midwestern lakes than to that of the original river (Li et al. 1987). In the Colorado River, virtually the entire native big-river fauna is endangered in part because they cannot complete their life cycles in reservoirs in the face of competition and predation from introduced fishes.

Dams also block the access of migratory fishes to upstream areas, not only eliminating the fish and their young as part of the upstream community but preventing recolonization of stream segments when local populations of resident fishes become extinct from natural or human-caused events. For example, in several streams draining the western slope of the southern Sierra Nevada, California, populations of an endemic cyprinid, *Lavinia symmetricus*, have become extinct in tributaries upstream of reservoirs (Moyle and Nichols 1974). Natural recolonization from sources either below the impoundment or in adjacent drainages is no longer possible.

Dams not only reduce downstream flows, but also reduce the amount of nutrients (e.g., detritus, silt, coarse organic material) flowing into downstream areas. This nutrient reduction eventually may affect the productivity of aquatic communities within the river, estuary, or adjacent ocean. Reduced productivity of fisheries in the Mediterranean and Caspian seas following construction of dams on the Nile and Volga rivers are in large part the result of decreased nutrient inputs (Welcomme 1985). Although commercial fisheries in Nassar Reservoir behind the Aswan Dam are productive, the total fish harvest is small compared to the amount lost in the Mediterranean Sea.

Siltation. Siltation from erosion has adverse effects on fish by covering spawning sites with sediment, destroying benthic food sources, and reducing water clarity for visual-feeding fish. In Sri Lanka, sedimentation from logging practices on steep rain-forest slopes and streamside mining has been a major contributor to the decline of the endemic fishes (Senanayake and Moyle 1982). In South Africa, sedimentation of estuaries has contributed to the decline of many native estuarine fishes (Skelton 1987).

Degradation of Wetlands. Wetlands are highly variable, complex, and productive systems that serve important functions to fish communities (e.g., maintaining water quality, regulating hydrologic patterns, controlling erosion, recharging groundwater, supporting the food chain, and providing habitat to various species and life history stages of fish). During flooding, riverine wetlands function to improve water quality by acting as sinks for nutrients such as nitrogen and phosphorus (Whigham, Chitterling, and Palmer 1988). In addition, terrestrial nutrients flushed into floodplain wetlands and forests significantly increase fish habitats and food organisms by promoting growth of emergent plants and trees. For example, Goulding (1980) showed that at least fifty species of fish in the Amazon basin move into inundated forests to feed on fruits, seeds, invertebrates, and detritus. Not only do these fishes depend on this food for growth and survival, but some tree species may depend on the fish for seed dispersal. Fishes that depend on flooded forest are some of the most abundant in the Amazon, constituting as much as 75 percent of the commercial fish catch (Goulding 1980).

Worldwide, wetlands are being lost rapidly to drainage, mainly to increase agricultural production (Figure 6.8). Already as much as 50 percent of the world's wetlands have been lost (IIED and WRI 1987). In the lower United States, 35 to 50 percent of the original 60 to 75 million ha of wetlands no longer exist (Mitsch and Gosselink 1986), and in arid states like California the losses may be as great as 90 percent. Of the 740 significant tropical wetlands listed by Scott and Carbonell (1986), 65 percent are probably threatened with degradation in the near future.

Pollution

Pollution of aquatic environments is pervasive. The U.S. Environmental Protection Agency (EPA) (1989) has listed 17,365 waterways within the United States that are "impaired" by either point or nonpoint source discharges of pollutants. These waterways require controls to restore water quality.

Pollution can affect fish species through direct mortality at any life stage or by sublethal effects influencing predation, foraging, and reproduction (Figure 6.9). Therefore, heavily polluted waters contain many

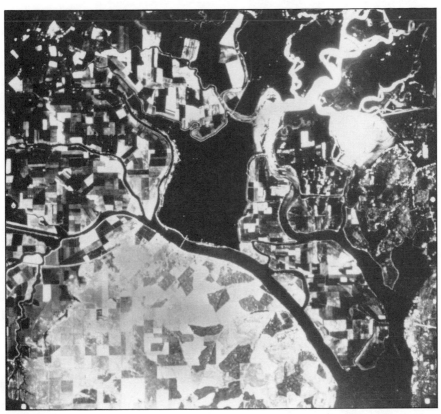

Figure 6.8
High-altitude photograph of agricultural lands surrounding the Sacra-
mento–San Joaquin River Delta. The vast majority of wetland losses are
the result of conversion to agricultural uses. Populations of several native
fish species that occur in the Delta have experienced dramatic declines in
recent years. Photography by P. Cotter.

fewer fish species than unpolluted waters. Chronic, sublethal pollution
(i.e., acid rain) often is a bigger problem than pollution that causes big
fish kills. Sublethal pollutants can cause fish to disappear gradually, as
they produce fewer young, grow more slowly, or die from stress-related
diseases. This presumably is one of the factors contributing to the decline
of big-river fishes throughout the world, as they often carry fairly heavy
loads of heavy metals and pesticides in their tissues. Fishes also must
survive reduced oxygen levels in rivers caused by sewage and other organic pollution that may be particularly hard on large, active fishes. Following we discuss several of the more prevalent forms of pollution affecting aquatic systems.

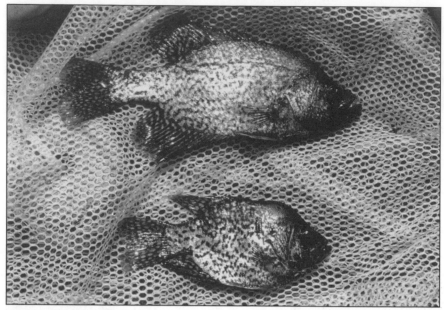

Figure 6.9
*A deformed (bottom) and normal (top) male black crappie (*Pomoxis ni-romaculatus*), both the same age, captured on their spawning grounds in Clear Lake, California. Such deformities are common in waters that are highly altered or polluted. It is unlikely that the deformed fish is able to reproduce, despite its maturity.*

Municipal and Industrial Point Source Pollution. The variety and volume of contaminants discharged directly into surface waters from municipal and industrial sources is staggering. Although there is evidence that some improvement has been made in the treatment of traditional pollutants (e.g., phosphorus, fecal matter, lead) in some industrialized nations, contamination of aquatic habitats by other substances continues to be a growing problem. For example, it is estimated that in the United States, approximately 160,000 industrial and commercial facilities discharge 92,000 metric tons of hazardous wastes into publicly owned treatment plants annually (U.S. EPA 1986). However, as much as 18 percent of this waste passes untreated to receiving waters, the result being that water quality standards are exceeded by up to 65 percent for some toxic substances (U.S. EPA 1986). Recently, U.S. EPA (1989a) listed more than 600 waterbodies in the United States as having excessive concentrations of toxic wastes. The effects of pollution are increasingly affecting marine environments as well. For example, roughly 50 percent of the deep waters of the Baltic Sea (100,000 km²) is devoid of oxygen and marine life as the result of 1.7 million metric tons of sewage being dumped into it each year (Larson, Rosenberg, and Elder 1986).

Fish populations show remarkable abilities to recover from devastating pollution events provided there is a source of colonists nearby. Olmsted and Cloutman (1974) recorded a pesticide spill that eliminated all individuals of twenty-nine species from a small creek. Within a year, the populations of all species had recovered. A similar spill in the isolated Pajaro River, California, however, eliminated one native species from the drainage because there were no refuges from the contaminated water or a source of colonists (Moyle 1976).

Agricultural pollutants. Agricultural runoff has adversely affected fish abundance and diversity in at least 47 percent of the stream kilometers in the United States (Juday 1982) as well as in numerous lakes. Its effects are the result of the combination of sediments, pesticides, fertilizers, and animal wastes. Excess sedimentation covers gravel beds needed for spawning, fills pools, and reduces food quality and quantity. Suspended sediment also can have direct adverse effects on fish and other aquatic organisms by damaging gills and other structures. Nutrients from agricultural fertilizers and animal wastes can stimulate blooms of noxious blue-green algae.

Detrimental effects of pesticides and other toxic substances on fish from agricultural runoff and other sources have been well documented (U.S. EPA 1979; Clark, Haversamp, and Chapman 1985). A recent review of 300 surveys (encompassing 35,000 fish representing over 100 families and 600 species) for trends in contamination by PCBs, DDT, dieldrin, chlordane, and other chlorinated pesticides in U.S. coastal waters indicates contamination to be widespread (Mearns et al. 1988). Freshwater fish generally are more contaminated than marine and estuarine fish.

Acid Rain. Emissions from the combustion of fossil fuels of nitrogen oxides (NO_x), sulfur dioxide (SO_2), heavy metals, and organic contaminants may be transported great distances before being deposited as rain, snow, dust, or fog. Acid rain is well documented for eastern North America and Europe, and evidence has accumulated for its occurrence in western North America, China, the Soviet Union, and South America as well (Schindler 1988). Acid rain causes significant changes in the chemistry and biota of lakes and streams. Adverse effects of the acidification of natural waters on fishes include direct mortality, changes in metabolism, reduced growth rates, and increased reproductive failure (Mills and Schindler 1986).

Acidification also may lead to changes in fish community composition and structure through a combination of interacting factors. Such factors may include (1) reduction in genetic variability by the elimination of isolated populations; (2) creation of conditions favorable to colonization by acid-tolerant species; (3) elimination of opportunities for recolonization following natural or human-caused extinction of local populations; (4) alterations to the food chain, especially at lower trophic levels; and (5) mobilization of toxic metals in the soil (e.g., aluminum and mercury), which may be released into aquatic systems during periods of heavy runoff (U.S. EPA 1986; Schofield 1988). Haines (1986) concluded that evidence exists for adverse effects on fish communities in lakes in Maine, Vermont, and New York as well as for fish in streams in Pennsylvania and Nova Scotia. Schofield (1988) found that approximately 40 percent of small, headwater lakes in Adirondack Park, New York, exhibited pH levels that were too low for fish survival. Similarly, Matuszek and Beggs (1988) showed that lakes affected by acid rain in Ontario, Canada, exhibit on the average a 40 percent reduction in fish species.

Introduction of Species

Introduction of nonnative fishes and invertebrates to "improve" local fishing has been a worldwide phenomenon for well over a century (Courtenay and Stauffer 1984, Moyle 1986). Introduced fishes occasionally replace native species in natural habitats through competition or predation, but most replacement occurs in altered environments that provide the introduced fishes an ecological advantage.

One way that introduced species eliminate native species is through the introduction of diseases. For example, the red shiner (*Notropis lutrensis*) has become a carrier of an Asiatic tapeworm, a parasite that came to North America with Chinese grass carp (*Ctenopharyngodon idella*). The shiner, an abundant and widely distributed species in eastern North America, apparently has been able to adapt to the effects of the tapeworm. Introduction of the shiner into the Virgin River, Utah, as a bait fish has corresponded with the decline of the endangered woundfin (*Plagopterus argentissimus*). The suspected mechanism behind this decline is the competitive superiority of red shiners to woundfins weakened by tapeworm infestations (Deacon 1988).

Introduced species also confound estimates of species diversity if this measure is calculated without regard to which species are present. For example, Clear Lake, California, originally contained twelve species of fish. Today the total is twenty-three species, including seven of the original natives, of which only four are common. The sixteen introduced species are abundant in many places besides Clear Lake, but two of the native species, Clear Lake splittail (*Pogonichthys ciscoides*) and thicktail chub (Gila *crassicauda*) are extinct (Moyle 1976). Thus, Clear Lake now has a much richer fish fauna, although it is uncertain whether the current fauna is a stable one. The world, however, has lost two species, both of which were once important food fishes for Native Americans.

Commercial Exploitation

About 9,000 of the 21,000 known fish species are currently exploited, and of these only 22 species (primarily herrings, cods, mackerels, jacks, and redfishes) are harvested in excess of 100,000 metric tons annually (WRI 1988). The United Nations Food and Agriculture Organization (FAO) (1987) estimates the maximum sustained yield for all world fisheries at 100 million metric tons annually, a level that is probably already exceeded. Exploitation has been a factor in creating endangered species mainly in large, slow-growing species such as sturgeon (*Acipenser* spp.) that also have high economic value. These species may have such a high value that they will continue to be exploited even when populations are small.

Exploitation may also affect smaller species as the value of wild-caught specimens for the aquarium trade rises. Collection of rare species for the aquarium trade is a factor in the decline of some native fishes of Sri Lanka (Senanayake and Moyle 1982). Much worse is the use of sodium cyanide in the Philippines for the collection of reef fishes for salt-water aquariums, because this chemical kills not only fish not collected, but the reefs themselves (Figure 6.10). Even the fish collected and put in clean water may eventually die from the effects of the chemical after they have been sold to an unwitting aquarist!

Cumulative Impacts on Aquatic Biodiversity

Cumulative impacts are the incremental effects of disturbances of many different types. The effects of a single disturbance may interact and multiply with the effects of other disturbances (Harris 1988; Preston and Bedford 1988). For example, California's Modoc sucker (*Catostomus microps*) started on its path to endangered status in the early 1900s when heavy grazing by livestock reduced riparian vegetation that served as cover in its small streams. Grazing also increased erosion of the drainages, filling pools with silt (Figure 6.11). Sucker populations were reduced further as water was diverted for irrigation of pastureland and when predatory brown trout (*Salmo trutta*) were introduced into the streams, probably in the 1930s. In the 1960s many of the stream reaches the sucker inhabited were channelized, not only eliminating pools that form the sucker's primary habitat but also allowing the invasion of another much

Figure 6.10
*Commercial fishing practices such as bottom trawling and the use of sodium cyanide destroy reef structure and alter trophic relationships. Visible on this Caribbean reef are blackbar soldier fish (*Myripritis jacobus*). Photography by P. Cotter.*

more common sucker species that eliminates Modoc suckers through hybridization and competition (Moyle 1976). All these factors worked cumulatively to make the Modoc sucker a federally endangered species. For most endangered or threatened fish species, one or two principal causes of their distress are emphasized, but their status is typically the result of multiple, cumulative long-term effects (Ono et al. 1983).

PROTECTING AQUATIC BIODIVERSITY

Ultimately, protection of the world's aquatic biota will depend on humans adopting the "spaceship Earth" attitude—that is, everything on this planet is interconnected, and the welfare of humans is intimately tied to maintaining healthy ecosystems everywhere. Only about 3 percent of the Earth's area is protected within parks, preserves, or refuges (Western 1989). It is obvious that the Earth's biodiversity cannot be maintained within this system of preserves, even if the total area of preserves were

Figure 6.11
*Dutch Flat Creek, California, is one of the few streams in which the endangered Modoc sucker (*Catostomus microps*) is found. Sixty to seventy years ago, the stream channel was probably only slightly lower than the meadow on the upper left. As a result of attempts to drain the meadow to make it more accessible for livestock and heavy use of the stream banks by livestock, the stream has cut down to bedrock and is very exposed, leaving few places for fish to live.*

doubled or tripled. This is largely because the majority of protected areas are islands of habitat surrounded by human-dominated landscapes. These islands, if too isolated from one another, are subject to high rates of extinction due to natural causes, with little hope of recolonization by individuals from other islands. At the same time they are susceptible to outside pressures such as edge effects and pollution that cross protected boundaries (Harris and Eisenberg 1989). The linear nature of streams and rivers and the long interface between oceans, lakes, and coastlines make it extremely difficult to contain entire drainages or systems within a preserve. For this reason, management of natural and altered landscapes outside the boundaries of formal preserves is necessary to protect aquatic biodiversity even for the short term. Realistically, however, the future of much aquatic life, especially fishes, will depend on having preserves in both inland and marine waters.

Preserves in Inland Waters

Few preserves have been established specifically to protect aquatic life, but many lakes and streams are protected as a result of their inclusion within terrestrial parks. If the protected region does not include all of the drainage to which a stream or lake belongs, however, protection for the fauna will be incomplete. Unprotected upstream areas may allow pollutants and sediments to wash into protected areas, and unprotected downstream areas may permit the invasion of unwanted species. For example, Moyle and Williams (1990) found that most native California fishes had very little of their habitat under formal protection. Exceptions to this finding were either formally listed endangered species that occupied limited habitats (e.g., pupfish in Death Valley National Monument, California/Nevada) or species of sport or commercial importance for which habitats for critical life-history stages are often protected.

Moyle and Sato (1991) place inland waters into six classes according to the degree of protection offered to native biota. Class 1 waters are pristine or near-pristine waters in which the native biota is still intact and for which the entire watershed is protected or undisturbed. Such waters are exceedingly rare in most regions and generally are small headwater streams or lakes. Larger class 1 waters are present in isolated regions of northern Canada and Alaska, where biodiversity is low, and in some tropical rain-forest areas. Class 2 waters are waters that are still in reasonably good condition and consequently contain most of the original native biota. Usually they are part of a national park or forest or some protected terrestrial area. Such waters currently are the principal hope for the long-term protection of native faunas, because they are fairly numerous. However, they do not include many important aquatic habitats, especially lowland habitats. For example, the many streams in the Great Smoky Mountains National Park, Tennessee/North Carolina, offer protection to small stream fishes, but not to the big-river species that occur in the Little Tennessee River that borders the park. The river has been disrupted by dams and reservoirs. Unfortunately, the isolation of tributary streams from one another has led to the extinction of several small-stream fishes (Sheldon 1988), one reason why they are classified as class 2 streams.

Class 3 waters may look natural but have been severely altered by human activity. The native biota has been changed, with many local species extinct and introduced species added. These waters thus may serve to protect only a fraction of the native fauna, with no guarantee of its long-term survival. Examples are large reservoirs, reaches of stream with altered flow and temperature regimes, and channelized or chronically polluted streams. Class 4 waters are refuges created entirely for the purpose of protecting endangered native biota in habitats that mimic native habitat, but are outside it and therefore need continuous management. For example, the outflow of Chimney Hot Springs, Nevada, was modified to create a spring pool refuge similar to that occupied by the endangered Railroad Valley springfish (*Crenichthys baileyi*) elsewhere (J. Williams, pers. com.). Class 5 waters are artificial refuges in which no attempt is made to recreate natural conditions (Moyle and Sato 1990, 47). Typically they are temporary holding facilities for endangered species for which the natural habitat is no longer suitable: aquaria, fish hatcheries, ponds, and so on. . Finally, class 6 waters are museum jars of preserved specimens in which the DNA of extinct species is more or less preserved. Obviously it would be best if all species of aquatic organisms were protected in class 1 or 2 waters, but it is likely that future protection will depend on all six classes. This means, unfortunately, that many fish species will be extinct and many natural communities disrupted.

Preserves in Estuarine and Marine Waters

Of the approximately 300 biosphere reserves that currently exist worldwide, roughly 85 (28%) are partially or entirely coastal marine or estuarine habitats (Ray and Gregg 1991). Preserves for marine organisms are limited largely to sections of shallow-water marine environments where fish, corals, and other organisms are visible to scuba divers; to intertidal areas where the organisms are extremely vulnerable to collecting; and to wetland areas used as nursery grounds by economically important fishes. One of the main motivations for protecting coral reefs and other marine areas is to provide places where large fishes and invertebrates, otherwise likely to be harvested, can be viewed by divers. Many tropical countries have found that protected coral reefs are of great economic benefit because they are enormously attractive to tourists.

In coastal areas of North America, approximately nineteen coastal biosphere reserves have been established that include submerged areas (Ray and Gregg 1991). Randall (1982) lists seventeen units of the U.S. National Park Service that include protection for underwater areas. The state of Florida has one large park that is primarily underwater, Pennecamp Coral Reef State Park, as well as the Everglades Biosphere Reserve. Probably the largest underwater sanctuaries have been established in Australia, where the Great Barrier Reef Marine Park Authority has the responsibility for setting aside and managing segments of the Great Barrier Reef. Other areas with significant coastal biosphere reserves include the United Kingdom and portions of Europe. Many other countries with coral reefs on their fringes also have established protected areas, but they are scattered and small and often exist only on paper (Randall 1982).

The degree of protection that marine reserves enjoy varies, however. In many, fishing is still permitted despite indications that removal of large predatory fishes may significantly alter the structure of the entire biotic community (Goeden 1982). Even nonconsumptive users of the reefs can harm them, because boat anchors break up corals, and scuba diving and snorkeling change the behavior of fishes. Many reef sanctuaries are also threatened by pollution from nearby cities and agricultural areas or oil spills from ships.

The biggest problem with marine sanctuaries is that they are fragmented pieces of the larger seascape. Increasingly, small isolated patches of protected marine habitat are unable to maintain a high level of diversity. This is not only because of the general problems associated with "island" preserves (Moyle and Sato 1991), but also because most of the organisms have pelagic larvae. These larvae are often swept away from the parental reef to colonize more distant areas while larvae from other areas colonize the original reef. Lack of protected reef areas of sufficient size or numbers is likely to mean the gradual disappearance of many species if connecting larval links among reefs are broken.

A major problem facing marine preserves located in bays is the invasion of exotic species that arrive in ballast water of ships and by other means. Carlton (1989) examined the flora and fauna of South Slough National Estuarine Research Reserve in Coos Bay, Oregon. He found that ninety species were either introduced or probably introduced, including the eel grass (*Zostera japonica*) that dominates the shoreline. The preserve looks natural, but the community of organisms is a mixture of native and introduced species, with no certainty that the remaining natives will be able to persist.

RESTORATION OF AQUATIC HABITATS

Many degraded aquatic habitats have the potential to be at least partially restored to more natural conditions. For example, revegetation of a disturbed riparian corridor or the elimination of livestock grazing may benefit native stream fishes through reduced streamside erosion and sedimentation of spawning area; increased shading resulting in lower water temperatures; and an improved food supply through increased insect populations. Restoration is not, however, a panacea for protecting aquatic biodiversity. Restoration may be most appropriate in maintaining fish biodiversity in relatively undisturbed to moderately disturbed habitats (Moyle and Sato's class 2 and 3 waters) where small restoration efforts may minimize the effects of past habitat degradation or halt trends toward additional habitat degradation.

CONCLUSIONS

If trends in fish faunas are any indication, we are losing the aquatic biota of Earth at an ever-accelerating rate. This loss is occurring because aquatic ecosystems are the sumps for terrestrial habitats inhabited by humans and thus tend to be highly altered and polluted. Therefore, the protection of aquatic biodiversity presents major challenges to conservation biologists. Not the least of these challenges is finding ways to design interconnected systems of protected areas to accommodate such natural events as larval drift and fish migration. As aquatic habitats become increasingly fragmented and isolated, preservation of genetic diversity and prevention of local extinctions become major problems (Meffe and Vrijenhoek 1988). In inland areas, disruption of big rivers worldwide already has resulted in the extinction and endangerment of river fishes, mollusks, and other organisms and the isolation of tributary drainages from one another. The best conservation strategy here is to find ways to maintain the remaining populations of big-river specialists on a species-by-species basis while attempting community-level protection for the fauna of tributaries. Sheldon (1988) suggested that the best protection strategy is to focus conservation efforts on the largest tributaries that still contain much of their original fauna, in as many regions as possible. Moyle and Sato (1990) suggested a general protocol for establishing a system of aquatic preserves for the arid western United States; McDowall (1984), Maitland (1985) and O'Keefe (1989) made suggestions for New Zealand, Great Britain, and southern Africa, respectively. The continued existence of many aquatic organisms will depend on the worldwide establishment of regional systems of aquatic natural areas managed specifically for their resident and migratory inhabitants. It is quite likely that such preserves would protect not only the aquatic fauna but water quality, valuable fisheries, and many terrestrial riparian organisms as well. To be effective in the long term, of course, the terrestrial systems surrounding and feeding such preserves would have to be managed appropriately, perhaps entailing major changes in human life-styles, especially in developed countries.

If the loss of biodiversity in fresh waters is halted or slowed, it may be possible to prevent similar losses from occurring in marine habitats. Protection of aquatic life not only has great economic and health benefits for humans but would also enable people to experience the diverse and beautiful world of fish. As Rupert Brooke expressed so well in his poem "The Fish,"

> In a cool curving world he lies
> And ripples in dark ecstasy.

LITERATURE CITED

Aleem, A.A. 1972. Effects of river outflow management on marine life. *Marine Biol.* 15:200–8.

Allen, G.R. 1987. Freshwater fishes of Australia—an annotated checklist. In *Proceedings of the conference on Australian threatened fishes*, ed. J.H. Harris, 35–43. Sydney: Division of Fisheries.

Arthington, A.H., D.A. Milton, and R.J. McKay. 1983. Effects of urban development and habitat alterations on the distribution and abundance of native and exotic freshwater fish in the Brisbane region, Queensland. *Austr. J. Ecol.* 8:87–101.

Balon E., and A.G. Coche. 1974. *Lake Kariba: A man-made tropical ecosystem in central Africa.* The Hague: Junk.

Brown, L.B., and P.B. Moyle. 1991. Native fishes of the San Joaquin Drainage: Status of a remnant fauna. In *Endangered and sensitive species of the San Joaquin Valley, California*, ed. D.L. Williams, T.A. Rado, and S. Byrne. Sacramento, Calif.: California Energy Commission.

Bussing, W.A. 1987. *Peces de las aquas continentales de Costa Rica.* San Jose: Editorial de la Universidad de Costa Rica.

California Advisory Committee on Salmon and Steelhead Trout. 1988. *Restoring the balance: 1988 annual report.* Sausalito, Calif.: California Advisory Committee on Salmon and Steelhead Trout.

Carlton, J.T. 1985. Transoceanic and interoceanic dispersal of coastal marine organisms: The biology of ballast water. *Ocean. Mar. Biol. Ann. Rev.* 23:313–71.

———. 1989. Man's role in changing the face of the ocean: Biological invasions and implications for conservation of near-shore environments. *Cons. Biol.* 3:265–73.

Carpenter, S.R., and J.F. Kitchell. 1988. Consumer control of lake productivity. *BioScience* 38:764–69.

Clark, E.H., J.A. Haversamp, and W. Chapman. 1985. *Eroding soils: The off-farm impacts.* Washington, D.C.: The Conservation Foundation.

Coad, B.W. 1981. Environmental change and its impact on the freshwater fishes of Iran. *Biol. Cons.* 19:51–80.

Cohen, D.M. 1970. How many recent fishes are there? *Proc. Calif. Acad. Sci.* 37:341–46.

Courtenay, W.R., and J.R. Stauffer, eds. 1984. *Distribution, biology, and management of exotic fishes.* Baltimore: Johns Hopkins Press.

Cross, F.B., and R.E. Moss. 1987. Historic changes in fish communities and aquatic habitats in plains streams of Kansas. In *Community and evolutionary ecology of North American stream fishes*, ed. W.J. Matthews and D.C. Heins, 155–65. Norman: University of Oklahoma Press.

Culliton, T.J., M.A. Warren, T.R. Goodspeed, D.G. Remer, C.M. Blackwell, and J.J. McDonough III. 1990. *Fifty years of population change along the nation's coasts 1960–2010.* Brancy, Md.: NOAA, Oceans Assessments Division, Strategic Assessment.

Deacon, J. 1988. The endangered woundfin and water management in the Virgin River, Arizona. *Fisheries* 13:18–29.

Ehrlich, P., and A. Ehrlich. 1981. *Extinction: The causes and consequences of the disappearance of species.* New York: Random House.

Foster, M.S., and D.R. Schiel. 1985. *The ecology of giant kelp forests in California: A community profile.* U.S. Fish Wildl. Ser. Biol. Rept. 85(7.2).

Glade, A.A., ed. 1987. *Libro rojo de los vertebrados terrestres de Chile.* Santiago: Corporacion Nacional Forestal.

Goeden, G.B. 1982. Intensive fishing and a 'keystone' predator species: Ingredients for community instability. *Biol. Cons.* 22:273–81.

Gosselink, J.G., and L.C. Lee. 1987. Cumulative impact assessment in bottomland hardwood forests. Report LSU-CEI-86-09. Baton Rouge: Center for Wetland Resources. Louisiana State University.

Goulding, M. 1980. *The fishes and the forest. Explorations in Amazonian Natural History.* Berkeley: University of California Press.

Haines, T.A. 1986. Fish population trends in response to surface water acidification. In *Acid deposition: Long term trends,* ed. C.M. Atad, Washington, D.C.: National Academic Press.

Harris, J.E., ed. 1987. *Proceedings of the conference on Australian threatened fishes.* Sydney: Division of Fisheries, Department of Agriculture.

———. ed. 1988. *Australian threatened fishes—1988* supplement. Report of the Australian Society of Fish Biology.

Harris, L.D. 1988. The nature of cumulative impacts on biotic diversity of wetland vertebrates. *Environ. Manage.* 12(5):675–93.

Harris, L.D., and J. Eisenberg. 1989. Enhanced linkages: Necessary steps for success in conservation of faunal diversity. In *Conservation for the twenty-first century,* ed. D. Western and M.C. Pearl, 166–81. New York: Oxford University Press.

Hesse, L.W. 1987. Taming the wild Missouri River: What will it cost? *Fisheries* (Bethesda) 12:2–9.

Hesse, L.W., G.R. Chaffin, and J. Brabander. 1989. Missouri River mitigation: A systems approach. *Fisheries* (Bethesda) 14:11–15.

Hughes, N.F. 1986. Changes in the feeding biology of the Nile perch, *Lates nilotica* (L.) in Lake Victoria, East Africa, since its introduction in 1960 and its impact on the native fish community of the Nyasa Gulf. *J. Fish. Biol.* 29:521–48.

Interagency Task Force on Floodplain Management. 1989. *A status report on the nation's floodplain management activity.* Washington, D.C.: L.R. Johnston.

International Institute for Environment and Development and World Resources Institute. 1987. *World resources 1987.* New York: Basic Books.

Juday, R.D. 1982. *National fisheries survey.* Report USFWS/OBS-84/06. Washington, D.C.: U.S. Government Printing Office.

Karr, J.R. 1981. Assessment of biotic integrity using fish communities. *Fisheries* (Bethesda) 7:2–8.

Kaufman, L. 1988. Marine biodiversity: The sleeping dragon. *Cons. Biol.* 2:307–8.

Lanly, J.P. 1982. *Tropical forest resources.* Food and Agriculture Organization Forestry Paper 30. Rome: Food and Agriculture Organization.

Larsson, U., R. Rosenberg, and L. Edler. 1986. Eutrophication in marine waters surrounding Sweden: A review. Solna, Sweden: National Swedish Environmental Protection Board.

Leidy, R.A. 1984. Distribution and ecology of stream fishes in the San Francisco Bay drainage. *Hilgardia* 52:1–175.

Leidy, R.A., and P.L. Fiedler. 1985. Human disturbance and patterns of fish species diversity in the San Francisco Bay drainage. *Biol. Cons.* 33:247–68.

Lelek, A. 1987. *The freshwater fishes of Europe. Vol. 9: Threatened fishes of Europe.* Wisbaden, W. Germany: Aula-Verlag.

Li, H.W., C.B. Schreck, C.E. Bond, and E. Rexstad. 1987. Factors influencing changes in fish assemblages of Pacific Northwest streams. In *Community and evolutionary ecology of North American stream fishes*, ed. W.J. Matthews and D.C. Heins, 193–202. Norman: University of Oklahoma Press.

Lovejoy, T.E. 1980. A projection of species extinctions. In *The Gobal 2000 report to the president: Entering the twenty-first century*, ed. G.O. Barney, 328–31. Washington, D.C.: Council on Environmental Quality.

Maitland, P.S. 1985. Criteria for the selection of important sites for freshwater fish in the British Isles. *Biol. Cons.* 31:335–53.

Manire, C.A., and S.H. Gruber. 1990. Many sharks may be headed toward extinction. *Cons. Biol.* 4:10–11.

Matusek, J.E., and G.L. Beggs. 1988. Fish species richness in relation to lake area, pH, and other abiotic factors in Ontario lakes. *Can. J. Fish. Aqua. Sci.* 45:1931–41.

Mearns, A.J., M.B. Matta, D. Simecel-Beatty, M.F. Buchman, G. Shigenaka, and W. A. Wert. 1988. PCB and chlorinated pesticide contamination in U.S. fish and shellfish: A historical assessment report. National Oceanic and Atmospheric Administration Tech. Memor. NOS OMA 39. Rockville, Md.: National Oceanic and Atmospheric Administration, Department of Commerce.

Meffe, G.K., and R.C. Vrijenhoek. 1988. Conservation genetics in the management of desert fishes. *Cons. Biol.* 2:159–69.

McCabe, C.A., H.D. Hall, and R.C. Barkley. 1981. Yazoo area pump study, Yazoo backwater project, Mississippi. Unpublished report. Jackson, Miss.: U.S. Fish and Wildlife Service.

McDowall, R.M. 1984. Designing reserves for freshwater fish in New Zealand. *J. Royal Soc. N.Z.* 14:17–27.

Micklin, P.P. 1988. Desiccation of the Aral Sea: A water management disaster in the Soviet Union. *Science* 241:1170–76.

Miller, R.R., J.D. Williams and J.E. Williams. 1989. Extinctions of North American fishes during the past century. *Fisheries* (Bethesda) 14:22–38.

Mills, K.H., and D.W. Schindler. 1986. Biological indicators of lake acidification. *Water Air Soil Pollut.* 30:779–89.

Mitsch, W.J., and J.G. Gosselink. 1986. *Wetlands.* New York: Van Nostrand Reinhold.

Moyle, P.B. 1976. *Inland fishes of California.* Berkeley: University of California Press.

———. 1986. Fish introductions into North America: Patterns and ecological impact. In *Biological invasions of North America and Hawaii*, ed. H.A. Mooney and J.A. Drake, 27–43. New York: Springer-Verlag.

———. 1991. Ballast water introductions. *Fisheries* (Bethesda) 16:4–6.

Moyle, P.B. and J.J. Cech, Jr. 1988. *Fishes: An introduction to ichthyology*, 2nd ed. Englewood Cliffs, N.J.: Prentice Hall.

Moyle, P.B., and R. Nichols. 1974. Decline of the native fish fauna of the Sierra Nevada foothills, central California. *Amer. Midl. Nat.* 92:72–83.

Moyle, P.B., and G.M. Sato. 1991. On the design of preserves to protect native fishes. In *Battle against extinction: Native fish management in the American West*, ed. W.L. Minckley and J.E. Deacon. Tucson: University of Arizona Press.

Moyle, P.B., and J.E. Williams. 1990. Biodiversity loss in the temperate zone: Decline of the native fish fauna of California. *Cons. Biol.* 4:475–84.

Moyle, P.B., J.E. Williams, and E.D. Wikramanayake. 1989. Fish species of special concern of California. Sacramento: California Department of Fish and Game.

Myers, N. 1980. *Conversion of tropical moist forests*. Washington, D.C.: National Academy of Sciences.

———. 1983. Conservation of rain forests for scientific research, for wildlife conservation, and for recreation and tourism. In *Tropical rain forest ecosystems, structure and function*, ed. F.B. Golley, 325–34. Amsterdam: Elsevier.

Nehlsen, W., J.E. Williams, and J.A. Lichatowich. 1991. Pacific salmon at the crossroads: Stocks at risk from California, Oregon, Idaho, and Washington. *Fisheries* (Bethesda) 16:4–21.

O'Keeffe, J.H. 1989. Conserving rivers in southern Africa. *Biol. Cons.* 49:255–74.

Olmsted, L.L., and D.G. Cloutman. 1974. Repopulation after a fish kill in Mud Creek, Washington County, Arkansas, following pesticide pollution. *Trans. Amer. Fish. Soc.* 103:79–87.

Ono, R.D., J.O. Williams, and A. Wagner. 1983. *Vanishing fishes of North America*. Washington, D.C.: Stone Wall Press.

Pathiyagoda, R. 1991. *Freshwater fishes of Sri Lanka*. Colombo: Wildlife Heritage Trust of Sri Lanka.

Petrere, M. Jr. 1989. River fisheries in Brazil: A review. *Regulated Rivers* 4:1–16.

Power, M., W. Matthews, and A.J. Stewart. 1985. Grazing minnows, piscivorous bass, and stream algae: Dynamics of a strong interaction. *Ecology* 66:1448–56.

Preston, E.M., and B.L. Bedford. 1988. Evaluating cumulative effects of wetland functions: A conceptual overview and generic framework. *Environ. Manage.* 12:675–93.

Randall, J.E. 1982. Tropical marine sanctuaries and their significance in reef fisheries research. In *The biological basis for reef fishery management*, ed. G.R. Huntsman, W.R. Nicholson, and W.W. Fox, 167–78. National Oceanic and Atmospheric Administration Tech. Memor. NMFS-SEFC-80. Rockville, Md.: National Oceanic and Atmospheric Administration.

Ray, G.C. 1988. Ecological diversity in coastal zones and oceans. In *Biodiversity*, ed. E.O. Wilson, 128–50. Washington, D.C.: National Academy Press.

Ray, G.C., and W.P. Gregg, Jr. 1991. Establishing biosphere reserves for coastal barrier ecosystems. *BioScience* 41:301–9.

Robison, H.W., and T.M. Buchanan. 1988. *Fishes of Arkansas*. Fayetteville: University of Arkansas Press.

Rozengurt, M.M., and J.W. Hedgepeth. 1989. The impact of altered river flow on the ecosystem of the Caspian Sea. *Rev. Aquatic Sci.* 1:337–62.

Rozengurt, M.M., M.J. Hertz, and S. Feld. 1987. *The role of water diversions in the decline of fisheries of the delta—San Francisco Bay and other estuaries.* Tiburon Center for Environmental Studies Tech. Rep. no. 87-8. Tiburon, Calif.: Tiburon Center for Environmental Studies.

Ryder, R.A. 1978. Fish yield assessment of large lakes and reservoirs—a prelude to management. In *Ecology of freshwater fish production*, ed. S.A. Gerking, 402–23. London: Blackwell Scientific.

Schindler, D.W. 1988. Effects of acid rain on freshwater ecosystems. *Science* 239:149–57.

Schofield, C.L. 1988. Lake acidification in wilderness areas: An evaluation of impacts and options for rehabilitation. In *Ecosystem management for parks and wilderness*, ed. J.K. Agee and D.R. Johnson, 135–44. Seattle, Washington: University of Washington Press.

Scoppetone, G.G., M. Coleman, and G.A. Wedemeyer. 1986. Life history and status of the endangered cui-ui of Pyramid Lake, Nevada. U.S. Fish and Wildlife Service Fish, Wildl. Res. 1. Reno, Nev.: U.S. Fish and Wildlife Service.

Scott, D.A., and M. Carbonell, compilers. 1986. *A directory of neotropical wetlands.* Cambridge and Slimbridge, U.K.: International Union for the Conservation of Nature and Natural Resources and International Waterfowl Research Bureau.

Scudder, T. 1989. River basin projects in Africa. *Environment* 31:4–32.

Senanayake, F.R., and P.B. Moyle. 1982. Conservation of the freshwater fishes of Sri Lanka. *Biol. Cons.* 22:181–95.

Sheldon, A.L. 1988. Conservation of stream fishes: Patterns of diversity, rarity, and risk. *Cons. Biol.* 2:149–56.

Sigler, W.F., W.T. Helm, P.A. Kucera, S. Vigg, and G.W. Workman. 1983. Life history of the Lahontan cutthroat trout, *Salmo clarki henshawi*, in Pyramid Lake, Nevada. *Great Basin Nat.* 43:1–29.

Simberloff, D. 1986. Are we on the verge of a mass extinction in tropical rain forests? In *Dynamics of extinction*, ed. D.K. Elliot, 1165–80. New York: John Wiley & Sons.

Simpson, P.W., J.R. Newman, M.A. Kevin, R.M. Matter, and R.A. Guthrie. 1982. Manual of stream channelization impacts on fish and wildlife. U.S. Fish and Wildlife Service, Office of Biological Services, Rep. 82/24. Washington, D.C.: U.S. Fish and Wildlife Service.

Skelton, P.H. 1987. *South African red data book—fishes.* South African National Scientific Programmes Rep. 137.

Smith, F.E. 1987a. *Water development and management in the Central Valley of California and the public trust.* Sacramento: U.S. Department of Interior, Fish and Wildlife Service.

————. 1987b. *The changing face of California's central valley fish and wildlife resources and the bay—delta hearings.* Sacramento: U.S. Department of Interior, Fish and Wildlife Service.

Smith, M.M., and P.C. Heemstra, eds. 1986. *Smith's sea fishes.* Johannesburg: Macmillan.

Terborgh, J. 1988. The big things that run the world—a sequel to E.O. Wilson. *Cons. Biol.* 2:402–3.

Tiner, R.W. 1984. *Wetlands of the United States: Current status and recent trends.* National wetlands inventory. Washington, D.C.: U.S. Fish and Wildlife Service.

United Nations Food and Agriculture Organization. 1987. *Review of the state of the world fishery resources.* Rome: United Nations.

U.S. Congress Office of Technology Assessment. 1987. *Waters in marine environments.* Washington, D.C.: U.S. Government Printing Office.

U.S. Environmental Protection Agency. 1979. *Fish kills caused by pollution, fifteen year summary 1961–1975.* U.S. EPA Rep. 440/4-78-011. Washington, D.C.: U.S. Government Printing Office.

———. 1986. *Report to Congress on the discharge of hazardous wastes to publicly owned treatment works.* U.S. EPA Rep. 530-SW-86–004. Washington, D.C.: U.S. Environmental Protection Agency.

———. 1989. *List of waters in the United States impaired by toxic and nontoxic pollutants.* Report to Congress, Clean Water Act, Section 304(l) Summary.

U.S. Fish and Wildlife Service. 1989. *Endangered and threatened wildlife and plants.* Reprint of 50 C.F.R., 17.11 and 17.12. Washington, D.C.: U.S. Government Printing Office.

Welcomme, R.L. 1985. *River fisheries.* FAO Tech. Paper 262.

Western, D. 1989. Conservation without parks: Wildlife in the rural landscape. In *Conservation for the twenty-first century,* ed. D. Western and M.C. Pearl, 158–65. New York: Oxford University Press.

Wharton, C.H., W.M. Kitchens, E.C. Pendleton, and T.W. Sipe. 1982. *The ecology of bottomland hardwood swamps of the Southeast: A community profile.* U.S. Fish and Wildlife Service, Biological Services Program. FWS/OBS-81/37.

Whigham, D.F., C. Chitterling, and B. Palmer. 1988. Impacts of freshwater wetlands on water quality: A landscape perspective. *Environ. Manage.* 12:663–74.

White, G.F. 1988. The environmental effects of the high dam at Aswan. *Environment* 30:5–40.

Williams, J.E., J.E. Johnson, D.A. Hendrickson, S. Contreras-Balderas, J.D. Williams, M. Navarro-Mendoza, D.E. NcAllister, and J.E. Deacon. 1989. Fishes of North America, endangered, threatened, or of special concern: 1989. *Fisheries* (Bethesda) 14:2–20.

Wilson, E.O., ed. 1988. *Biodiversity.* Washington, D.C.: National Academy Press.

World Resources Institute. 1988. *World resources 1988–89: An assessment of the resource base that supports the global economy.* New York: Basic Books.

CHAPTER

7

Threats to Invertebrate Biodiversity: Implications for Conservation Strategies

JOHN E. HAFERNIK, JR.

ABSTRACT

Traditionally, conservation efforts have been directed at saving large vertebrates and their habitats. There is increasing awareness, however, of threats to invertebrates and concern about detrimental effects of loss of invertebrate diversity. I summarize the geographic distribution of extinct and threatened U.S. invertebrates, discuss reasons for recent extinctions, and comment on the role invertebrates should play in conservation strategies. Areas of great endemism are also areas of greater observed and potential extinctions. These include (1) Hawaii, (2) California, (3) large eastern river systems, and (4) subterranean caves. Several examples of successful efforts to conserve endangered invertebrates are provided. The preservation of small remaining parcels of native habitat even in urban areas can help maintain invertebrate diversity. The chapter concludes with suggestions for ways that studies of invertebrate diversity can contribute to theories of reserve design and habitat restoration.

INTRODUCTION

Healthy biological communities depend largely on interactions among small or nonspectacular organisms—mostly invertebrates, plants, and microbes (Wilson 1987; Majer 1987; Ehrlich 1988). Comparative studies of secondary production by insects and vertebrates invariably show insects to be the greater producers of biomass and conduit of energy through communities (Price 1984). Leaf cutter ants, not large vertebrates, are the principal herbivores in neotropical forests (Wilson 1987). In the temperate zone, studies of old-field trophic structure (Wiegert and Evans 1967; Odum, Connell, and Davenport 1962) indicate that insects are the most important herbivores, not only consuming more of the primary production than birds and rodents do, but also more efficiently converting it to herbivore biomass, thus allowing support of a larger number of trophic levels. Even spittlebugs, a seemingly insignificant group of Homoptera, ingest more than mice or sparrows do (Wiegert and Evans 1967). Many other inconspicuous invertebrates also play major roles in ecosystem function because of their importance in soil aeration and drainage, litter decomposition and nutrient cycling, pollination, seed distribution and survival, and herbivory, and as predators and food sources for predators (Majer 1987). In fact, the first evidence for the importance of keystone species in some communities came from studies of marine invertebrates (Paine 1966).

Yet despite the ecological importance of invertebrates, their conservation has received little attention. Rather, conservation strategies often have been based more on emotional reaction rather than scientific or societal goals. Mostly, efforts and funding have been directed at maintaining or increasing the abundance of large birds and mammals. I believe the neglect of invertebrates is largely the result of lack of knowledge and consequent lack of appreciation of invertebrate diversity not only by the general public but also often by conservation biologists. Conservation biology as a discipline has its academic lineage rooted in wildlife biology, vertebrate ecology, and resource management. Many conservation biologists have little more than superficial knowledge of invertebrate diversity. A quick look at the index of Western and Pearl (1989), the most recent conservation biology book I could find, demonstrates the continuing emphasis on vertebrates. Index citations for vertebrates outnumber those for invertebrates and plants by a factor of 10, and few of those citations refer to efforts at invertebrate conservation.

Although concern about decline in invertebrate populations can be traced to the early nineteenth century (Pyle, Bentzien, and Opler 1981), the milestone publications by the International Union for the Conservation of Nature and Natural Resources (IUCN) of Red Data Books on invertebrates (Wells, Pyle, and Collins 1983) and swallowtail butterflies (Collins and Morris 1985) provides the first compilations of threats worldwide to a large variety of invertebrates. So far, however, most positive results in invertebrate conservation have come as unplanned consequences of efforts to preserve vertebrate or plant species. As the emphasis in conservation biology turns more to preservation of natural biological diversity, specific efforts must be directed at invertebrates. While to most people the words "birds" and "mammals," and the collective term "animal" are used as virtual synonyms, conservation biologists must be aware that animal biodiversity and invertebrate diversity are the true virtual synonyms.

The great diversity of invertebrates and their numerical abundance has contributed to their neglect. Jones (1987, 169) expresses a view that is probably common even among conservationists. That view is that invertebrates such as arthropods and mollusks are so diverse that the loss of any one species would be of less ecological importance than the disappearance of a fish, bird, or mammal. Although it is true that the effects of extinction of a species on its associated community vary greatly depending on the species involved (Westman 1990), the presence of a backbone per se is not a sufficient predictor of great impact.

In this chapter, I analyze threats to invertebrate diversity primarily in the United States, assess characteristics of extinction-prone invertebrates, and suggest ways that conservation efforts aimed at maintenance of invertebrate diversity can complement traditional vertebrate-centered approaches. I present these from the view of an entomologist with the implicit assumption that invertebrates are critical to ecosystem function and that they are at least the equal aesthetically and otherwise of vertebrates.

HISTORICAL RECORD OF THREATS TO U.S. INVERTEBRATES

The historical record of extinctions and changes in distribution and abundance of U.S. vertebrates is relatively well documented. For invertebrates, the record is much poorer. Historical and present distributions are known in detail for only a few well-studied species (Opler 1987). Many species remain to be described, and many others are known from one or a few collection records. Until recently, extinctions probably have gone mostly unnoticed (Pyle et al. 1981; Opler 1976).

In contrast, the most prominent invertebrates are usually those relative few whose numbers and distribution have expanded as the result of human activities—the species we often recognize as agricultural, horticultural, and household pests. Despite the enhanced success of these few, it is clear that many other invertebrates have declined in numbers during the last century, and an unknown but substantial number have gone extinct, probably mostly as the result of human activities.

This pattern of decline in diversity is well illustrated for temperate-zone invertebrates by changes in abundance of British butterflies over recent decades. Of the fifty-five U.K.-resident butterflies, three species have increased in abundance and eight have held their own, while forty-four have declined (Thomas 1984). Some declines have been extremely rapid (Thomas 1983). Moore (1981) reports a similar pattern of decline for British dragonflies and damselflies. Declines in butterflies also have occurred over much of Europe (Heath 1981). Data are more fragmentary for U.S. butterflies, but the pattern for urban areas appears similar. There have been numerous local extinctions, and several taxa have been lost, especially in California (Pyle et al. 1981). Historical records indicate that fifty-seven species have been recorded from San Francisco, California. Presently, fewer than forty persist, many in small, fragmented populations (Reinhard n.d.).

Recently the U.S. Fish and Wildlife Service (USFWS) published an updated list of extinct, endangered, and threatened U.S. invertebrates (Dunlop 1989). This tabulation lists eighty-three U.S. insects and other invertebrates as extinct (category 3a) and ninety-two species of insects that have not been recorded in the last twenty-five years. Using the guideline of Diamond (1989), these latter species should be considered extinct until proven otherwise. Another eighty-three species (50 mollusks and 33 arthropods) are federally listed as endangered or threatened (Endangered Sp. Bull. 16:5. 1991). These numbers probably underestimate threats to invertebrate diversity by at least an order of magnitude.

The rapid rise in invertebrates that are candidates for federal listing is probably more indicative of the magnitude of threat to invertebrates. By early 1989, 1,042 invertebrates, most of them insects and mollusks, were category 1 or 2 candidates for federal listing as threatened or endangered species (Dunlop 1989). This represents an increase of 310 candidate species since the previous comprehensive list (Arnett 1984).

Despite the rapidly increasing number of invertebrate candidates for listing, the pace of evaluating and listing invertebrate species, especially insects, has been exceedingly slow compared to that of plants and vertebrates (Figure 7.1). The fragmentary knowledge of many invertebrates,

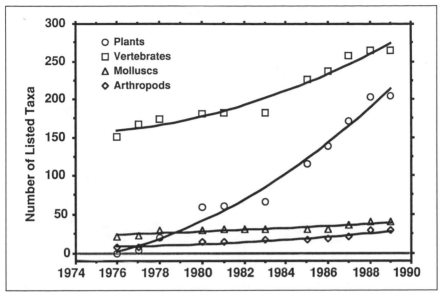

Figure 7.1
Listing history of various taxa under the U.S. Endangered Species Act.

even relatively well-known taxa, makes evaluation of candidates difficult and contributes to their slow rate of evaluation. But the primary cause is lack of adequate staff, resources, and commitment to preservation of invertebrates, especially at higher levels of the federal bureaucracy.

The current list of extinct and threatened invertebrates is far from complete and is biased in its taxonomic composition to snails, freshwater mussels, and a few well-studied insect groups. These relatively well-studied taxa can serve as indicators of change in invertebrate diversity, however, and provide a basis for assessing characteristics of extinction-prone taxa. The pattern suggests that at least some invertebrate groups have suffered recent extinction rates that are the equal of rates for birds and mammals.

Whereas large vertebrates are usually wide-ranging generalists, not specifically associated with areas of high endemism or diversity, many invertebrates are characterized by restricted distributions, movements, and associations with unique habitats. Most extinct invertebrates were naturally restricted in distribution as are most invertebrates listed or proposed for listing. This should not be taken as a blanket indication of the safety of wide-ranging species, however, since data on patterns of abundance for wide-ranging invertebrates are lacking. Declines are likely to go unnoticed in these species until drastic reductions have occurred.

The recent listing of the American burying beetle (*Nicrophorus americanus*) as endangered is a case in point (Endangered Sp. Bull. 14:8. 1989). This species once fed widely on carrion in thirty-two states, the District of Columbia, and three Canadian provinces. It now is apparently restricted to two widely separated localities, one an island off the coast of New England and the other in eastern Oklahoma (Amaral and Morse 1990). The cause for its decline is unknown. Furthermore, a number of other candidates for listing have not been observed in large parts of their range in recent years.

For warm-blooded vertebrates, the probability of extinction is greater for larger species (Eisenberg and Harris 1989). No such correlation is obvious for invertebrates, at least in the temperate zone. Rather, most recent extinctions have been of moderate- to small-sized species.

Geographically, within the United States, areas of greatest apparent loss of invertebrate species are Hawaii, California, and river systems of the eastern states, especially those of Tennessee and Alabama (Figure 7.2). The Rocky Mountains and Great Plains states have had few recorded losses. Geographic distributions of listed and candidate species show a

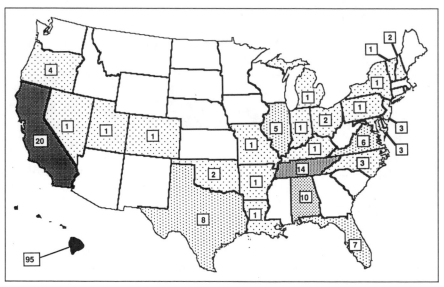

Figure 7.2
Historrical distribution by state of invertebrates listed as extinct (USFWS category 3a) or for which there are no records in the last twenty-five years (USFWS category 2*). Boxes indicate the number of taxa from each state; density of stipling increases with the number of extinctions. (Source: Dunlop 1989)

similar pattern (Figures 7.3 and 7.4). All states, except Alaska, have at least one candidate species. The low apparent threat to invertebrates in the Great Plains states should be viewed with caution because butterflies and other insects have undergone serious declines in prairie habitats (Opler 1981; Johnson 1986).

The fate of the diverse freshwater molluscan fauna of eastern river systems has received intensive scrutiny. This once-rich fauna has been decimated by dam construction, channel alteration, and chemical pollution (Palmer 1986). Twenty-eight species of mussels and clams are believed to be extinct (Palmer 1986; Dunlop 1989), thirty-nine are federally listed, and approximately sixty others are candidates for listing. The status of mussels in the Cumberland and Tennessee rivers demonstrates the magnitude of threat to surviving species. Here, fourteen of the eighteen mussel species are federally listed as endangered, at least in part, because of habitat deterioration related to coal mining (Biggins 1989). Other river systems also have seen a dramatic decline in mollusk populations in the last decade. This pattern is repeated for gastropods (Palmer 1986) and probably is typical for at least some aquatic insects that doubt-

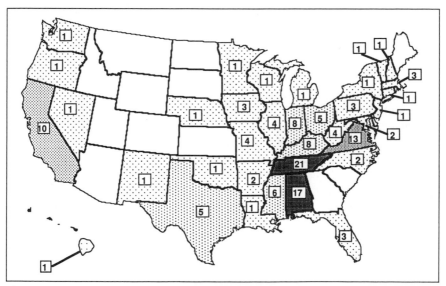

Figure 7.3
Distribution of invertebrates listed as endangered or threatened under the U.S. Endangered Species Act. Historical distributions are depicted for all species except the wide-ranging American burying beetle, whose current distribution is plotted. Boxes indicate the number of taxa for each state; the density of stipling increases with the number of listed taxa.

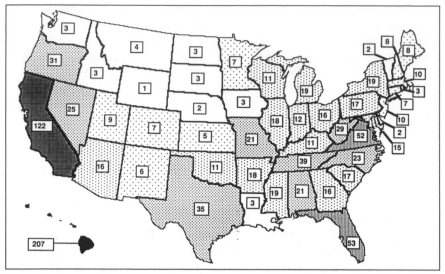

Figure 7.4
Distribution by state of invertebrates listed as candidates (USFWS categories 1, 2) for listing as endangered or threatened species. Boxes indicate the number of candidates known from each state; the density of stipling increases with the number of candidates. (Source: Dunlop 1989)

less have also undergone at least local extinctions and rapid change in faunal composition. In this regard, extinctions have been recorded for two mayflies, a stonefly, and a caddisfly from the eastern U.S. An additional four mayflies and three caddisflies have not been recorded in the last quarter century, and another six mayflies and seventeen caddisflies are candidates for listing (Dunlop 1989).

The Hawaiian Islands have experienced a cataclysmic extinction of species unequaled in historic time. The fate of birds is well known, as at least 50 percent of birds once found on the islands are extinct and 41 percent of the existing fauna is federally listed as endangered or threatened (Stone 1989a). Less well known is that losses of comparable magnitude have been suffered by some invertebrates. At least 95 species apparently either are extinct or have not been recorded for the last twenty-five years (Figure 7.2), while 207 species are candidates for listing (Figure 7.4).

Lack of knowledge of the systematics of most Hawaiian invertebrates leaves the fate of most species unknown (Gagné 1989). Groups that are comparatively well studied, however, present a bleak picture. Hawaii once had perhaps the world's most spectacular molluscan fauna, with 1,061 named endemics. This magnificent fauna has been devastated by deforestation and introduced predators. Extinction has been documented for a number of genera (Palmer 1986), and many others have seriously declined in numbers. For land snails in the Oahu genus *Achatinella*, approximately 50 percent of the forty-one known species are extinct, and the few remaining strong colonies face probable extirpation (Hart 1978; Palmer 1986). The genus is now federally listed as an endangered "species" and as such is the only currently federally listed Hawaiian invertebrate, a fact that is symbolic of the low priority invertebrate conservation is given at the national level.

Hawaiian yellow-faced bees of the genus *Nesoprosopis* were once a spectacular example of species radiation. Sadly, eighteen species are now extinct, thirty-five have not been recorded in the last quarter century, and another four are candidates for listing. None is listed as threatened or endangered, apparently because of insufficient data to warrant listing!

For terrestrial invertebrates from the continental United States, California is the site of the most extinctions (20) as well as listed (10) and candidate (122) species, most of which are terrestrial. The first recorded extinctions for U.S. insects were from California, and the first U.S. insects listed as endangered or threatened were seven California butterflies. California's flora and fauna are isolated by mountain ranges and deserts from the rest of the continental biota and populate a mosaic of diverse, often small, patches of habitats. These factors contribute to the high degree of endemism shown by California's biota and the susceptibility of a number of taxa to natural and human-caused extinction.

Worldwide, caves and cavelike habitats are also areas of great invertebrate endemism. Moreover, cave environments are fragile and easily upset by pollution, disturbance, and infilling. More than 150 cave invertebrates are candidates for listing, and most recent federal listings of arthropods (9) have been of cave associates.

CAUSES OF RECENT EXTINCTIONS

Based mostly on his study of vertebrates, Diamond (1989) identified an "evil quartet" of mechanisms responsible for most extinctions. These are (1) habitat destruction and fragmentation, (2) introduced species, (3) chains of extinction, and (4) overkill. Let us briefly consider the impact each of these overlapping and intertwined mechanisms has on invertebrate diversity.

Habitat Destruction and Fragmentation

The rate of recent animal and plant extinctions in the United States has roughly paralleled patterns of human population growth (Opler 1976).

The primary cause for this is the expanding exploitation of natural habitats for human living space, food production, and recreation. Fragmentation of once-widespread habitats and destruction of all or part of naturally fragmented habitats are the most serious threats to invertebrate biodiversity.

Urban development of California's San Francisco Bay and Los Angeles Basin regions and its effect on insect diversity, especially as exemplified by butterflies, illustrates the common pattern. The beginning of urbanization of the San Francisco peninsula during the mid-nineteenth century quickly led to the extinction of the endemic satyr butterfly *Cercyonis sthenele sthenele*, the first U.S. invertebrate whose extinction was noted, and the near-extinction of the endemic xerces blue butterfly, *Glaucopsyche xerces*. In 1875, Henry Behr, a famous nineteenth-century lepidopterist, wrote to his colleague Henry Strecker describing what was to be an increasingly common fate for invertebrates in urban areas (letter in Field Museum Archives quoted in Pyle et al. 1981, 241). "*Glaucopsyche xerces* is now extinct, as regards the neighborhood of San Francisco. The locality where it used to be found is converted to building lots, and between German chickens and Irish hogs no insect can exist besides louse and flea." In fact, the xerces blue survived that first wave of habitat destruction, only to succumb in the 1940s to continuing development, especially of sand dune habitats. Fragmentation of grassland and sand dune habitat and housing development on serpentine areas also caused the local extinction in San Francisco of a number of other butterfly species by the 1970s. Included on the list are the callippe silverspot butterfly (*Speyeria callippe callippe*) and the pheres blue butterfly (*Icaricia icarioides pheres*), both candidates for federal listing, and the bay checkerspot butterfly (*Euphydryas editha bayensis*), a recently listed taxon. Housing development of much of the remaining San Francisco hill habitats in the late 1960s (with urban renewal funds!) eliminated most of the remaining habitat in the city of the mission blue butterfly (*Icaricia icarioides missionensis*), a listed taxon.

Besides the xerces blue butterfly, a number of other insects associated with sand dunes have been especially hard hit (Arnold 1983, 1985; Andrews, Hardy, and Giuliani 1979; Powell 1981). The best studied area is the Antioch dunes. These are a small, ancient wind-deposited dune system of the San Joaquin Delta region near San Francisco. Farming, industrial development, and sand mining have reduced the original 200 acres of sand hills to 55 acres of highly altered habitat (Arnold 1985). This fragment supports the only surviving population of Lange's metalmark butterfly (*Apodemia mormo langei*) and several endemic plants. It is also the type locality for twenty-seven insect taxa, eight of which are known only from the Antioch dunes (Powell 1981). In an attempt to preserve its highly endemic fauna and flora, this remnant recently was acquired as a National Wildlife Refuge. Protection has come too late for some species, notably the Antioch dunes shieldback katydid (*Nebduba extincta*), a species whose specific epithet speaks to its fate (Rentz 1977).

Draining wetlands for housing and industry, as well as channeling creeks for flood control, has greatly reduced and altered aquatic habitats in the San Francisco Bay Area. These activities contribute to high extinction rates for populations of species such as the San Francisco forktail damselfly (*Ischnura gemina*) (Figure 7.5) (Hafernik 1989).

The cause of extinction of a particular invertebrate may seem as simple as the complete replacement of a species' habitat by a condominium complex, but often the causes are more subtle and the result of a series of historical events. Murphy and Weiss (1988a) summarize factors contributing to the decline of the bay checkerspot butterfly. This butterfly was probably once widely distributed in grassland habitats of the Bay Area. Replacement of native perennial grassland by introduced annual grasses likely restricted the distribution of the butterfly primarily to frag-

Figure 7.5
*A mating pair of the San Francisco forktail damselfly (*Ischnura gemina*).
This species has the most restricted range of any western odonate and is
a category 2 candidate for listing as an endangered species. Its existence
is threatened by the development and alteration of wetlands.*

mented grasslands associated with serpentine soils. Housing development and freeway construction in the last few decades have further reduced the size of these fragments. Furthermore, surviving patches vary widely in habitat quality, especially in drought years and as the result of livestock grazing. Because local extinctions and recolonizations are common, Murphy and Weiss suggest that the long-term success of this butterfly will require maintaining not only currently utilized fragments, but also fragments that periodically are colonized and can act as stepping stones for movement of butterflies and thus their genes within the overall metapopulation. A detailed discussion of evidence supporting a metapopulation model for the distribution of the bay checkerspot butterfly can be found in Harrison, Murphy, and Erlich (1988). An analysis of environmental factors that influence population viability of the butterfly is reported in Murphy, Freas, and Weiss (1990).

Many threatened invertebrates are probably characterized by a metapopulation structure similar to that of the bay checkerspot butterfly, in which frequent extinction of demes, dispersal, and recolonization allow regional persistence of a species. The San Francisco forktail damselfly and the curve-footed hygrotus diving beetle (*Hygrotus curvipes*) (Figure 7.6) appear to fit this pattern (Hafernik 1989). For the damselfly, colonies are associated with ponds and sluggish streams that are not filled with cattails or shaded by willows. Fluctuating water levels cause frequent local extinctions. Many sites have been channeled for flood control, preventing natural scouring of vegetation by occasional flooding and leaving the survival of many populations dependent on the cleanout and dredging cycle of these channels. The beetle occupies vernal pools and temporary wetlands. Some demes may inhabit a single drying puddle in a creek bed. Increasing use of wells for irrigation and for water for livestock and the subsequent lowering of water tables threatens to eliminate or reduce the time that suitable habitat is available. In both species, maintaining sufficient habitat for dispersal and recolonization is critical to their long-term survival.

Introduced Species

The intentional and unintentional introduction of nonnative species by humans has also been an important cause of invertebrate extinctions, especially on islands. Nowhere are such extinctions better documented than for Hawaii. Approximately 870 nonnative plants have naturalized in Hawaii, a number that almost equals the estimates of native plant richness at the time of Captain Cook's arrival in 1778 (Smith 1989). At least 2,000 species of alien invertebrates have become established on the islands (Howarth and Medeiros 1989), along with 81 species of verte-

Figure 7.6
*A male curved-foot hygrotus diving beetle (*Hygrotus curvipes*) at the water's surface. This small predaceous beetle is known mainly from ponds and vernal wetlands from a few localities in western Contra Costa and Alameda counties, California. Its status as a federal category 2 candidate for listing has fostered study of its ecology and distribution.*

brates (Stone 1989b). The adverse effects on the native land snail fauna by cattle, pigs, and rats were noted over a hundred years ago (Baldwin 1887, cited in Palmer 1986). Extinction of the snail genus *Carelia* from Kaui probably was caused by cattle trampling forests and by pigs eating the snails (Palmer 1986). Vegetation changes have contributed to a reduction in Hawaiian native insect diversity. Destruction by European hares of vegetation on Laysan, one of the leeward Hawaiian Islands, caused the extinction of several endemic noctuid moths (Opler 1976). Replacement of native vegetation with nonnative species has largely destroyed the native herbivore fauna of lowland areas and increasingly threatens other locations (Smith 1989).

Attempts to control earlier introductions have sometimes backfired, leading to additional extinctions. The predatory snail *Euglandina rosea* was introduced from Florida to control the giant African snail (*Achataina fulica*) but proved ineffective (Howarth and Medeiros 1989). Unfortunately, it was very effective in eliminating native land snails. Hart (1975) estimated that one-third of *Achatinella* land snail species were driven to extinction by *Euglandina rosea*, whose effects were exacerbated by destruction of vegetation by pigs, goats, and sheep. Attempts to control exotic introductions have also negatively affected some endemic Hawaiian insects. Mosquito fish have been widely introduced to control introduced mosquitos that are vectors of avian malaria, a major threat to native birds. Unfortunately, this has restricted distribution of the once widespread damselfly *Megalagrion pacificum*, itself a potential mosquito predator, to two isolated streams (Simon et al. 1984). The damselfly has also been threatened by the introduced long-legged ant (*Anoplolepis longipes*). It was once one of the most common odonates, and its imminent demise and that of its 27 congeners threaten a cascade of extinctions throughout the native stream biota (Stone 1989c).

On the U. S. mainland, cattle grazing probably contributed to extinction of the atossa fritillary butterfly (*Speyeria adiaste atossa*) and to local extinction of demes of the bay checkerspot butterfly. Here, as elsewhere, negative effects of introduced species are often one of several factors that combine to threaten native species. As mentioned above, introduction of European annual grasses contributed to fragmentation of the range of the bay checkerspot butterfly. This fragmentation was affected further by urban development and overgrazing by an introduced vertebrate, the cow, contributing to local extinctions during drought. For the mission blue butterfly, urban development has reduced and fragmented its habitat, and that process has been augmented by enlarging stands of introduced gorse and broom species (Arnold 1985).

Chains of Extinction

From the preceding examples, it should be apparent that the extinction of a species or a drastic change in its abundance or range can have ripple effects through native biotas, leading to a potential cascade of further extinctions (Diamond 1989). Studies of invertebrates have been important in understanding the dynamics of such phenomena. The mutualism between plants and their insect pollinators is often a keystone association that, if broken by extinction of important pollinators, can lead to the decline or extinction of the plant, of associated herbivores, and of predators and parasites of the herbivores (Gilbert 1980).

Chains of extinction sometimes begin with the introduction of an exotic organism and do not always require the complete demise of a species for the chain to be initiated. A classic example from eastern North America involves insects that specialized on American chestnut (*Castanea dentata*). Mature American chestnut plants were mostly destroyed by chestnut blight, a disease caused by an introduced pathogen. As a result, at least five moth species, including the chestnut borer, were driven to extinction, even though their host plant survived (Opler 1979).

The history of the decline of the large blue butterfly in England provides a second example. Despite forty years of active conservation efforts, the large blue butterfly became extinct in England in 1979 (Thomas 1981). The final causes for extinction point out the critical need for understanding any threatened organism's ecological requirements if appropriate management decisions are to be made. Female large blues lay eggs on thyme (*Thymus praecox*), on which the young larvae feed. Older larvae, however, are carried by red ants (*Myrmica* spp.) into their underground nests where the larvae feed for nine months on the ant's brood. Only one species of English ant (*Myrmica sabuleti*) was a suitable host for the blue's larvae. In ungrazed habitats it was replaced by another red ant species, but one that was unsuitable. The last British colony of the butterfly declined to extinction because ant colonies were the most important limiting resource. Overcrowding of butterfly larvae in ant nests resulted in high larval mortality and a penultimate year in which only a few adults, unfortunately mostly males, survived, followed by a final year in which no successful mating occurred and all eggs were sterile. Here, the local decline of one species contributed to the regional extinction of another.

Overkill (Collecting)

Before the agricultural revolution, hunting probably was the main human-caused mechanism of extinction of vertebrates (Diamond 1989). Overkill by early humans may have caused the Pleistocene extinctions of many large mammals in the New World (Martin 1984). Today hunting has declined in importance for vertebrates but is still a major threat to the survival of a few spectacular species such as whales, elephants, and rhinoceroses.

Collecting for biological research, recreation, or for commercial purposes is the invertebrate analogue of hunting. Overcollecting of some snails for their beautiful shells may have caused the initial decline in populations of Hawaiian *Achatinella* (Hart 1978). Because of possible threats to its survival caused by continued collecting, Mitchell's satyr (*Neonympha mitchelli*), a butterfly restricted to a few wetlands in the eastern United States, recently has been added to the endangered species list.

Current evidence suggests that collecting is rarely an important factor in the survival of most arthropods, even those with restricted distributions, nor is it likely to be (Pyle et al. 1981). The high fecundity and rapid generation time of most arthropods, coupled with the tendency for males of many species to be more commonly collected than females, usually mitigate small losses due to collection. Legal restrictions on private collecting of endangered arthropods serve primarily a symbolic function. Because of the ethical considerations involved in collecting specimens from small populations, however, several entomological societies recently have established collecting guidelines (Pyle et al. 1981).

Commercial exploitation may threaten populations of economically valuable species such as tropical birdwing butterflies (*Ornithoptera*, *Troides*, and *Trogonoptera*) for which populations are already critically depleted by habitat destruction (Collins and Morris 1985). This activity needs to be closely monitored, and, to this end, birdwing butterflies and the European Apollo butterfly (*Parnassius apollo*) are Appendix II species under the Convention on International Trade in Endangered Species of Wild Fauna and Flora (CITES). Female tarantulas are now popular in the pet trade because of their longevity in captivity and exotic body form. Because most of these are collected from the wild rather than reared commercially, they also could be threatened by commercial collection.

The widespread use of pesticides to control insect pests and its potential effect on nontarget invertebrates provides another analogue to hunting. Despite the commonly expressed view that pesticide use has resulted in decreased insect diversity, pesticide use has not been linked to the extinction of any invertebrate (Pyle et al. 1981). So far, greater negative effects have been documented on nontarget vertebrates such as fish and birds of prey. Nevertheless, there are many examples of the adverse effects of pesticides on nontarget invertebrates, especially in agroecosystems, leading to local collapse of populations of naturally occurring biological control agents and pollinators and to the induction of secondary pest outbreaks (Metcalf 1982). Pesticide-caused catastrophic disruptions of agroecosystems indicate a potential for major changes in natural systems, resulting in chains of extinction and alterations of food webs and nutrient flow patterns.

THE ROLE OF INVERTEBRATE CONSERVATION IN STRATEGIES TO CONSERVE TEMPERATE BIODIVERSITY

Vertebrate-centered conservation strategies continue to dominate efforts to conserve temperate biodiversity. Money and energy are allocated mostly to efforts to save large charismatic species such as the California condor, grizzly bear, or mountain lion. Much less attention is given to rare or endangered invertebrates or plants. Large warm-blooded animals require large reserves to support minimum viable populations (Eisenberg and Harris 1989), and this has led to an emphasis on acquiring large plots and/or linking current reserves with corridors.

Because island biogeography theory predicts a general positive relationship between area and biodiversity, large reserves would be expected to contain large numbers of invertebrate and plant species, and in this way efforts directed at large vertebrates should enhance invertebrate conservation as well. Unfortunately, the distribution of threatened large vertebrates is poorly correlated with the distribution of threatened invertebrates or plants. Large vertebrates, especially predators, have been eliminated from most temperate habitats. Those populations that remain are mainly in relatively pristine wilderness areas. Few threatened invertebrates occur in these areas. For instance, in the coterminous United States, the grizzly bear is restricted to areas in Wyoming, Idaho, and Montana. These states have no listed invertebrates and the lowest number of invertebrate candidate species of any U.S. region.

Most threatened invertebrates are associated with fragmented habitats, either naturally so or as the result of recent human activities. These fragments, though too small to support viable populations of most vertebrates, can sustain viable populations of many species of invertebrates (Wilson 1987) and plants.

Many threatened invertebrates are urban or suburban species whose habitats survived the first wave of development that eliminated most large vertebrates. A conservation strategy that includes as a priority preservation of these relatively small expanses of habitat can help to conserve a number of otherwise doomed taxa. Preservation will also help maintain a diversity of nonlisted invertebrates and plants, as well as some smaller vertebrates. Moreover, such a strategy helps call attention to the importance of conserving biodiversity in urban areas as well as wildlands.

A major impediment to the conservation of invertebrate diversity is the poor public image of most invertebrates, especially insects. The media constantly present us images of ways to live in a bug-free world through the use of pesticides and reinforce the common image of insects as disgusting creatures that eat our food, sting us, or transmit diseases, characteristics true of a very small percentage of species. Entomologists contribute to this image by generally referring to insect predators and parasitoids as "beneficial insects," ignoring the important role insects play at all trophic levels.

Conservation efforts for invertebrates can serve an educational function and help to dispel common misconceptions. By calling attention to unique and threatened organisms in a particular area, emphasizing the importance of invertebrates in ecosystem function and the intricate natural histories of many of these species, conservation biologists can help the informed citizen develop a better appreciation of these creatures.

Successful efforts in this regard are beginning to accumulate. Despite the fact that beetles generally have a poor public image (Kellert and Berry 1980), citizens of Plainfield, New Hampshire, adopted the rare cobblestone tiger beetle (*Cicindela marginipennis*) as their official town insect. A neighborhood group in San Francisco, California, used the occurrence of the San Francisco forktail damselfly in a city park as an example of the importance of such open space to maintaining biodiversity in urban settings. The citizens' group convinced the city of San Francisco to fund a biological inventory of the park that includes invertebrates.

Butterflies, because of their great beauty and apparent fragility, are one of the few insect groups with a positive image among the average citizenry (Kellert and Berry 1980). This explains in part why more efforts have been directed at their conservation than that of other insects. Because of the dependence of butterflies on plants for larval and adult food, their interaction with other animals, and their response to often subtle habitat changes, they have been suggested as ecological indicators of endangered habitats (Arnold 1983), for which they can serve as "umbrella" species (Murphy et al. 1990).

The listing of butterflies as endangered species already has helped to preserve a number of habitats that would otherwise have been destroyed (Arnold 1985). I will mention only two examples from California. The occurrence of the listed Lange's metalmark butterfly on the Antioch dunes of California led to the purchase of the remaining dune remnants as a national wildlife refuge, preserving not only the butterfly's habitat but also that of a number of other endemic insects and plants. The Antioch Dunes National Wildlife Refuge was the first refuge acquired by the U.S. Department of Interior for the protection of insects and plants. The mission blue butterfly (Figure 7.7) has become a conservation symbol in the San Francisco Bay Area. Its largest surviving population inhabits San Bruno Mountain, a mountain just south of San Francisco that is surrounded by a sea of urban development and that supports the largest remaining remnant of Franciscan flora and fauna. The mission blue butterfly's endangered status contributed to a long battle over the preservation of the mountain that eventually resulted in a compromise that preserved much of it as a state and county park. Although elements of the compromise remain controversial, without the attention focused on the mission blue and other endemic insects, greater habitat loss would probably have been sustained.

The listing of invertebrates sometimes has led to confrontation between real estate developers and conservationists, as was true in the San Bruno Mountain controversy. In that case, political efforts by developers contributed to modifications of the Endangered Species Act that affect vertebrates as well. Section 10a of the act now provides a mechanism for

Figure 7.7
*A female mission blue butterfly (*Icaricia icariodes missionensis*). This butterfly is endemic to a few localities within or near San Francisco, California, and was one of the first U.S. insects to be listed as an endangered species. Its listing promoted the conservation of its habitat and that of several other endangered insects and plants.*

the granting of permits to individuals for incidental take of a threatened or endangered species, providing such actions do not jeopardize the long-term survival of the species. As a result, habitat conservation plans that outline steps to minimize and mitigate negative impacts on endangered species are now commonly proposed by developers and their consultants. The habitat conservation plans approved so far vary widely in quality.

The listing of invertebrates has not always led to confrontation, however. Rather, it sometimes has encouraged cooperation among government agencies, industry, and conservationists. Chevron Corporation has cooperated in the enhancement of a small remnant of the El Segundo dunes located on one of their refinery sites in Los Angeles, California. This 0.6 ha site is one of two remaining localities for the endangered El Segundo blue butterfly (*Euphilotes battoides allyni*) and supports other dune species (Arnold and Goins 1987). Waste Management of California, Inc., has entered into a multifaceted conservation plan to protect the bay checkerspot butterfly (Murphy 1988). Both corporations have highlighted their actions in television commercials focusing on conservation of the butterflies. Pacific Gas and Electric Company has cooperated in the enhancement of Lange's metalmark butterfly habitat under power lines adjoining the Antioch Dunes National Wildlife Refuge.

Conservation efforts directed at invertebrates have also stimulated interest in the population biology of rare or restricted species. Notable in this regard are the continuing studies of Paul Ehrlich and his group on checkerspot butterfly populations (Ehrlich and Murphy 1987a, 1987b; Weiss, Murphy, and White 1988; Murphy and Weiss 1988a, 1988b; Murphy et al.1990) and the work of Richard Arnold (1983) on various species of threatened insects. The population structure of the San Francisco forktail damselfly (Garrison and Hafernik 1981) has made it an excellent system for testing predictions of mating system theory (Hafernik and Garrison 1986). More such research is needed on a broader array of invertebrates to provide the information needed for informed conservation decisions.

Invertebrates have already played a major role in the development of ecological and island biogeography theory. As debates continue about the best criteria for reserve design; the relative importance of genetic, demographic, and environmental factors in causing extinction; and so on, studies of invertebrates will be important. The small areas needed to support invertebrate populations make them relatively easy subjects for manipulation experiments. In addition, their generally fast generation times allow results to be gathered quickly. The classic experiments of Simberloff and Wilson (1969, 1970) on colonization and extinction rates of islands illustrate how the manipulation of invertebrate populations can be used to test theory.

Time is running out. The world faces the extinction, in our lifetimes, of untold numbers of species, resulting in unknown ecological consequences. In the future, experimental field studies of invertebrates must play a major role in the development of theories of reserve design and restoration ecology. Invertebrates are perfect subjects for reintroduction projects designed not only to enhance the range of restricted species but also, through replicated natural experiments, to investigate factors affecting successful reestablishment. Not only is the potential for rapid establishment of breeding colonies good, if natural resources are sufficient, but also the funding needed per project is modest. If a fraction of the money now being spent on the potential reintroduction of such species as the California condor were invested in reestablishment and enhancement of populations of threatened invertebrates, conservation biology could significantly expand its number of successes and gain a broader understanding of the diverse factors influencing success.

CONCLUSIONS

1. Human activities increasingly threaten invertebrate biodiversity. A number of extinctions have been documented, especially for Hawaii, and many more species are currently at risk of extinction.
2. Public and private conservation policy emphasizes the conservation of large vertebrates. In most cases, however, invertebrates are more crucial to ecosystem function.
3. Invertebrates usually interact with their environments on a finer scale, require smaller patches for survival, and thus are good indicators of unique and unusual habitats.
4. A modest redirection of priorities and resources could have a major positive impact on invertebrate conservation efforts. Relatively small, cost-effective reserves often can preserve significant invertebrate diversity.
5. Invertebrate conservation can help focus attention on unique remnants of natural habitats in urban and suburban areas. The few federally listed insects have resulted in significant successes in habitat conservation and public education.
6. Invertebrates have played and must continue to play an important role in the development of ecological theory as it applies to conservation biology.

ACKNOWLEDGMENTS

Peggy L. Fiedler, Johnnie Johnson Hafernik, Subodh Jain, and V. Thomas Parker reviewed earlier drafts of this chapter and provided many helpful suggestions.

LITERATURE CITED

Amaral, M., and L. Morse. 1990. Reintroducing the American burying beetle. *Endangered Sp. Tech. Bull.* 15:3.

Andrews, F.G., A.R. Hardy, and D. Giuliani. 1979. The coleopterous fauna of selected California sand dunes. Tech. Rep. U.S. Bureau of Land Management.

Arnett, G. R. 1984. Endangered and threatened wildlife and plants; review of invertebrate wildlife for listing as endangered or threatened species. *Federal Register* 49:21664–75.

Arnold, R.A. 1983. Ecological studies of six endangered butterflies (Lepidoptera, Lycaenidae): Island biogeography, patch dynamics, and the design of habitat preserves. *Univ. Calif. Pub. Entom.* 99:1–161.

————. 1985. Private and government-funded conservation programs for endangered insects in California. *Natural Areas J.* 5:28–39.

Arnold, R.A., and A.E. Goins. 1987. Habitat enhancement techniques for the El Segundo blue butterfly: An urban endangered species. In *Integrating man and nature in the metropolitan environment,* ed. L.W. Adams and D.L. . Leedy, 173–81. Columbia, Md.: National Institute for Urban Wildlife.

Biggins, D. 1989. Coal mining and the decline of freshwater mussels. *Endangered Sp. Tech. Bull.* 14:5.

Collins, N.M., and M.G. Morris. 1985. *Threatened swallowtail butterflies of the world. The IUCN Red Data Book.* Gland, Switzerland: International Union for the Conservation of Nature and Natural Resources.

Diamond, J. 1989. Overview of recent extinctions. In *Conservation for the twenty-first century,* ed. D. Western and M. Pearl, 37–41. New York: Oxford University Press.

Dunlop, B.N. 1989. Endangered and threatened wildlife and plants; Animal notice of review. *Federal Register* 54:554–79.

Ehrlich, P.R. 1988. The loss of diversity: Causes and consequences. *Biodiversity,* ed. E.O. Wilson and F.M. Peter, 21–27. Washington, D.C.: National Academy Press.

Ehrlich, P.R., and D.D. Murphy. 1987a. Conservation lessons from long-term studies of checkerspot butterflies. *Cons. Biol.* 1:122–31.

————. 1987b. Monitoring populations on remnants of native vegetation. *Nature conservation: the role of remnants of native vegetation,* ed. D. Saunders, G.W. Arnold, A.A. Burbidge, and A.J.M. Hopkins. Chipping Norton, New South Wales: Surrey Beatty and Sons.

Eisenberg, J.F., and L.D. Harris. 1989. Conservation: A consideration of evolution, population, and life history. In *Conservation for the twenty-first century,* ed. D. Western and M. Pearl, 99–108. New York: Oxford University Press.

Gagné, W.C. 1989. Native terrestrial invertebrates. In *Conservation biology in Hawai'i,* ed. C.P. Stone and D.P. Stone, 77–81. Honolulu: University of Hawaii Cooperative National Park Resources Study Unit.

Garrison, R.W., and J.E. Hafernik. 1981. Population structure of the rare damselfly, *Ischnura gemina* (Kennedy) (Odonata: Coenagrionidae). *Oecologia* 48: 377–84.

Gilbert, L.E. 1980. Food web organization and the conservation of neotropical diversity. *Conservation Biology.* ed. M.E. Soulé and B.A. Wilcox, 11–33. Sunderland, Mass.: Sinauer.

Hafernik, J.E. 1989. Surveys of potentially threatened Bay Area water beetles and the San Francisco Forktail damselfly. Tech. Rep. U.S. Fish and Wildlife Agency.

Hafernik, J.E., and R.W. Garrison. 1986. Mating success and survival rate in a population of damselflies: Results at variance with theory?. *Amer. Nat.* 128:353–65.

Harrison, S., D.D. Murphy, and P.R. Ehrlich. 1988. Distribution of the bay checkerspot butterfly, *Euphydryas editha bayensis*: Evidence for a metapopulation model. *Amer. Nat.* 132:360–82.

Hart, A.D. 1975. Living jewels imperiled. *Defenders* 50:482–86.

————. 1978. List of probably extinct, possibly extinct, and extremely rare *Achatinella*. Washington, D.C.: U.S. Fish and Wildlife Service Tech. Rep.

Heath, J. 1981. *Threatened Rhopalocera (butterflies) in Europe*. Strasbourg: Council of Europe Nature and Environment Series, no. 23.

Howarth, F.G., and A.C. Medeiros. 1989. In *Conservation Biology in Hawai'i*, ed. C.P. Stone and D.P. Stone, 82–87. Honolulu: University of Hawaii Cooperative National Park Resources Study Unit.

Johnson, K. 1986. Prairie and plains disclimax and disappearing butterflies in the central United States. *Atala* 10–12: 20–30.

Jones, G.E. 1987. *The conservation of ecosystems and species*. London: Croom Helm.

Kellert, S.R., and J.K. Berry. 1980. Knowledge, affection and basic attitudes toward animals in American society, Phase III. Report to U.S. Fish and Wildlife Service.

Majer, J.D. 1987. The conservation and study of invertebrates in remnants of native vegetation. In *Nature conservation: The role of remnants of native vegetation*, ed. D.A. Saunders, G.W. Arnold, A.A. Burbidge, and A.J.M. Hopkins, 333–35. Chipping Norton, New South Wales: Surrey Beatty and Sons.

Martin, P.S. 1984. Prehistoric overkill: The global model. In *Quaternary extinctions: A prehistoric revolution*, ed. P.S. Martin and R.G. Klein, 354–403. Tucson: University of Arizona Press.

Metcalf, R.L. 1982. Insecticides in pest management. In *Introduction to pest management*, ed. R.L. Metcalf and W.H. Luckmann, 217–77. New York: John Wiley & Sons.

Moore, N.W. 1981. The conservation of Odonata in Great Britain. *Atala* 7:64–67.

Murphy D.D. 1988. The Kirby Canyon conservation agreement: A model for resolution of land-use conflicts involving threatened invertebrates. *Environ. Cons.* 15:45–57.

Murphy, D.D., K.E. Freas, and S.B. Weiss. 1990. An environment-metapopulation approach to population viability analysis for a threatened invertebrate. *Cons. Biol.* 4:41–51.

Murphy, D.D., and S.B. Weiss. 1988a. Ecological studies and the conservation of the bay checkerspot butterfly, *Euphydryas editha bayensis*. *Biol. Cons.* 46:183–200.

————. 1988b. A long-term monitoring plan for a threatened butterfly. *Cons. Biol.* 2:367–74.

Odum, E.P., C.E. Connell, and L.B. Davenport. 1962. Population energy flow of three primary consumer components of old-field ecosystems. *Ecology* 43:88–96.

Opler, P.A. 1976. The parade of passing species: Extinctions past and present. *Sci. Teach.* 43:30–34.

————. 1979. Insects of American chestnut: Possible importance and conservation concern. In *The American chestnut symposium*, ed. W. McDonald, 83–85. Morgantown: West Virginia University Press.

————. 1981. Management of prairie habitats for insect conservation. *Natural Areas Assoc. J.* 1:3–6.

————. 1987. Invertebrate surveys in North America are necessary. *Wings* 12:8–10.

Paine, R. 1966. Food web complexity and species diversity. *Amer. Nat.* 100:65–75.

Palmer, S. 1986. Some extinct molluscs of the U.S.A. *Atala* 13:1–7.

Powell, J.A. 1981. Endangered habitats for insects: California coastal sand dunes. *Atala* 6:41–55.

Price, P.W. 1984. *Insect ecology.* New York: John Wiley & Sons.

Pyle, R., M. Bentzien, and P. Opler. 1981. Insect conservation. *Ann. Rev. Entomol.* 26:233–58.

Rentz, D.C. 1977. A new and apparently extinct katydid from Antioch sand dunes. *Entomol. News.* 88:241–45.

Simberloff, D., and E.O. Wilson. 1969. Experimental zoogeography of islands: The colonization of empty islands. *Ecology* 50:278–96.

————. 1970. Experimental zoogeography of islands: A two-year record of colonization. *Ecology* 51:934–37.

Simon, C.M., W.C. Gagné, F.G. Howarth, and F.J. Radovsky. 1984. Hawaii: A natural entomological laboratory. *Bull. Entomol. Soc. Amer.* 30:8–17.

Smith, C.W. 1989. Non-native plants. In *Conservation biology in Hawai'i,* ed. C.P. Stone and D.P. Stone, 60–69. Honolulu: University of Hawaii Cooperative National Park Resources Study Unit.

Stone, C.P. 1989a. Native birds. In *Conservation biology in Hawai'i,* ed. C.P. Stone and D.P. Stone, 96–102. Honolulu: University of Hawaii Cooperative National Park Resources Study Unit.

————. 1989b. Non-native land vertebrates. In *Conservation biology in Hawai'i,* ed. C.P. Stone and D.P. Stone, 88–95. Honolulu: University of Hawaii Cooperative National Park Resources Study Unit.

————. 1989c. Hawaii's wetlands, streams, fishponds, and pools. In *Conservation Biology in Hawai'i,* ed. C.P. Stone and D.P. Stone, 125–36. Honolulu: University of Hawaii Cooperative National Park Resources Study Unit.

Thomas, J.A. 1981. Insect conservation in Britain: Some case histories. *Atala* 6:31–36.

————. 1983. The ecology and conservation of *Lysandra bellargus* in Britain. *J. Appl. Ecol.* 20:59–83.

————. 1984. The conservation of butterflies in temperate countries: Past efforts and lessons for the future. In *The biology of butterflies,* ed. R.I. Vane-Wright and P.R. Ackery, 333–53. London: Academic Press.

Weiss, S.B., D.D. Murphy, and R.R. White. 1988. Sun, slope, and butterflies: Topographic determinants of habitat quality for *Euphydryas editha. Ecology* 69:1486–96.

Wells, S.M., R.M. Pyle, and N.M. Collins. 1983. *The International Union for the Conservation of Nature and Natural Resources Invertebrate Red Data Book.* Cambridge, U.K.: International Union for the Conservation of Nature and Natural Resources.

Western, D., and M. Pearl, eds. 1989. *Conservation biology for the twenty-first century.* New York: Oxford University Press.

Westman, W.E. 1990. Managing for biodiversity. *BioScience* 40:26–33.

Wiegert, R.G., and F.C. Evans. 1967. Investigations of secondary productivity in grasslands. In *Secondary productivity in terrestrial ecosystems*, ed. K. Petrusewicz, 499–518. Warsaw: Inst. Ecol. Polish Acad. Sci.

Wilson, E.O. 1987. The little things that run the world (the importance and conservation of invertebrates). *Cons. Biol.* 1:344–446.

CHAPTER 8

Forest Fragmentation and the Conservation of Biological Diversity

LARRY D. HARRIS

and GILBERTO SILVA-LOPEZ

ABSTRACT

Definitions and clarifications of terms and concepts relevant to the effects of habitat fragmentation on biological diversity are presented. Habitat fragmentation differs from habitat patchiness, and it is illustrated how forest fragmentation is distinguished from a series of forest fragments and a single tract of insular forest. Five types of fragmentation are illustrated, and land use in and around the Ocala National Forest illustrates how these are relevant to management decisions. The full impacts of fragmentation cannot be appreciated unless the concept of wildlife is distinguished from that of native fauna because faunal relaxation and faunal collapse are measurable only against the backdrop of native biota. Faunal collapse occurs when sufficient levels of disturbance cause fundamentally different intensities of ecological processes to prevail. Habitat fragmentation effects cannot be gauged independent of the scale of evaluation, and again, the case of the Ocala National Forest is used to illustrate the issues.

INTRODUCTION

Over 100 years ago the French ecologist de Candolle observed that "the breakup of a large landmass into smaller units would necessarily lead to the extinction or local extermination of one or more species and the differential preservation of others" (de Candolle 1855, in Browne 1983, 44). This observation, perhaps the first to note the effects of habitat fragmentation, has held profound implications for present-day ecologists and resource managers. Along with the complete and permanent removal of large expanses of forest, habitat fragmentation is occurring at an alarming rate and is the principal cause of current problems in conservation biology, many of which are only now emerging.

There is increasing concern among conservation biologists that habitat fragmentation ranks among the most serious causes of the erosion of biological diversity. In the words of Wilcox and Murphy (1985, 884), it is "the most serious threat to biological diversity, and the primary cause of the present extinction crisis." This chapter is written as a primer to the habitat fragmentation phenomenon, its nature and its consequences. Although all of the relevant literature cannot be reviewed here, we provide an entry to the scientific papers on the subject. Our emphasis is focused on fragmentation of forest habitats, and, in order to provide data-based examples, we focus on the north-central Florida region that is dominated by the Ocala National Forest (Ocala NF).

DEFINITIONS AND CLARIFICATIONS·

Biodiversity, short for *biological diversity*, "is the variety and variability among living organisms and the ecological complexes in which they occur" (Office of Technology Assessment 1987, 37; see Glossary for additional relevant definitions). The hereditary component of this diversity is propagated at the molecular level and is referred to as *genetic diversity*. But all individuals and higher levels of biological diversity such as the community are molded and sculpted in an environmental context. Thus *phenotypic diversity* constitutes genotypic diversity as well as the many complex morphological combinations molded by the environmental context within which the genome occurs. Natural selection operates on phenotypes, and populations consist of phenotypes. For example, in Florida and throughout the tropics, reptiles and amphibians commonly dominate the animal communities. Yet the gender of such stereotypic species as alligator (*Alligator mississippiensis*), rattlesnake (*Crotalus* spp.), gopher tortoise (*Gopherus polyphemus*), marine turtles (Cheloniidae), and bull-frogs (*Rana catesbeiana*) is dictated by incubation temperature of the eggs, not genetics. Thus, species, and even individuals, represent much more diversity than can be measured in their genes (e.g., Lewontin, Rose, and Kamin 1984).

It is because of the importance of environmental context that bio-diversity conservation is so critically affected by how humans impact the environment. The previously described phenomenon, referred to as tem-perature-dependent sex (TDS) (Bull 1980), means that not only is the sex ratio of this year's offspring governed by environment, but also that the sex ratio of present populations of long-lived marine turtles and crocod-ilians was determined by the environmental temperatures that prevailed decades ago. As long as the diversity of nesting sites, year classes, and sources of population recruits was great, there was virtually no prospect that imbalanced sex ratio would be a problem. But as we eliminate the diversity of regional sources of recruits, nesting conditions, and year classes within the population, we court disaster.

The issue of old-growth forests is of intense current interest among conservation biologists. Existing old-growth forests are the outcome of centuries of environmental and community forces that have operated on the genes of seeds that germinated from 500 to 1,000 years ago. Although it is obvious that tree shape and form (e.g., bonsai) is controlled by environment, it is less appreciated that the abundance of cavities and the habitat structure utilized by all arboreal species also results from forces such as fungal attack and rot and the population levels of cavity-exca-vating woodpeckers. The presence of such obligate cavity-nesting birds as bluebird (*Sialia sialis*), wood duck (*Aix sponsa*), kestrel (*Falco spar-varius*), and screech owl (*Otus asio*) is controlled by community inter-actions. Diversity begets diversity. The suggestion that an old-growth forest can be duplicated by the planting of genetically improved nursery stock represents the height of naïveté. Even if a replacement stand were cultured to a very old age, this action presumes that

1. old age is an adequate representation of old growth (which it is not),
2. a stand of planted trees will develop into a forest (which it might or might not), and
3. trees that are planted in the greatly altered landscape milieu of today might somehow come to resemble the forest that was regenerated by natural processes in a very different environment of 500 years ago.

Even if a stand of trees were equal to a forest, if the genome planted now was the same as that removed, and if all the attendant organisms were reintroduced along with the seedlings, there remains an infinitesimal prospect that the forces shaping the seedlings that are planted in a cutover landscape are similar to those that shaped the development of the previous old growth. The full complement of biological diversity represented by an old-growth forest ecosystem is only vaguely known and simply cannot be represented by our present vocabulary or mathematical index and cannot be regenerated by present technology.

The *Oxford English Dictionary* defines *fragment* as a portion broken off or otherwise detached from a whole — comparatively small detached portions of anything; *fragmentation* is the breaking or separation into fragments. Therefore, in the context of biological conservation we define *forest fragmentation* as an unnatural detaching or separation of expansive tracts into spatially segregated fragments.

Fragmented forests are not the same as patchy forests. All naturally regenerated forests are patchy in the sense that the trees and associated organisms do not occur in uniform patterns. Areas of forest dominated by one species customarily grade into areas dominated by another; tree falls and light gaps are created by insect and fungal infestations and/or blowdowns; colonizing vines may so dominate the trees as to obscure their existence at all. But importantly, whereas ecologists would dispute the difference between oak-hickory (*Quercus-Carya*) and beech-maple (*Fagus-Acer*) forest types, few could agree on the location of a specific boundary between them. In natural systems, one patch generally grades into another; forest types are intergraded.

A large scientific literature deals with ecological patchiness, namely, the "discontinuities in environmental character states . . . that have biological significance to the organism . . . and are revealed in a nearly endless spectrum of spatial scales" (Wiens 1976, 83). Thus, a patchy forest is one where the heterogeneity and/or the discontinuities associated with natural gradations in the environment are coupled with natural regenerative and colonizing processes of the forest organisms. Researchers may very well establish *sampling plots* within the forest, but the boundaries are totally human-made (Figure 8.1A).

On the contrary, a fragmented forest refers to a landscape that was formerly forested but now consists of forested tracts that are segregated and sometimes isolated in a matrix of non forested habitat. As the degree of fragmentation progresses, a naturally patchy forest is transformed into a *fragmented forest*, then to a number of *forest fragments*, and finally a single *insular tract* (Figure 8.1). As discussed later, the consequences of these different stages are distinct.

Dominance by trees is a useful criterion for designating and mapping forests. But although trees are necessary to the definition of a forest, they

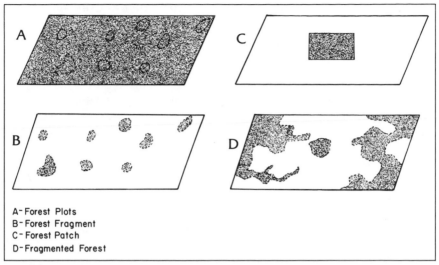

Figure 8.1
Schematic illustration of the different degrees of forest fragmentation. Panel A depicts a closed-canopy forest with designated plots representing 10% of the area. Panel B depicts a series of forest fragments that remain after 90% of the original forest has been removed. Panel C represents a single tract of insular forest habitat remaining after 90% of the forest has been removed, and Panel D depicts a fragmented forest.

certainly are not sufficient. For example, foresters consider that a tract of cleared forest that will be replanted to trees should still be classified as forest acreage (USDA 1978). Thus, even though a formerly forested landscape that is 90 percent barren of trees may seem like a fragmented forest to a conservation biologist, it technically qualifies as nonfragmented forest from the standpoint of forest statistics. To be effective, the definition of forest must also make reference to the structural characteristics that occur within the designated ecosystem. Just as orchards consisting of fruit or nut trees are not considered forests, neither are vast stretches of citrus-covered landscape generally included in forest statistics. The consequences of grove fragmentation are not discussed here. Similarly, the point at which a landscape covered with equally spaced and manicured Christmas trees grades into the typical conifer plantation, or the point at which the latter deserves consideration as a forest, is yet to be resolved. Thus, to assess the consequences of forest fragmentation, a conservation biologist would evaluate parameters such as

1. the amount, composition, and distribution of residual forest;
2. the abruptness of gradation between remaining patches;

3. the continuity or disruption of the distribution and movement of native organisms;
4. the composition and structure of the vegetation that now constitutes the landscape matrix; and
5. the compositional pattern of the overall landscape.

Forest fragmentation may be independent of forest loss. It is commonly assumed that forest fragmentation is simply one aspect of the overall loss of forest or that habitat loss follows necessarily from forest fragmentation. Neither is the case. Although the two phenomena commonly occur simultaneously, they can be independent, and it is important for researchers to distinguish between the two. For example, a forestry enterprise might remove ten 50-ha cuts from an existing 5,000-ha closed-canopy forest and simultaneously purchase an adjacent 500-ha tract of forest or perhaps replant an equal acreage to forest. In either case, we are left with the same amount of forest, 5,000 hectares. On the other hand, if the cuts were interspersed throughout the tract, fragmentation effects would seriously alter the overall habitat quality even though total acreage remained the same. Similarly, several state and federal agencies in the United States have habitat acquisition programs that allow annual purchase of wildlife habitat, or they provide incentives for the reforestation of agricultural land. In this case, the level of forest fragmentation may be increased simultaneously with an increase in acreage rather than with a decrease in forest acreage. Future research must be directed toward distinguishing between the effects of habitat loss, those of habitat fragmentation, and the effects of interactions between the two phenomena.

From a traditional natural resource agency viewpoint, one might conclude that a 90 percent loss of a state's forest habitat would result in the loss of 90 percent of the wildlife. This might be couched in terms of population size (i.e., 90% of the individuals), species, or simply in terms of "wildlife." In any event, the question rapidly reduces to how many individuals and which species will be lost. But only when the remaining tracts occur as a series of fragments (as in Figure 8.1B), and only to the degree that measured responses are different from the 90 percent prediction, can fragmentation effects be distinguished from the effects of forest loss (as in Figure 8.1D).

On the other hand, principles of biogeography (Darlington, 1957; Preston 1962a, 1962b; MacArthur and Wilson 1967) might lead to totally different null hypotheses. Habitat islands of tenfold greater size are expected to support twice as many species as the reference patch, whereas those of one-tenth the size are expected to support half as many species (see Harris 1984). Thus, if 90 percent of the habitat is lost, but the remaining 10 percent occurs as a single patch, then one might expect only a 50 percent reduction in species. It is clearly of some significance whether the effects of fragmentation are tested by comparing forest habitats as represented by Figure 8.1A versus 8.1B, 8.1A versus 8.1D, or 8.1B versus 8.1D.

Fragmented habitat is not the same as insular habitat. When land development encroaches on a tract of forest and completely severs it from the original tract or completely surrounds and isolates it, then the tract becomes a forest habitat island, and it is referred to as insular habitat (Figure 8.1C). In other words, as long as organisms and ecological processes that are characteristic of forests continue to dominate, we refer to it as a fragmented forest (Figure 8.1D). As the surrounding matrix of landscape changes from one that is dominated by flows of energy and matter characteristic of forests to one that is not, then it ceases to be a fragmented forest and is at best a series of forest fragments. When a tract of forest becomes functionally isolated from other, larger tracts of forest, it is referred to as a *forest island.* In other words, we define *insular habitat* as that which occurs when the flows of energy and matter across the habitat island edge become totally dominated by flows to or from a non-forest landscape matrix. A forest fragment becomes insular habitat when it is sufficiently isolated by geographical space and by differentiation from the landscape matrix to allow the interactions with the matrix to overwhelmingly dominate the interactions with former or related tracts of forest.

Fragmentation effects depend on scale and the normal movement pattern of organisms in question. Habitat fragmentation must be viewed as a multiscale, multidimensional problem. What constitutes an isolated fragment to a specialist species may be neither isolated nor a fragment to the wide-ranging generalist species. Conversely, a road that carries a heavy traffic stream may constitute a much more serious fragmenting force to a wide-ranging generalist species that must move across it (and get killed on it) than it does to a sedentary specialist (Harris and Gallagher 1989).

Because most species possess a unique need to move and because they express different tolerances to habitat disturbance, the intervening distances that determine whether the landscape is a fragmented forest, a set of forest fragments, or an insular tract of habitat must be judged against species-specific data. For example, cougars (*Felis concolor*) are habitat generalists, and a single dominant male may range over 100,000 ha. A now–predominantly cleared but formerly forested landscape that contains one residual 10 ha tract of forest per 10,000 ha (0.1%) would contain as many as ten tracts within the home range of a single cougar.

Therefore, from the standpoint of cougars and other wide-ranging generalist species such as black bears (*Ursus americanus*) or elk (*Cervus canadensis*), the landscape might appear to contain widely dispersed forest fragments.

Eastern wild turkeys (*Meleagris gallopavo*) are more reliant on forest habitat and have home ranges typically less than a few thousand hectares. If they occurred in the same landscape, their centers of activity would be tied to a single forest island, and they would rarely come in contact with two or more fragments in the course of their day-to-day movements. Thus, from the standpoint of turkeys, the landscape would constitute a coarse-grained or patchy forest landscape.

The northern parula warbler (*Parula americana*) establishes breeding territories of a hectare or two, and thus three or more pair could breed within a 10-ha tract. As a consequence, their distribution in such a landscape would be clumped. Ecological interactions between birds within a tract would probably be intense, but interactions between the birds of separate patches would be weak at best.

Populations of walking sticks (Orthoptera: Phasmatidae) would occur in each of the insular habitats, and each could exist for numerous generations entirely within a single tract. Because the distance between tracts would be very great relative to their normal movement, interactions between subpopulations would occur very rarely. Thus, any given tract could be described as insular habitat relative to walking sticks. The distinction between what appears as a forested universe, a fragmented forest, and a forest fragment is clearly dependent on one's perspective. The subsequent evaluation of consequences will be markedly different.

STRUCTURAL CHARACTERISTICS AND DISPERSION OF TREES

The preceding discussion has focused on situations where the composition of the patches of forest differs greatly from that of the surrounding matrix. It must also be recognized that the within-tract structure of forests differs a very great deal as does the degree of distinction between the forest tract and the surrounding matrix (Figure 8.2C). Consider the difference between a structurally complex old-growth forest that supports only about twenty-five very large Douglas fir (*Pseudotsuga menziesii*) trees per hectare but thousands of tons of dead wood and understory woody plants, and a woodland that has the exact number of live mature trees but little if any understory. If both were surrounded by the same clear-cut matrix, it seems improbable that the consequences of fragmentation would be similar. Not only the structural complexity of the relictual

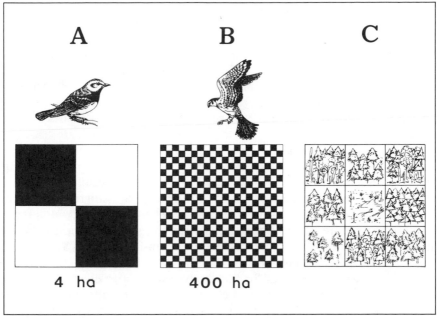

Figure 8.2
Important aspects of the habitat fragmentation issue involve fragment size, dispersion across the landscape, and contrast between adjacent units. A 400-ha (1,000-acre) landscape that is divided into 1-ha blocks would seemingly appear quite different to a warbler (as in panel A) than it would to a wide-ranging species (as in panel B). Fragmentation effects may be reduced by affecting the structural qualities of different fragments (as in "the new forestry," panel C).

forests must be considered; the structure of and degree of distinctness of the surrounding matrix is equally relevant.

Fragmenting Forces May be Encroaching, Enveloping, Divisive, or Intrusive. Different processes lead to forest fragmentation, and notably different landscape patterns result. Unfortunately, habitat fragmentation research has not progressed sufficiently for the distinctions between the types of fragmentation, the resulting landscape patterns, and the consequences for biological diversity to be differentiated in the scientific literature. Clarification of the differences in patterns of fragmentation will be necessary before major advances to understanding and predicting fragmentation effects can be made.

Regressive fragmentation results when forest clearing occurs from a single direction and the frontal edge of the forest is simply pushed back by successive bouts of cutting. This commonly occurs as coastal areas are settled and human population expands inland or as cutting progresses from lower to higher elevations. Urban population growth that pushes toward a large forest tract and causes wilderness and wildlife species to regress ahead of the development is also of this type (Figure 8.3A). Importantly, there is always a hinterland, and no matter how severe the barrier effects on the frontal edge may be, immigration and emigration may still be possible from some source pool in the opposite direction of the fragmenting force.

Enveloping fragmentation occurs when clearing, developmental pressures, or both surround the perimeter of a tract and cause a contraction of the forest area from all directions (Figure 8.3B). Because there are fewer, if any, opportunities for immigration and/or emigration into a contiguous forest, the consequences are believed to be very different. Because the nature of the interactions with the surrounding habitat is also different, contraction of a forest area in a nonforest landscape matrix results in more severe effects than if a tract is simply severed from an expansive forest. One reason for this derives from edge effects. Whereas the severed tract will manifest some effects due to the altered edges and partially different matrix, insular tracts that are created by enveloping fragmentation will suffer the effects of reduced habitat size and increased isolation, and both of these consequences will be amplified by edge effects from the totally surrounding matrix.

Divisive fragmentation (Figure 8.3C) occurs when an intrusive force such as a turnpike, a road with attendant strip development, powerlines, or railroads bisects an expansive tract such that the movement of organisms between the bisected parts is significantly obstructed. Effects of this type of fragmentation are believed to be proportional to the magnitude of the dividing force and the severity of the separation of the two parcels.

Intrusive fragmentation occurs when forest habitat is removed or greatly altered from within, such as would result from placing clear-cuts or food-plot cuttings within an existing tract of forest (Figure 8.3D). Whereas the first three types of fragmentation directly impact the surrounding matrix (the context) and indirectly affect internal forest structure, this form directly affects the structural integrity of the forest from within—that is, it directly impacts the content as opposed to the context. Although it would seem that the ultimate effect on forest pattern might be the same whether the openings began in the interior and worked their way outward or vice versa, it seems intuitive that the consequences for relictual biological diversity would be very different.

Encroaching fragmentation is distinguished from both regressive and enveloping forms inasmuch as the forest landscape is commonly removed from either side of a linear gallery forest, but the riparian or gallery forest

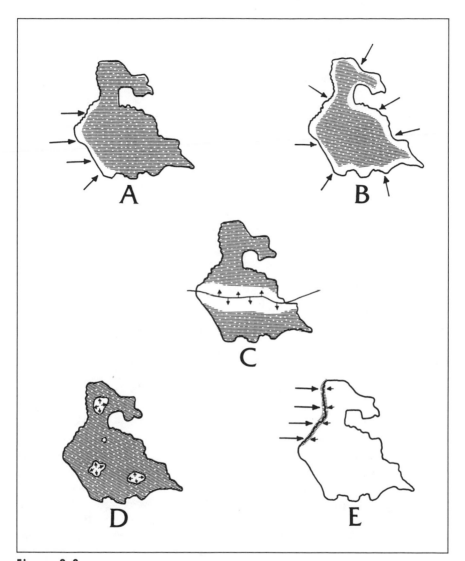

Figure 8.3
Different types of fragmentation processes depicted against a silhouette of the Ocala National Forest. Panel A depicts the regressive form as intense human land use approaches to the western edge. Panel B depicts enveloping fragmentation as the entire tract is surrounded by converted land use. Panel C depicts divisive fragmentation as results from highway development bisecting a tract. Panel D depicts the nature of intrusive developments that occur from within the forest tract, and Panel E illustrates how a river corridor such as the Oklawaha may remain connected at either end but become isolated by land use developments on either side.

remains connected to larger patches of forest on either end. As a result, energy, matter, and organisms can continue to move freely along the length of the fragment but are totally isolated except for the corridor effect (Figure 8.3E).

Two or more of these types of fragmentation may impact a single forest tract, and we will show in a following section how several of the different types are operative simultaneously. Nonetheless, a clear understanding of fragmentation effects will not be possible until which time as the various types and degrees of fragmentation have been distinguished by research.

DESCRIPTIVE ASPECTS OF FRAGMENTATION EFFECTS

When a tract of mature forest is cut and removed, one new habitat is introduced and a second new habitat is created. This is for two reasons. The site where mature forest existed is no longer mature forest; it is a new habitat type, and the overall composition of the forest landscape has been changed. Many natural forces such as hurricanes, tornadoes, crown fires, and insect outbreaks created gaps in the natural forest. These gaps were a natural stage in the forest regeneration process and are thus referred to as the *gap phase* of forest regeneration. The opportunistic species that evolved to inhabit or depend on these gaps are referred to as *gap phase species.*

Although timber harvests are ecologically very different from natural gap-forming processes, timber cuts can be made to approximate natural gaps as long as the individual cuts are small relative to the total expanse of forest, and as long as the total acreage of cutover forest does not exceed 50 percent of the total forest acreage. In other words, as long as small cuts constitute a minority percentage of the total forest acreage and occur in a *mature forest matrix*, native forest species may be able to persist in fragmented forests. The percentage of the landscape remaining in mature forest and its connectivity are important considerations, however, because it is the mature forest species that have intrinsically low dispersal capacities as compared to the early successional species that have intrinsically high dispersal capacities. Relictual tracts of mature or old-growth forest need to be large enough to maintain interior forest conditions for species that require such conditions.

When the proportion of the forest that is cutover comes to dominate the landscape and the relictual forest fragments that remain come to exist in a *matrix of cutover landscape*, the faunal relations shift dramatically. Species that evolved to exist in a forest matrix, albeit a patchy one, are suddenly confronted with the prospect of existence in a cutover landscape. Moreover, as the matrix of second growth increases, *common, opportunistic species* colonize the new habitat and exert amplified levels of competition, parasitism, and predation upon the remaining forest species. The full range of effects of habitat fragmentation that have been recorded in the literature cannot be understood without first introducing the concept of fauna.

Fauna versus Wildlife

Since Linnaeus used it in the eighteenth century, the term *fauna* refers to the assemblage of animals that is sufficiently characteristic of a particular area or era that it is distinguishable from the animal life of other areas and/or eras. In addition to the actual species present, characteristics such as unique or endemic taxa, primitive versus recent taxa, life form of the animals (e.g., fossorial, cursorial, arboreal), mean body size (e.g., large plains game vs. smaller forest species), the ratio of foraging guilds (e.g., grazers vs. browsers vs. granivores vs. frugivorous vs. insectivores), and trophic relations (e.g., trophic pyramids) can be used to distinguish one fauna from another. For example, tundra, forest, plains, and desert faunas are so distinctly different as to not be confused with one another. In order to change from one fauna to another, significant alterations in both the presence or absence and the relative abundance of species must occur. For example, simply exterminating a species or two from the desert fauna does not change it into a plains fauna. Conversely, the invasion of a fauna by one or a few exotic species is not, in and of itself, cause for redesignation. But if a significant loss of native species is coupled with a significant invasion by exotic species, and these events ultimately lead to major changes in all or most of the systematic and ecological parameters identified earlier, then a change in the native fauna has occurred. This is precisely what happens when a formerly forested landscape is transformed into either a fragmented forest or a collection of forest fragments.

The names of P.L. Sclater, A. Wallace, C. Hart Merriam, and J. Allen stand out as prominent contributors to the delineation of regional faunas. Not uncommonly, biotic provinces were differentiated at the scale of a region or an area the size of a few states (e.g., Dice 1943). Biogeographers of the nineteenth century believed that one characteristic of native faunal assemblages was that of balance or harmony as illustrated in the following quotes from Baur (1897, 217):

> Continental islands, therefore, may be composed of two floral and faunal elements: first, an original (endogenous) one; and second, a secondary (exogenous) one. Oceanic islands, however, will only contain a secondary

(exogenous) floral and faunal element. .... The flora and fauna of the first group [continental] will be more or less harmonic,—that is to say, the islands will be like satellites of the continent from which they developed, and the whole group comparable to a planetary system. The flora and fauna of the second group [oceanic islands] will be disharmonic, —that is to say, it will be composed of a mixture of forms which have been introduced accidentally from other places.

Contrary to the term fauna, the term *wildlife* is of very recent coinage and has its first official usage associated with the formation of The Wildlife Society in 1937 (Meine 1988; Hunter 1990). It refers to all free-ranging plants and animals of an area and does not distinguish between game versus nongame, common versus rare, or native versus exotic (Harris 1988a). Therefore, at the same time that the effects of forest fragmentation on native fauna might be catastrophic, it may not be possible to convince a nonecologist that there have been any effects on wildlife other than a change in species abundance. Indeed, in many cases, fragmentation leads to increases in both species diversity and abundance. Because the term wildlife is so general, and because all species are assumed to be of equal intrinsic value, discussions quickly degrade into assertions about human biases and favoritism.

Faunal relaxation refers to the process whereby the number of species inhabiting an area approaches some dynamic equilibrium between rates of colonization and local extinction (Diamond 1972, 1973; Terbourgh 1974). Although originally applied to true islands, the concept was rapidly applied to habitat islands and tacitly expanded to embrace the loss of species with certain characteristics and the increasing dominance of species with different characteristics. Harris (1988a) broadened the term still further to include more diverse changes in the native fauna in response to habitat fragmentation.

As large expanses of closed-canopy forest are cut and fragmented, the original plant and animal communities "relax" to a new galaxy of species; as the patches of habitat become smaller, the proportion of edge-tolerant and opportunistic species continues to increase (Diamond 1972; Terbourgh 1974; Whitcomb et al. 1981; Anderson and Robbins 1981; Lynch and Whigham 1984; Robbins, Dawson, and Dowell 1989; Wilcove, Mclellan, and Dobson 1986).

Consistent with the original formulation, we use *faunal relaxation* to refer to the collective set of faunal responses to reduced habitat area and/or fragmentation and/or insularization. These responses include (1) loss of certain native species and an increase in abundance of others; (2) colonization by alien species; (3) inbreeding depression resulting from small population size of sequestered species; and (4) dramatic shifts in the overall species frequency distribution that results from the preceding.

For example, forest fragmentation typically results in a series of fragments that are too small and/or too separated to provide adequate habitat
for the occurrence or successful reproduction of some species. These are
referred to as *area sensitive species* because they require large territories
or foraging or ranging areas, and they are negatively impacted by reductions in habitat area. Even when the total area of mature forest is kept
constant but fragmenting forces intervene, these species frequently experience unsustainably high levels of mortality as they move from one
fragment to another.

A second group of organisms inhabits only the interior portions of
relatively large tracts; these are referred to as *interior species* because their
habitat exists only some distance removed from the forest edge. Whereas
all interior species appear to be area-sensitive to a greater or lesser degree,
species such as black bears and barred owls (*Strix varia*) are area-sensitive
species but are not interior species.

Unless the species benefits from openings and edges, forest fragmentation that is accompanied by reduced acreage of forest must reduce the
populations of even those species that are tolerant of fragmentation effects. This holds important consequences because most utilitarian values
that derive from wildlife (including ecological services such as pollination) hinge on large, viable populations. Thus, the mere reduction of a
species to rarity may detract a great deal from its utilitarian values. Moreover, such reduced populations are subject to a new set of uncertainties
not shared by larger populations, and thus they also may dwindle and
become locally extinct. Uncertainties associated with small population
size are classed as (1) *demographic*, such as uncertain sex ratio, number
of offspring, or age of death; (2) *environmental*, such as aberrancies in
weather or food supply; and (3) *genetic*, such as the uncertainties associated with inbreeding and genetic drift (Soulé 1987; Brussard and Gilpin
1989).

The collective result is that endemic and/or characteristic species,
and most certainly the characteristic assemblage of forest species, are lost.
But more is lost. It is not uncommon to observe increases in species
richness in association with fragmentation. This is because even though
some native forest species are lost, many new species are attracted to the
second growth that now constitutes the landscape matrix. Because gaps
and second growth are less abundant, less predictable, and more patchy
in the presettlement forest, second-growth species have evolved high dispersal and colonization abilities. They colonize quickly and thrive wherever second growth occurs. And, unlike mature forest specialists that have
narrower geographic ranges, second-growth generalists have broad distributions and are quick to broaden them still farther. Thus, one of the
greatest concerns associated with the habitat fragmentation issue is that

the identity and distinctiveness of all of North America's native faunal regions are being eroded and replaced by a homogenized collection of wildlife. It would seem that homogenization of regional faunas and floras represents nearly as serious a threat to North America's biological diversity resource as does the loss of a few old-growth species (Samson and Knopf 1982; Noss 1983; Knopf 1986).

Species-Area Relations for Insular Forest Tracts

The species-area relation is frequently described by this equation:

$$S = cA^z \tag{8.1}$$

When the exponent (z) values are positive, this relation predicts that larger areas will contain more species than comparable but smaller-sized areas. Moreover, this relation has been shown to apply generally to both *enumeration data*, when one counts each and every species present, and to *sampling data* derived by counts of the number occurring in sample plots of various sizes. Roughly the same relation, but different parameter values, holds whether dealing with sample plots of different sizes drawn from a larger universe or from discrete habitats such as islands (i.e., Figure 8.1A vs. 8.1B). Additional properties of this relation also follow.

When the value of exponent z is between 0.0 and 1.0 (far and away the most common), any additional unit area added to plot size or island size contributes an ever-smaller number of new species to the cumulative total. An important but commonly overlooked corollary of this is that species density (i.e., the number of species per unit area) declines with increasing area.

It is commonly implied (and sometimes stated) that equation (8.1) produces a curve that "levels off"—that is, has an inflection point at some loosely defined section of its range. This is not true, however. A necessary condition for any function $y = f(x)$ to have an inflection point at $x = x_0$ is for:

$$d^2y / dx^2 = 0, \tag{8.2}$$

that is, at an inflection point of $y = f(x)$ the second derivative is zero.

The species-area curve (8.1) with $0 < c$, $0 < z < J$ has a second derivative:

$$d^2S / dA^2 = cz (z-1)A^{z-2} \tag{8.3}$$
$$\text{with} \quad cz(z-1) < 0, \quad -2 < (z-2) < -J.$$

That is,

$$d_2S / dA^2 < 0, \text{ all } 0 < A. \tag{8.4}$$

In other words, d^2S/dA^2 is never 0 and $S = cA^z$ has no inflection point. The meaning of this in the real world is that there is a continuously decreasing rate of increase in cumulative species added in response to increasing sample plot size and/or increasing habitat island size. Conversely, if one continues to reduce the size of the area, there is no point at which species loss becomes notably accentuated. Therefore, whether one is losing habitat as in reducing the size of forest islands, or whether one is deliberating on some magical minimum critical size necessary for a protected area, there is no "breaking point" where the number of species added is notably diminished or accelerated. Any threshold of forest fragment size or habitat island size that is chosen as a decision point is subjectively, and not mathematically, defined. Any perceived "leveling off" that is often given as a justification for such decisions is simply an artifact of scale. For the purposes of demonstration we provide the following example.

If one were to collect representative species-occurrence data from appropriate protected areas throughout a region, then it becomes a straightforward exercise to derive the best-fit equation (e.g., least-squares equation) for explaining the species richness that occurs in the different areas. For the purpose here we use this equation:

$$S = 16.3 \, A^{0.16} \tag{8.5}$$

This equation could then be used to predict the number of species likely to occur in any comparable natural area within the sampling universe. San Felasco Hammock State Preserve is a 2,400-ha protected area in Alachua County, Florida. St. Marks National Wildlife Refuge constitutes a second important protected area in north Florida that is approximately ten times as large (26,000 ha), and the Ocala National Forest constitutes yet a third important protected area that is nearly ten times larger still (i.e., 182,000 ha).

It will be quickly noted from Panel A of Figure 8.4 that the San Felasco Hammock Preserve lies well out onto the flattened portion of the projected species-area curve. Similarly, it will be noted from Panel B of Figure 8.4 that the St. Marks National Wildlife Refuge also lies well onto the flattened portion of its species-area plot. And likewise, the Ocala National Forest falls well onto the flattened area of the Panel C species-area plot. Each of these curves is produced with exactly the same equation, and the graphs are different only in the scale used along the axes.

Now, by contrast, note what an utterly different impression is gained by looking at the relative position of the three areas when they are placed on the same graph in Figure 8.5. Recall that the three panels of Figure 8.4 and the single graph of Figure 8.5 all depict the same numerical relationship, namely that of equation (8.5).

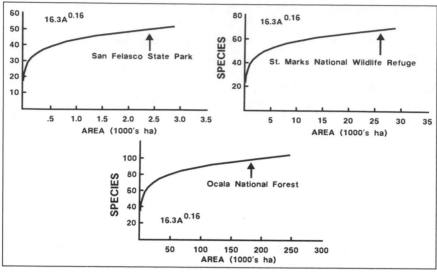

Figure 8.4
*Three species-area curves generated from an empirically derived equation S = 16.3 A * 0.16. Note that in all three panels, the position of the designated protected area occurs to the right on the "flattened" part of the species-area curve.*

We may deduce a powerful conservation principle from this exercise: Whenever the deliberation involves the value of incremental acreage to be added to or subtracted from a conservation area, the judgment of value is subjective and biological, *not* mathematical. Moreover, perceptions from a graph of the species-area curve will as likely be misleading as convincing. And most importantly, equal increments (or decrements) of area will always appear more critical to smaller areas, that is, those that occur on the left of the graph compared to larger areas that occur to the right. Moreover, acreage increments or decrements will always appear less consequential to areas that occur toward the right hand-side of the graph. Or, said another way, it will always appear less threatening to slightly diminish the acreage of a tract *if* it is positioned near the right-hand margin of the graph, and it will always appear more fruitful to add area to a tract that is positioned near the left margin of the graph. If one wishes to make a specific case regarding a specific area, one simply needs to choose the comparisons prudently.

Ecological Processes and Forest Fragmentation

Previous sections described a series of definitional issues and structural changes associated with habitat fragmentation. We now describe changes

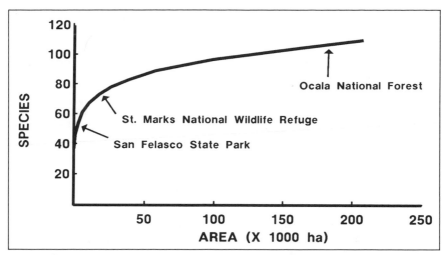

Figure 8.5
The same species-area curve as depicted in Figure 8.4 but with all three designated areas identified on the same curve. Note how it would now appear that an increment of area added to the smaller area would be immensely more valuable than an equal area added to the larger area.

in ecological relations of animal populations inhabiting fragmented forest landscapes. Competition for limited forest resources such as dead wood and nesting cavities can be critical to the maintenance of forest biodiversity. Forest fragmentation processes lead to not only reduced levels of standing dead wood but also an increase in the abundance of aggressive cavity competitors. One characteristic of old-growth forests is the common occurrence of dead wood. Fungal invasion leads to the frequent occurrence of rotting branches, and when coupled with large woodpecker populations that excavate nesting cavities there seems to be no shortage of nest sites for obligate secondary cavity nesters (i.e., those that cannot excavate cavities), ranging from bluebirds to wood ducks and gray squirrels (*Scirus carolinensis*).

But unlike old-growth forests, cleared landscapes, silvicultural clearcuts, second-growth forests, and especially conifer plantations do not harbor large amounts of dead wood. Because all woodpeckers in the Southeast, other than the red-cockaded woodpecker (*Picoides borealis*), excavate cavities only in dead wood, cavity competition in landscapes dominated by these land uses can be severe.

Only two species of woodpeckers have increased in absolute abundance and/or relative dominance of the woodpecker community in Florida; these are the red-bellied and pileated woodpeckers (Harris and Frederick 1990). The red-bellied woodpecker seems to be the more aggressive and ubiquitous competitor with cavity nesters such as the endangered red-cockaded woodpecker (Jackson 1978).

Numerous researchers have analyzed the role of cavity competition in the Southeast (see Thompson 1971; Wood 1983). The consensus is that activities that either reduce the availability of standing dead wood or increase the density of cavity competitors near red-cockaded woodpecker colonies will have detrimental effects. In addition to competition for existing cavities, cavity entrance enlargement by pileated woodpeckers further increases the spectrum of potential competitors. Jackson (1978) reported that 69 percent of the cavities he examined in Mississippi and Georgia had been enlarged by pileated woodpeckers. Harlow and Lennartz (1983) demonstrated an inverse correlation between snag density and the proportion of cavities that had been enlarged. Avifaunal diversity throughout the Southeast is impacted by forest fragmentation that not only directly reduces the abundance of cavities but also brings open-landscape competitors into close proximity to remaining forest-nesting species. We believe that species such as the red-cockaded woodpecker are threatened not just by loss of habitat, but by altered ecological relations. Simply providing more habitat such as presently exists will not solve the problem of maintaining these species well into the next century.

Predation equals or exceeds the force of competition in driving processes of natural selection and maintenance of biological diversity. Amplified levels of nest predation are a particularly important force in fragmented forests. Researchers such as Wilcove (1985), Andren et al. (1985), Andren and Angelstam (1988), Yahner and coworkers (Yahner 1988; Yahner and Cypher 1987; Yahner and Scott 1988; Yahner and Wright 1985), and Reese and Ratti (1988) have documented high levels of predation in and around forest openings, as well as how nearness to forest edge amplifies predation effects on forest species.

Many breeding organisms face a double jeopardy because they are attracted to nest or to reproduce near the forest edge, precisely where predation rates are the highest. Gates and Gysel (1978) coined the term *ecological trap* to refer to this phenomenon, and researchers such as Temple and Cary (1988) and Robinson (1988, 1990) have demonstrated how highly fragmented forests may have negative net production of nesting birds. For example, Robinson (1991) "found little evidence of successful reproduction of forest birds that winter in the tropics but breed in Illinois (neotropical migrants). Most nests found in the area failed because of high predation rates (80% of all nests) and brood parasitism by cowbirds (76% of the nests of neotropical migrants). . . . Taken together, these data strongly suggest that populations of most neotropical migrants do not produce enough young to replace themselves in the Shelbyville area."

Two distinct groups of predators are involved. Mammals such foxes (*Vulpes fulva, Urocyon* sp.), skunks (*Mephitis* spp. and *Spilogale* spp.), raccoons (*Procyon lotor*), and domestic dogs (*Canis familiaris*) and cats (*Felis domesticus*) are primarily predators on ground nests. Because they are cursorial, are predominantly nocturnal, and hunt by use of scent, they have limited effect on aerial nests. Birds of the family Corvidae—ravens, crows, and jays—are diurnal, hunt by use of sight, and prey on any open nests, whether on the ground or above it. Both groups, the middle-sized mammal predators and the Corvidae, abound in human-dominated landscapes and exert extraordinarily high predation pressure on nesting organisms in fragmented forest environments.

Damage caused by predation is often amplified when the predator is an alien species. Savidge (1987), for example, found that the introduced brown tree snake (*Boiga irregularis*) was responsible for the extinction or decline of ten species of native forest birds on the island of Guam. In particular, the Philippine turtle dove (*Streptopelia bitorquata*) that nests in vegetational remnants was found to be impacted greatly by snake predation (Conry 1988).

Parasitism is predicted to increase in stressed systems, whether organisms or communities (Odum 1985). One form of parasitism that has been reported to seriously impact the faunal community of North American fragmented forests is nest parasitism by brown-headed cowbirds (*Molothrus ater*). Naturally rare in eastern North America, the species has proliferated throughout the East following human settlement and forest fragmentation (Mayfield 1977; Brittingham and Temple 1983). According to Mayfield (1977), the effect of the cowbird is particularly insidious because it uses many different host species, and therefore its parasitism of any single species is not reduced even though the host population is reduced to near-extinction. A second species, the shiny cowbird (*M. bonariensis*), is more tropical in distribution but is rapidly expanding its range northward through the Antilles (Wiley 1988) and has now colonized the southern tip of Florida. It poses a serious threat to the avifauna of Florida.

The proportion of forest bird nests parasitized is strongly correlated with nearness to edges between forest and clearings such as those introduced by clear-cutting, energy transmission lines, or game-management food plots. Numerous eastern forest species are now threatened with extinction due to nest predation and parasitism because of wildlife-management practices that have inadvertently maximized habitat fragmentation under the guise of maximum interspersion. Kirtland's warbler (*Dendroica kirtlandii*) hangs on the verge of extinction, and its close relative Bachman's warbler (*D. bachmanii*) may well have already been exterminated by these forces (Harris 1988b).

Decomposition rates are suspected to be affected by forest fragmentation, higher rates of decomposition being associated with intact forest and lower rates being associated with the smaller forest islands (Klein 1989). In many forest ecosystems, especially tropical ones, available nutrients are highly limiting, and thus reduced decomposition directly affects the nutrient cycling rate with possible productivity implications. In the study by Klein, the species composition and abundance of both scavenging and dung-rolling beetles (Coleoptera) were reduced in forest fragments, and possibly it was the reduction of these organisms that caused the suspected decline in decomposition rates. Much more research is required before a pattern can be developed.

In total, it can now be demonstrated that forest fragmentation not only directly affects the presence and absence of species, but it also alters the fundamental ecological processes that shape and govern the nature of ecological communities (Table 8.1). When sufficient alteration of ecological process has occurred, the passive process of faunal relaxation may accelerate to become active *faunal collapse*. In other words, entire native faunas may collapse because of altered ecological mechanisms that create domino effects that ripple through the entire ecological community.

Focus on North-Central Florida and the Ocala National Forest— Habitat Isolation at the Regional Scale

Florida's human population of 12 million is growing at an annual rate of 4 percent and is resulting in the rapid loss of Florida's rural character and a rapid restriction of native faunal populations to a system of protected areas. Although aggressive conservation land–acquisition programs exist, the protected areas are quickly becoming habitat islands in a sea of human development. Processes at work in and around the Ocala NF exemplify the consequences of enveloping fragmentation and isolation at the regional scale.

Up until 200 years ago Florida supported eleven species of mammals larger than 5 kg in size. Of these eleven, the monk seal (*Monachus monachus*) is now globally extinct, and the red wolf (*Canis rufus*) and bison (*Bison bison*) are locally extinct. The manatee (*Trichechus manatus*), Florida panther (*Felis concolor coryi*), and key deer (*Odocoileus virginiana clavium*) are federally listed as endangered. Three other species, black bear, bobcat (*Felis rufus*), and otter (*Lutra canadensis*), are either listed as threatened by the state of Florida or are listed by the International Convention on Trade in Endangered Species (CITES). In other words, of the eleven large mammals native to Florida, nine have been extirpated,

Table 8.1
Modified Ecological Functions that Result from Fragmentation of Florida Habitats.

I. Amplified levels of competition for nesting cavities
   A. Result from weedy species (e.g., red-bellied woodpecker, European starling)
   B. Impact 22 species of obligate cavity nesters (e.g., red-cockaded woodpecker, eastern bluebird)
II. Amplified levels of open-nest parasitism
   A. Results from nuisance species (e.g., brown-headed cowbird)
   B. Impact open-nesting species (e.g., Bachman's warbler (extinct?), Swainson's warbler)
III. Amplified levels of ground-nest predation
   A. Results from unrestricted mid-sized mammals (e.g., raccoon)
   B. Impact over 150 ground-nesting species (e.g., turkey, gopher tortoise)
IV. Amplified levels of aerial nest predation
   A. Result from increased populations of anthrophilic species (e.g., crows and jays)
   B. Impact both ground- and aerial-nesting species (e.g., game birds and colonial nesting waders)
V. Amplified levels of parasites and infectious disease
   A. Result from unrestricted populations of mammals (e.g., raccoons, foxes, coyotes, and domestic dogs)
   B. Carry disease and parasites (e.g., heartworms, rabies)

are endangered, or are thought to be threatened; only the mainland white-tailed deer (*Odocoileus virginiana*), and the raccoon are doing demonstrably well statewide. Many other species of mammal such as the Everglades mink (*Mustela vison*) are endangered, and other species are being lost to extinction (e.g., Anastasia beech mouse, *Peromyscus gossypinus anastasae* [Humphrey, Kern, and Ludlow 1988]). However, the toll on large species has been particularly great.

Various subpopulations of remaining large-bodied species are increasingly isolated and occur only in large federally owned protected areas such as the Ocala NF. This is an example of enveloping fragmentation operative at the regional scale. The three stochastic effects (environmental, demographic, and genetic) known to be operative on small sequestered populations are already seen to be at work in species such as the Florida panther, where 95 percent of the spermatozoa are found to be congenitally infertile, a likely consequence of inbreeding depression (U.S. Fish and Wildlife Service 1987).

Highways and Divisive Fragmentation

Much has been made of the distinction between true islands that are isolated by water, and terrestrial habitat islands that are isolated in a human-dominated landscape matrix. It is commonly implied that water constitutes a more serious barrier to immigration and emigration than do human artifacts such as cities and large highways. Just the opposite is true. To accommodate the phenomenal human population growth rate and a tourist industry involving 40 million tourists each year, the state has built hard-surface highways at a rate of over 6 km/day for the last fifty years. Whereas Florida's native species evolved in a setting surrounded by sea, none of them evolved in the presence of high-density, high-speed traffic. The consequence is devastating.

Mortality from collisions between motor vehicles and animals represents the primary known source of mortality for most of Florida's endangered large wildlife species, including the panther, black bear, key deer, manatee, American crocodile (*Crocodylus acutus*), and bald eagle (*Haliaeetus leucocephalus*). At minimum, high-volume traffic streams such as Florida State Road 40 (SR40) that bisects the Ocala NF (Figure 8.1C) have a major isolating effect on populations existing on either side of the road. In fact, the consequences are much worse when roads cross traditional animal movement corridors and no concessions are made for movement. For example, data are available for Florida State Road 46 (SR46) that runs parallel to but is located 50 km south of SR 40. Because no provisions were made for wildlife movement beneath the road, it is notorious as a principal black bear death trap in the state of Florida. A biologist with the state Department of Natural Resources has observed that for bears moving southward from the Ocala NF, "State Road 46 is basically functioning as a wildlife killing machine" (W. Thomson, biologist, Florida Dept. of Natural Resource, pers. com.).

Data are also available for US Highway 441 that runs through Paynes Prairie State Preserve 50 km to the north of SR40. An animal mortality survey of 3.25 km of this road yielded the estimate that 13,000 snakes belonging to twelve species with an estimated mass of 1.3 tons were killed during a 4.5-year period (Richard Franz, Florida Museum of Natural History, pers. com.). These situations are not unique to the Ocala NF nor to the state of Florida (see chapter 9). As protected areas become increasingly surrounded by roads and human development, the accentuated mortality rates due to vehicle collision will eliminate many species (see also Oxley, Fenton, and Carmody 1974; Mader 1984; Lalo 1987).

When Fragmented Forests Become Forest Fragments

The conversion of forests to agricultural land is common in north-central Florida, and Florida's annual rate of forest acreage loss is 1 percent per year, a rate much greater than pertains in countries such as Brazil (Harris and Frederick 1990). A study aimed at assessing effects of habitat insularization on avifauna (as in Figure 8.2B) and conducted on twelve forest fragments ranging from 0.4 ha to 30 ha in size revealed a typical species-area relation of increasing breeding bird species as a function of area (Harris and Wallace 1984). The number of species associated with the islands increased over the range of islands sampled, and generally the larger islands included species that occurred on the smaller islands along with additional new species. It seems that certain "assembly rules" (sensu Diamond 1975) were operative and that islands were colonized by forest-breeding species as their acreage exceeded known territory sizes. This phenomenon does not hold for raptors and edge species, however, because they use the forest island only for breeding while using large expanses of surrounding agricultural land for foraging.

Of forty-five species of birds that commonly breed in expansive tracts of north Florida forest, only twenty-four species (53%) used the twelve forest island fragments. The overwhelming majority of these twenty-four species are common throughout residential and agricultural areas of north Florida. The twenty-one species that did not breed in any of the forest fragments consist of area-sensitive species (see Table 8.2) and interior species (see Table 8.3).

The Matrix Surrounding Forest Fragments Is Not Simply Black or White

Isolated cypress (*Taxodium distichum*) wetlands are common throughout the Southeastern Coastal Plain. The surrounding flatwoods were historically dominated by pine of several species (*Pinus* spp.). These flatwoods

Table 8.2
Bird species that breed in the southeastern United States that are classified as "area-sensitive" (from Hamel et al. 1982).

Swallow-tailed kite	Solitary vireo
Mississippi kite	Red-eyed vireo
Red-shouldered hawk	Black-and-white warbler
Broad-winged hawk	Bachman's warbler
Ruffed grouse	Northern parula warbler
Wild turkey	Black-throated green warbler
Black-billed cuckoo	Yellow-throated warbler
Barred owl	Pine warbler
Pileated woodpecker	Scarlet tanager
Red-cockaded woodpecker	Summer tanager
Yellow-throated vireo	Rose-breasted grosbeak

Table 8.3
Bird species that breed in the southeastern United States that are designated as "interior" species (from Hamel et al. 1982).

Sharp-shinned hawk	Prothonotary warbler
Cooper's hawk	Worm-eating warbler
Hairy woodpecker	Swainson's warbler
Acadian flycatcher	Black-throated blue warbler
Winter wren	Ovenbird
Wood thrush	Louisiana waterthrush
Hermit thrush	Kentucky warbler
Swainson's thrush	Hooded warbler
Veery	

grade into the higher pine flatwoods that were formerly dominated by longleaf pine (*Pinus palustris*) and are now dominated by either naturally regenerated or planted slash pine (*Pinus elliottii*). When the surrounding landscape matrix consists of mature pine, then the gradation between cypress and pine is gradual. But when the pine is harvested by clear-cutting, the cypress suddenly becomes the only forest habitat islands in the cleared landscape, and what was formerly a gradual interface becomes a totally exposed vertical face (Figure 8.6).

Thorough breeding-bird surveys of isolated cypress wetlands that occurred in the clear-cut landscape were compared with comparable surveys of cypress islands occurring in a landscape matrix of mid-aged planted slash pine (McElveen 1977; Harris and McElveen 1981). A species-area curve was fitted to the data, and it is self-evident that the larger islands of cypress supported more breeding-bird species than did the smaller islands. However, there was also a significant difference in breeding bird species richness, with the cypress island isolates that stood naked in the clear-cut landscape supporting more species than the cypress islands that were surrounded by mid-aged planted pine (Figure 8.7). Extrapolation from the observed species-area relations suggest that cypress habitat islands of 40 ha or more would be necessary to extinguish the effect of surrounding habitat types (Harris 1988a). This result compares favorably with more recent results published by Robbins et al. (1989).

Because the study emphasized avian species dynamics in the cypress island edges, the data on edge effects are particularly strong. Without doubt, the sharp edges consistently supported greater species richness, species density, species diversity, and density of individuals than did the interfaces between cypress and surrounding pine. Moreover, stands of cypress isolated by clear-cutting consistently supported greater species richness, species density, species diversity, and, most notably, two times

Figure 8.6
Schematic illustration of tracts of cypress forest occurring in a planted slash pine landscape matrix (top) compared to cypress habitat islands occurring in a clear-cut landscape matrix (bottom). McElveen researched the effects of altering the landscape matrix within which the cypress occurred.

as many breeding individuals as did the cypress surrounded by planted pine (Table 8.4).

Recall that the cypress stands themselves were not directly impacted by the treatment and that there were no discernible differences between those that were surrounded by treatment and those that were not. Thus, a quick interpretation is that the cypress stands that were surrounded by clear-cut landscape served as refuges for breeding birds from the pinelands. However, several lines of evidence argue against this. First, six of the fourteen breeding bird species that occurred in the enhanced avifauna

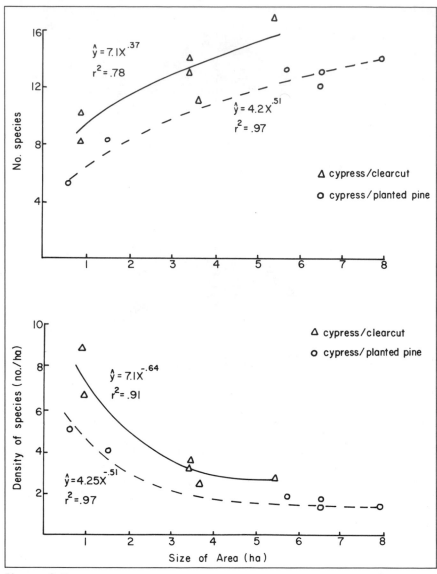

Figure 8.7
*A: Species-area curves for breeding birds occurring in cypress habitat is-
lands, some occurring in a clear-cut landscape matrix compared to others
occurring in a pine plantation matrix. B: Species density as a function of
size of area. If the species-area curves appear as in part A, then it is
axiomatic that the density of species will follow curves as depicted in part
B.*

Table 8.4

Average density of birds in a north Florida cypress-pine landscape where cypress islands were surrounded by mid-rotation planted slash pine and an ecotone existed between the two. Other comparable cypress islands were in a clear-cut landscape and a sharp edge occurred between the two.

Species	Pine Plantation	Ecotone	Sharp Cypress	Edge	Clear-Cut
Rufus-sided towhee	1	1.7	0.1	3	0.01
Carolina wren	0.1	0.6	0.7	1	
Cardinal	0.6	1.5	0.7	0.8	
Parula warbler	0.1	0.7	1.9	0.4	
White-eyed vireo		0.1	1.2	0.8	0.5
Great crested flycatcher	0.2	0.9	0.9	1.6	
Common yellowthr ·	0.1	0.3	0.4	2.2	
Tufted titmouse	0.1	0.2	0.1	0.1	
Red-bellied woodpecker		0.1	0.4	0.2	
Blue-gray gnatcatcher			0.8	0.8	
Yellow-throated warbler			0.1	0.1	
Prothonotary warbler		0.6			
Pileated woodpecker		0.1			
Yellow-billed cuckoo			0.03		
Yellow-throated vireo			0.03		
Bachman's sparrow	0.03	0.2			
Number of species	8	9	15	12	2
Bird density/ha	2.2	7.0 ·	7.7	11.1	0.2

were never observed in the pinelands or on the clear-cut landscape. Three of the species were new to the system and did not occur in the pine-surrounded cypress islands (Table 8.3). Second, Bierregaard (1989, 6) collected long-term data from a similar research design and reported that "after about 200 days the activity stabilized around the expected values for non-isolated reserves." The responses reported here were as great in the second breeding season after clear-cutting as in the first, and thus they showed no sign of being ameliorated by time. It is our interpretation that the enhanced edge effect resulting from the sharp and accentuated difference between contrasting systems carried over deep into the remaining cypress islands. Therefore, the predominance of increased species and individuals was the result of birds typical of secondary and fragmented landscapes.

Edge Effects or Area Effects

Reports and analyses of the negative aspects of edge effects are common in recent years (see Harris 1988c; Alverson, Waller, and Solheim 1988;

Temple and Cary 1988; Yahner 1988; and Reese and Ratti 1988 for reviews). Yet additional insight can be gleaned from the above-mentioned Florida study.

It follows from simple geometry that the area of a habitat is always a squared function of any linear measure of size. For example, L is a linear measure of the size of any square, and area is L squared; r is a linear measure of the size of circles, and πr^2 represents the area. On the other hand, perimeter is always a simple linear function of any linear measure of size (e.g., 4L, or $2\pi r$). This means that the amount of edge surrounding a cypress island is always a linear function of the square root of area ($\sqrt{a}$). Therefore, if the number of species of birds found to inhabit the cypress habitat islands was more strongly a function of edge than area, it should be reflected by analysis of the square root transformed data. Coefficients of determination (R) associated with least-squares analysis of the data can be used as an index to goodness of fit.

For the cypress islands surrounded by planted pines, linear least-squares analysis of S = f(a) produced an equal coefficient (i.e., 0.97) to that achieved from analysis of the log-log transformed data. We interpret this to mean that the number of breeding bird species inhabiting these cypress islands was no more a function of edge than of area. It was slightly different for cypress islands surrounded by clear-cut, where analysis for edge resulted in a slightly, but not significantly, better fit than the analysis for area (R = 0.70 vs. 0.68). Albeit weak, this is added evidence for the significance of the edge effect associated with the abrupt margins. Importantly, as postulated by Thomas (1979), the data show a clear relation between the magnitude of the edge effect and the size of the cypress islands (Figure 8.8).

Species Presence or Absence

The biogeography approach to conservation strategies has been criticized because of the heavy emphasis placed on simple tallies of the number of species present or the inferred local extinction of species on the simple basis of presence-absence data. Indeed, more scrutiny must be given to the issue of what constitutes a "resident species." For example, consider that a Florida panther might physically occupy a few square meters at any given time. Yet its territory encompasses 10^8 m^2. How frequently must the cat visit, or how much time must it spend in any given square meter in order to fall within the definition of "present"?

This is a significant problem in the real world of conservation biology. In the case of the now-extinct dusky seaside sparrow (*Ammospiza maritima nigrescens*), scientists monitored population status by using the conventional "territorial-male" census technique only to learn after the

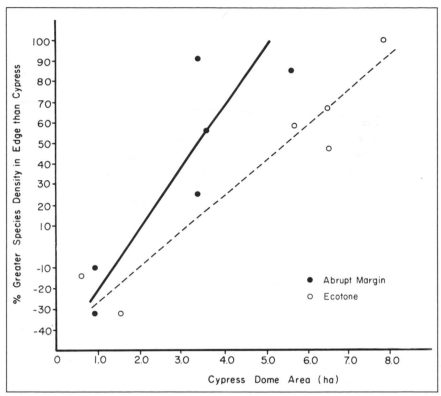

Figure 8.8
Measured increase or decrease of bird density in the edge surrounding cypress habitat islands. Not only is edge effect shown to be a function of the size of the habitat island involved, it is also shown to be a function of the sharpness of the edge involved.

fact that there were no remaining females (see Harris 1988b). Bluntly, how faithful to a site must an organism be in order to be tallied as present or to constitute being a resident?

Based on data from the breeding bird studies referred to previously, it is possible to shed light on this matter. For example, McElveen (Harris and McElveen 1981) mapped the territorial males in each of six cypress habitat islands on six separate counts in each of two successive breeding seasons. Let us consider the results of counting territorial males in any one of the islands such as the third of the sharp-edge treatments (e.g., E3). Because of year-to-year environmental variation, the number of breeding bird species differed between years (Figure 8.9a). Now, if the criterion for a species being considered present consists of territorial de-

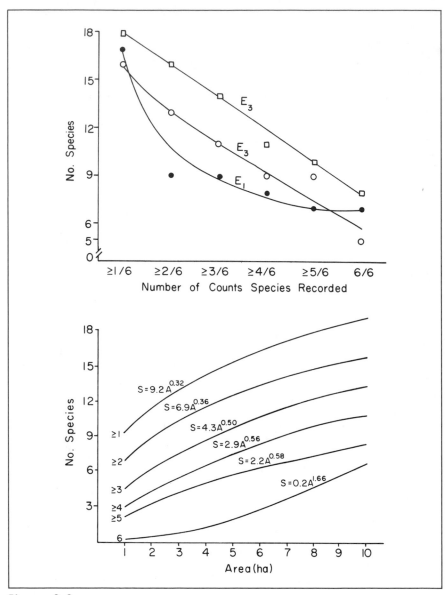

Figure 8.9
The number of species that is enumerated to occur in a designated habitat island is dependent on the criteria used for presence or absence. Part A illustrates how the inferred number of breeding bird species is dependent on the proportion of counts upon which the species is sighted. Part B illustrates how the shape of the species-area curve depends on the criterion applied.

fense on at least one of the six counts (the criterion usually cited in the published literature), then the number of species present in this island was either sixteen or eighteen, depending on the year. If the criterion for presence were changed to require that a species be present on at least two of the six counts (33%), then the number of species enumerated would drop to either sixteen or thirteen, depending on the year. If the criterion were very rigid and at least one territorial male were required to be singing on all six of the six counts before a species could be enumerated, then the number of species would be only five or eight, depending on the year. Clearly, as the rigor with which we define presence is increased, the number of species concluded to be present decreases.

Development of a species-area curve depends on data from sample plots or habitat islands of different size, and therefore the same exercise should be repeated for each year of all cypress habitat islands. When this is done it is clear that the form of the species-area curve changes in response to the rigor of the criteria for defining species presence (Figure 8.9b). This exercise should once again confirm that the interpretation of species presence-or-absence data and species-area relations should not be done casually.

Intrusive Fragmentation and the Colonization by Weedy Species

Weeds are commonly thought of as species growing where they are not wanted. And although animals are not commonly thought of as weeds, if conservation of biodiversity is the goal, then animal species that proliferate in human-dominated environments might also be considered weeds. Intrusive fragmentation is that which creates significant openings in the middle of the forest with the effects then radiating outward rather than progressing from outside the area inward. Clearings such as logging roads, quarries, homesteads, clear-cuts, and even food plots for game management have the net effect of fragmenting otherwise expansive tracts of closed-canopy forest and facilitating colonization by common, weedy species.

With approximately 7,000 "inholdings," the Ocala NF is perhaps the most urbanized forest in the National Forest System. About 10 percent of the area within the forest boundary is privately owned, with developments ranging from small homesteads with gardens, crops, and stock to recreational facilities and high-density residential developments. The U.S. Navy maintains a designated bombing range within the forest. The interspersion of human residents and their physical structures throughout the forest challenges both the skill and patience of forest managers.

About 33 percent of the plant species occurring in central Florida are exotic (R. Wunderlin, University of Southern Florida, pers. com.) and perhaps 200 of these species occur within the Ocala NF. This is no small addition to the plant species diversity of the forest. But as expressed previously, diversity in and of itself is not necessarily a good or laudable management objective. This is especially true if the biota consists of alien or exotic species, or even native species that occur commonly in areas or on sites where they did not occur in presettlement times. For example, a population of free-ranging Asian rhesus monkeys (*Macaca mulatta*) remains in the forest as a holdover from the filming of Tarzan movies earlier in this century (Wolfe and Peters 1987).

With homesteads, humans, and property developments interspersed throughout the forest, the use of critically important management practices such as prescribed burning becomes difficult if not impossible. Longleaf pine is the dominant species of one community that requires fire for regeneration and maintenance. Several oak species (*Quercus* spp.) are quick to invade when fire is excluded, and abundance of gray squirrels and flying squirrels (*Glaucomys volans*) increases in response to the increased mast availability. Thus, biological diversity may seemingly increase on the formerly pine-dominated sites. But this apparent increase is not worthy of designation as a goal or achievement because these species are already common elsewhere, and their negative interactions with rare pine-site species is, on balance, detrimental in that many species that occur only in the fire-maintained longleaf pine community are lost (Noss 1989).

For example, the red-cockaded woodpecker is a federally listed endangered species that evolved in occurrence with fire and depends on a plant community structure largely free of hardwood invasion. The red-cockaded woodpecker is both directly and indirectly impacted by the invasion by oaks and attendant species. Not only does the presence of oaks reduce habitat suitability for the red-cockaded woodpecker, but their presence also allows fires to burn physically higher into the midstory canopy, destroying the nesting cavities of the birds. Both species of squirrels, but especially the flying squirrel, use red-cockaded woodpecker cavities and pose serious threats by competing with, and preying upon, the endangered woodpecker (Dennis 1971).

Although data are few, the list of negative interactions caused by increasing numbers of alien and/or anthrophilic species is great. Brown-headed cowbirds have now colonized the Ocala NF and are seemingly increasing in abundance. Red-bellied woodpeckers are now the most common woodpecker in the region, and in conjunction with European starlings (*Sturnus vulgaris*) no doubt exert severe cavity competition on more characteristic and rare pineland species. With elimination of top carnivores and great reductions in fur trapping and hunting for middle-sized furbearers, the generalist mammals such as opposum, raccoon, and gray fox (*Urocyon cinereoargenteus*) are abundant. All of these prey on ground-nesting or small ground-dwelling animal species, including several that are threatened or endangered. Alien species such as armadillo (*Dasypus novimcinctus*), red fox (*Vulpes fulva*), and coyote (*Canis latrans*) have invaded the region, and feral populations of domestic dogs and cats add to the natural levels of competition, predation, and disease (e.g., distemper and rabies).

Predator-Prey Reversal

Being habitat generalists and omnivores, raccoons do not seem to be affected negatively by habitat fragmentation. Not only do they profit from most forestry and wildlife management practices that promulgate fragmentation, their populations are actively encouraged by many humans that live within the urbanized and fragmented forest. Without the panther and red wolf as native large mammal predators, and with reduced numbers of the formerly widely distributed large alligators to serve as predators, raccoon numbers have exploded. For example, Sanderson (1988, 484) states that "A continent wide population explosion began with the 1943 breeding season . . . it is conservatively estimated that today there are 15–20 times as many raccoons in North America as there were during the 1930's."

Raccoon predation on the nests of game species such as bobwhite (*Colonus virginiana*), turkey, and wood duck; threatened and endangered species such as the gopher tortoise, indigo snake (*Drymarchon corais couperi*), gopher frog (*Rana areolata aesopus*), and one-toed amphiuma (*Amphiuma pholeter*); and over 100 additional species that nest on or near the ground surface surely must stand as a primary threat to the native biological diversity of north-central Florida. Indeed, inasmuch as the alligator itself must nest on dry mounds, its nests are seriously depredated by raccoon predation. Thus the raccoon, formerly a prey item held to low numbers by habitat configuration and native large predators has escaped its former controls, exists in greatly inflated numbers and now exerts a heavy predation pressure (via egg predation) on scores of native species including its former predator, the alligator.

LANDSCAPE MANAGEMENT IMPLICATIONS OF FRAGMENTATION

The Ocala NF was chosen as an example because it so graphically illustrates the multilevel nature associated with the habitat fragmentation issue. At the highest level of analysis, that of the region, survival of the large, wide-ranging, generalist species such as panther and black bear is totally dependent on the existence of large federal land holdings such as the Ocala NF. Because they are generalists, their populations are probably not adversely affected by forest fragmentation at the level of individual square kilometers. But because they are physically large and their densities are low, it is difficult if not impossible to maintain a viable population within the bounds of the protected area.

Landscape linkages that run southward down the Wekiva River basin, northward up the Oklawaha River basin, and connecting into the cross-Florida barge canal lands are essential to the regional existence of these species in the future. Indeed, if the boundaries of the Ocala NF are expanded eastward to the St. Johns River and if the Oklawaha River is restored suitable for free migration by aquatic organisms, then a major conservation advance will have been made. The manatee is one of North America's largest and longest-distance migrant mammal species. Silver Springs on the southwest corner of the Ocala NF is one of Florida's largest artesian springs (820 ft^3/sec). Manatees used the Oklawaha and Silver Rivers historically, and if the Oklawaha is restored, this could be the case in the future. Thus, with only minor restorative measures, the Ocala NF will become the only national forest to serve as habitat to an endangered marine mammal.

The distribution of black bears in the eastern U.S. is almost totally limited to tracts of large, publicly owned protected areas such as the national parks and national forests (Pelton 1986). Still, most of these are too small to maintain viable populations. Without landscape corridors to interconnect these large tracts with nearby protected areas and provide a full spectrum of habitat types from river bottoms to arid uplands, there is little hope for maintaining the black bear over long periods of time. Because of the prospect of developing river-basin corridors that connect the Ocala NF with surrounding protected areas, this area is a strategic key to maintaining eastern North America's southernmost population of bears. Statutory conversion of the cross-Florida barge canal lands into one of America's first greenway parks will serve just such a purpose.

At a level below that of the region, several major highways serve to fragment the Ocala NF into five distinct sectors unless the retrofitting of highway underpasses is achieved. SR40, SR314, and SR316 sever the area into four north-south parcels, and SR19 divides the area into east and west sectors. As traffic loads increase, there will be increased pressure for bigger and faster roads and the road-induced mortality on terrestrial species will continue to increase. Roads are perhaps America's number-one fragmenting force.

A yet lower level of scale involves fragmentation induced by the major city-size inholdings that occur within the forest. It is erroneous to think of the Ocala NF as a single large protected area when so much human development occurs within its bounds. Without aggressive programs to purchase inholdings and guide developmental choices, the urbanized forest will cease to serve as a major native diversity-protected area.

At yet a lower level of scale comes the effect of silvicultural practices such as clear-cutting and the inadvertent policy of maximum fragmentation that derives from the game-management practice of maximum interspersion. Species such as the red-cockaded woodpecker have no future in the Ocala NF unless major changes in landscape patterns are rapidly implemented.

To perpetuate the notion that maximum interspersion maximizes species diversity is shortsighted and of questionable validity in a state where every acre of native habitat is needed to protect the native fauna. Florida and the Southeast do not need more of what is already common, and they do not need fewer of what is already rare.

Managing the effects of fragmentation at the lowest levels of scale, the individual patch cut or thinning operation, may require the most intense levels of analysis and monitoring. With over 200,000 ha to monitor, the task of evaluating the prudence of each and every intrusive practice may appear to be an impossible task. High-speed computers equipped for advanced geographical information systems analysis will be required. Yet, as one looks beyond the challenge, the prospect of actually managing an entire national forest to abate the consequences of forest fragmentation seems exciting.

ACKNOWLEDGMENTS

Monetary support for this work derives from the Ocala National Forest challenge grant program, the U.S. Forest Service Southeast Forest Experiment Station, and from the U.S. National Park Service, Southeast regional office in Atlanta. Dr. Dominic Dottavio of the Park Service and Benee F. Swindel, Jerry Clutts, and Laura Cockerham of the U.S. Forest Service facilitated our work.

LITERATURE CITED

Alverson, W., D. Waller, and S. Solheim. 1988. Forests too deer: Edge effects in northern Wisconsin. *Cons. Biol.* 2:348–58.

Anderson, S., and C. Robbins. 1981. Habitat size and bird community management. *Trans. N. Am. Wild. Nat. Res. Conf.* 46:511–20.

Andren, H., and P. Angelstam. 1988. Elevated predation rates as an edge effect in habitat islands: Experimental evidence. *Ecology* 69:544–47.

Andren, H., P. Angelstam, E. Lindstrom, and P. Widen. 1985. Differences in predation pressure in relation to habitat fragmentation: An experiment. *Oikos* 45:273–77.

Baur, G. 1897. New observations on the origin of the Galapagos Islands with remarks on the geological age of the Pacific Ocean. *Am. Nat.* 31:661–80.

Bierregaard, R., Jr. 1989. Avian communities in the understory of Amazonian forest fragments. In *Biogeography and ecology of forest bird communities*, ed. A. Keast, 1–11. The Hague: SPB Academic.

Brittingham, M., and S. Temple. 1983. Have cowbirds caused forest songbirds to decline? *BioScience* 33:31–35.

Browne, J. 1983. *The secular ark, studies in the history of biogeography*. New Haven: Yale University Press.

Brussard, P., and M. Gilpin. 1989. Demographic and genetic problems of small populations. In *Conservation biology and the black-footed ferret*, ed. U. Seal, E. Thorne, M. Bogan, and S. Anderson, 37–48. New Haven: Yale University Press.

Bull, J. 1980. Sex determination in reptiles. *Q. Rev. Biol.* 55:3–21.

Conry, P. 1988. High nest predation by brown tree snakes on Guam. *Condor* 90:478–82.

Darlington, P. 1957. *Zoogeography: The geographical distribution of animals.* New York: John Wiley & Sons.

Dennis, J. 1971. Utilization of pine resin by the red-cockaded woodpecker and its effectiveness in protecting roosting and nest sites. In *The ecology and management of the red-cockaded woodpecker*, ed. R. Thompson, 78–85. Atlanta: U.S. Fish and Wildlife Service.

Diamond, J. 1972. Biogeographic kinetics: Estimation of relaxation times for avifaunas of southwest Pacific islands. *Proc. Nat. Acad. Sci. U.S.A.* 69:3199–202.

———. 1973. Distributional ecology of New Guinea birds. *Science* 179:759–69.

———. 1975. Assembly of species communities. In *Ecology and evolution of communities*, ed. M. Cody and J. Diamond, 342–44. Cambridge: Harvard University Press.

Dice, L. 1943. *The biotic provinces of North America.* Ann Arbor: University of Michigan Press.

Gates, J., and L. Gysel. 1978. Avian nest dispersion and fledging success in field-forest ecotones. *Ecology* 59:871–83.

Hamel, P., H. LeGrand, Jr., M. Lennartz, and S. Gauthreaux, Jr. 1982. Bird-habitat relationships on southeastern forest lands. General Tech. Report SE-22. Asheville, N.C.: U.S.D.A. Forest Service.

Harlow, R., and M. Lennartz. 1983. Interspecific competition for red-cockaded woodpecker cavities during the nesting season in South Carolina. In *Proceedings of red-cockaded woodpecker symposium II*, ed. D. Wood, 41–43. Tallahassee: Florida Game and Freshwater Fish Commission.

Harris, L. 1984. *The fragmented forest: Island biogeography theory and the preservation of biotic diversity.* Chicago: University of Chicago Press.

————. 1988a. The faunal significance of fragmentation of southeastern bottom-land forest. In *Proceedings of the symposium: Forested wetlands of the southern United States*, ed. D. Hook and R. Lea, 126–34. General Tech. Report SE-50. Asheville, N.C.: U.S.D.A. Forest Service.

————. 1988b. The nature of cumulative impacts on biotic diversity of wetland vertebrates. *Environ. Manage.* 12:675–93.

————. 1988c. Edge effects and conservation of biotic diversity. *Cons. Biol.* 2:2–4.

Harris, L., and J. McElveen. 1981. *Effect of forest edges on north Florida breeding birds*. IMPAC Report 6 (no. 4). Gainesville: University of Florida School of Forest Resources and Conservation.

Harris, L., and R. Wallace. 1984. Breeding bird species in Florida forest fragments. *Proc. Ann. Conf. Southeast. Assoc. Fish Wildl. Agencies* 38:87–96.

Harris, L., and P. Frederick. 1990. The role of the Endangered Species Act in the conservation of biological diversity: An assessment. In *Integrated environmental management*, ed. J. Cairns, Jr., and T. Crawford, 99–117. Chelsea, Mich.: Lewis.

Harris, L., and P. Gallagher. 1989. New initiatives for wildlife conservation, the need for movement corridors. In *In defense of wildlife: Preserving communities and corridors*, ed. G. Mackintosh, 11–34. Washington, D.C.: Defenders of Wildlife.

Humphrey, S., W. Kern, Jr., and M. Ludlow. 1988. The Anastasia Island cotton mouse may be extinct. *Fl. Sci.* 51:150–56.

Hunter, M. 1990. *Wildlife, forest, and forestry: Principles of managing forests for biological diversity*. Englewood Cliffs, N.J.: Prentice Hall.

Jackson, J. 1978. Competition for cavities and red-cockaded woodpecker management. In *Endangered birds: Management techniques for preserving threatened species*, ed. S.Temple, 103–12. Madison: University of Wisconsin Press.

Klein, B. 1989. Effects of forest fragmentation on dung and carrion beetle communities in central Amazonia. *Ecology* 70:1715–25.

Knopf, F. 1986. Changing landscapes and the cosmopolitism of the eastern Colorado avifauna. *Wildl. Soc. Bull.* 14:132–42.

Lalo, J. 1987. The problem of road kill. *Am. For.* 50:47–49.

Lewontin, R., S. Rose, and L. Kamin. 1984. *Not in our genes*. New York: Pantheon.

Lynch, J., and D. Whigham. 1984. Effects of forest fragmentation on breeding bird communities in Maryland. *Biol. Cons.* 28:287–324.

MacArthur, R., and E. Wilson. 1967. *The theory of island biogeography*. Princeton, N.J.: Princeton University Press.

Mader, H. 1984. Animal habitat isolation by roads and agricultural fields. *Biol. Cons.* 29:81–96.

Mayfield, H. 1977. Brood parasitism reducing interactions between Kirtland's warblers and brown-headed cowbirds. In *Endangered birds: Management techniques for preserving threatened species*, ed. S. Temple, 85–91. Madison: University of Wisconsin Press.

McElveen, J. 1977. The edge effect on a forest bird community in north Florida. *Proc. Southeast. Game Fish Comm. Conf.* 31:212–15.

Meine, C. 1988. *Aldo Leopold, his life and work*. Madison: University of Wisconsin Press.

Noss, R. 1983. A regional landscape approach to maintain diversity. *BioScience* 33:700–6.

Noss, R. 1989. The longleaf pine landscape of the Southeast: Almost gone and almost forgotten. *Endang. Spec. Update* 5:1–7.

Odum, E. 1985. Trends expected in stressed ecosystems. *BioScience* 35:419–22.

Office of Technology Assessment. 1987. *Technologies to maintain biological diversity*. O.T.A.-F-330. Washington, D.C.: U.S. Government Printing Office.

Oxley, D., M. Fenton, and G. Carmody. 1974. The effects of roads on populations of small mammals. *J. App. Ecol.* 11:51–59.

Pelton, M. 1986. Habitat needs of black bears in the East. In *Wilderness and natural areas in the eastern United States: A management challenge*, ed. D. Kulhavy and R. Conner, 49–53. Nacogdoches, Tex.: Stephen F. Austin State University. Center for Applied Studies.

Preston, F. 1962a. The canonical distribution of commonness and rarity: Part I. *Ecology* 43:185–215.

———. 1962b. The canonical distribution of commonness and rarity: Part II. *Ecology* 43:410–32.

Reese, K., and J. Ratti. 1988. Edge effect: A concept under scrutiny. *Trans. N. Am. Wildl. Nat. Res. Conf.* 53:127–36.

Robbins, C., D. Dawson, and B. Dowell. 1989. Habitat area requirements of breeding birds of the middle Atlantic states. *Wildl. Monogr.* 103.

Robinson, S. 1988. Reappraisal of the costs and benefits of habitat heterogeneity for nongame wildlife. *Trans. N. Am. Wildl. Nat. Res. Conf.* 53:145–55.

———. 1990. Effects of forest fragmentation on nesting songbirds. Report no. 296. Champaign: Illinois Natural History Survey.

Samson, F., and F. Knopf. 1982. In search of a diversity ethic for wildlife management. *Trans. N. Am. Wildl. Nat. Res. Conf.* 47:421–31.

Sanderson, G. 1988. Raccoon. In *Wild furbearer management and conservation in North America*, ed. M. Novak, J. Baker, M. Obbard, and B. Malloch, 487–99. Toronto: Ontario Ministry of Natural Resources.

Savidge, J. 1987. Extinction of an island forest avifauna by an introduced snake. *Ecology* 68:660–68.

Soulé, M., ed. 1987. *Viable populations for conservation*. New York: Cambridge University Press.

Temple, S., and J. Cary. 1988. Modelling dynamics of habitat—interior bird populations in fragmented landscapes. *Cons. Biol.* 2:340–47.

Terbourgh, J. 1974. Preservation of natural diversity: The problem of extinction prone species. *BioScience* 24:715–22.

Thompson, R., ed. 1971. The ecology and management of the red-cockaded woodpecker. Proceedings of a symposium. Atlanta: U.S. Fish and Wildlife Service.

Thomas, J., technical ed. 1979. Wildlife habitats in managed forests, the Blue Mountains of Oregon and Washington. Agriculture handbook 553. Washington, D.C.: U.S.D.A. Forest Service.

U.S. Department of Agriculture Forest Service. 1978. *Forest statistics of the U.S., 1977*. Washington, D.C.: U.S. Government Printing Office.

U.S. Fish and Wildlife Service. 1987. *Florida panther* (Felis concolor coryi) *recovery plan*. Atlanta: U.S. Fish and Wildlife Service.

Whitcomb, R., C. Robbins, J. Lynch, B. Whitcomb, M. Klmkiewicz, and D. Bystrak. 1981. Effects of forest fragmentation on avifauna of the eastern deciduous forest. In *Forest island dynamics in man-dominated landscapes*, ed. R. Burgess and D. Sharpe, 125–206. New York: Springer-Verlag.

Wiens, J. 1976. Population responses to patchy environments. *Ann. Rev. Ecol. Syst.* 7:81–120.

Wilcove, D. 1985. Nest predation in forest tracts and the decline of migratory songbirds. *Ecology* 66:1211–14.

Wilcove, D., C. McClellan, and A. Dobson. 1986. Habitat fragmentation in the temperate zone. In *Conservation biology, the science of scarcity and diversity*, ed. M. Soulé, 237–56. Sunderland, Mass: Sinauer.

Wilcox, B., and D. Murphy. 1985. Conservation strategy: The effects of fragmentation on extinction. *Am. Nat.* 125:879–87.

Wiley, J. 1988. Host selection by the shiny cowbird. *Condor* 90:289–303.

Wolfe, L., and E. Peters. 1987. History of the freeranging rhesus monkeys (*Macaca mulatta*) of Silver Springs. *Fl. Sci.* 50:234–45.

Wood, D. 1983. Foraging and colony habitat characteristics of the red-cockaded woodpecker in Oklahoma. In *Proceedings of red-cockaded woodpecker symposium II*, ed. D. Wood. Tallahassee: Florida Game and Freshwater Fish Commission.

Yahner, R. 1988. Changes in wildlife communities near edges. *Cons. Biol.* 2:333–39.

Yahner, R., and B. Cypher. 1987. Effects of nest location on predation of artificial arboreal nests. *J. Wildl. Manage.* 51:178–81.

Yahner, R., and D. Scott. 1988. Effects of forest fragmentation on depredation of artificial avian nests. *J. Wildl. Manage.* 52:158–61.

Yahner, R., and A. Wright. 1985. Depredation on artificial ground nests: Effects of edge and plot size. *J. Wildl. Manage.* 49:508–13.

CHAPTER 9

Issues of Scale in Conservation Biology

Essay by REED F. NOSS

INTRODUCTION

So often, the solution to a problem is overlooked because it requires a fundamentally different way of looking at things. Once the problem is seen in a new light, the resolution is obvious, and we wonder how we could have ever been so stupid. Is it possible to identify "new lights" at the outset, and then proceed to determine which might best illuminate the problem at hand?

To see a problem in a new light is to change the scale of observation. Many of the most persistent controversies in ecology can be traced to different parties viewing a situation at different spatial or temporal scales (Allen and Starr 1982; O'Neill et al. 1986; Wiens et al. 1986). The scale at which Nature is viewed determines the patterns and processes detected. For example, is a forest in equilibrium with regard to species composition and habitat diversity? Certainly not if we look at an individual stand of a hectare or so. At this scale, even single tree falls disrupt stability. If we enlarge our view to a small watershed (say, 50 ha), we begin to see an equilibrium—until a wildfire or tornado arrives. But even very large disturbances can be "incorporated" into an equilibrium, of sorts, if we expand our landscape boundaries accordingly (Shugart and West 1981). As W.S. Cooper (1913, 43) observed on Isle Royale, the virgin forest is "a mosaic or patchwork which is in a state of continual change. The forest as a whole remains the same, the changes in various parts balancing each other." Bormann and Likens (1979) called this phenomenon a "shifting-mosaic steady state." Over a longer time span, on the order of centuries and millennia (or possibly much shorter, with greenhouse effects), a forest of any size never remains the same; species distributions and disturbance regimes shift with changing climate. Because species migrate at different rates, forest communities are ephemeral and unlikely to attain equilibrium (Davis 1981).

For any problem, spatial and temporal scale must both be carefully defined. Spatial and temporal scales are positively correlated in that processes at larger spatial scales are usually slower. For example, physico-chemical changes in a leaf occur at a faster rate than in a whole tree, which in turn changes faster than a forest (O'Neill et al. 1986). There is no "best" scale at which to study ecology; the appropriate scale depends on the research question at hand. Scale problems have been particularly troublesome in conservation. Many applied ecologists and land managers have demonstrated a narrow spatiotemporal perspective or have even failed to recognize that scale is an issue (Noss and Harris 1986). Because biotic impoverishment occurs at many different scales and levels of organization (i.e., from loss of alleles to global change), conservation must be pluralistic. To save the Florida panther, for instance, we must consider the genetics of inbreeding, the impacts of disease and prey scarcity on individual health and population viability, the pattern of patches and corridors across landscapes, and regional habitat changes that may result from climate change. And all of these considerations must be integrated into a single recovery plan! In the following pages I provide a few illustrative examples of scale problems in conservation and suggest some ways they might be resolved.

PROBLEMS IN SPACE

Biodiversity can be conceived of as a nested hierarchy of elements at several levels of biological organization. Familiar levels of organization are genetic, population-species, community-ecosystem, and landscape. Generally speaking, as level of organization ascends from gene to landscape (and beyond, to biosphere), so does the spatial scale at which these elements occur. The regional landscape (generally in the range of 10^3 to 10^5 km^2) is a convenient scale at which to integrate planning and management for multiple levels of organization. It is the scale of a constellation of national forests, parks, and surrounding private lands, or of a large watershed or mountain range (Noss 1983, 1987). The regional landscape is big enough to comprise numerous, interacting ecosystems; to incorporate large natural disturbances; and to maintain viable populations of large, wide-ranging animals. Yet it is small enough to be biogeographically distinct, and to be mapped in detail and managed by people who know the land well. It often has cultural significance and coherence as a "bioregion."

Many land-management decisions ignore the regional context of biodiversity. Practices that maximize habitat interspersion and edge effect are designed to enhance diversity at a local scale (10^2 to 10^3 ha), and often cause weedy species to proliferate (Samson and Knopf 1982; Noss 1983). Edge-oriented management may be applied to benefit game species, such as white-tailed deer (*Odocoileus virginianus*), or to provide esthetic, recreational, or interpretive opportunities. Toward such ends, some managers maintain seral meadows dominated by exotic and ruderal species in nature reserves. Sensitive species that inhabit forest interiors or that are easily disturbed by humans suffer from this kind of management and may disappear from landscapes where no large, unmanipulated natural areas remain (Noss 1983). Local enhancement of diversity can thereby lead to regional impoverishment—hence the need to consider biodiversity at a regional scale and to pay attention to species composition (identity) as well as diversity. Even if the management concern is strictly with a local area, regional processes (such as geographic dispersal) as well as local processes (such as adaptation to particular habitat conditions) must be taken into account (Ricklefs 1987). A failure to consider regional processes that control local biodiversity may result in the disruption of these processes, as when habitat fragmentation eliminates opportunities for species migration in response to changing climate.

Managing biodiversity at a regional scale still requires a broader context: ultimately, the biosphere. The Nature Conservancy has provided a framework for considering the status of "elements of diversity" (species and community-types) at different scales. Elements are ranked from 1 (most imperiled) to 5 (most common or secure) at state, national, and global scales. This framework allows for conservation to be tailored to the scale of concern of different agencies or governments, with the implicit understanding that the global ranking is biologically most critical.

Unfortunately, many states concentrate their conservation efforts on elements ranked high at a state scale, regardless of their global status. Thus, species at the edges of their ranges are favored over species that are more typical of a particular state and that may be rarer globally. Preservation of peripheral populations of species is a worthy goal, especially when such populations are genetically distinct, but not if this results in abandoned concern for the characteristic regional biota. Generally, too little effort has been made to distinguish the various kinds of species rarity, which are determined by three factors: geographic range, habitat specificity, and local population size (Rabinowitz, Cairns, and Dillon 1986). A global strategy that seeks to represent all species throughout their ranges, but with species-specific effort proportional to global endangerment, may help overcome provinciality. Conservation plans that identify and protect centers of species richness, endemism, and vegetative diversity (Scott et al. 1991) complement approaches based on qualities of individual species.

Many problems of spatial (and temporal) scale in conservation relate to disturbance. The area affected by a disturbance (patch size) is a primary consideration in studies of disturbance ecology (Pickett and Thompson 1978; Pickett and White 1985) and determines the grain of a landscape. Small disturbances such as individual tree falls create a fine-grained pattern, whereas large disturbances create a coarse-grained pattern. Different types of organisms undoubtedly perceive landscape grain very differently. Most landscapes are characterized by an overlaying of fine and coarse-grained patterns, reflecting a diverse history of disturbances. The range in disturbance patch sizes is almost infinite, from hoofprints of deer across a forest floor; to major fires, floods, and basalt flows that cover thousands or even millions of hectares; to glaciations that bulldoze large portions of continents; and finally to events, such as asteroid impacts, that may perturb the entire planet.

The most catastrophic of disturbances are capable of wiping out virtually everything in their paths; such cataclysms have been responsible for prehistoric extinctions, probably including some mass extinctions. But in general, species that exist in an area have evolved ways of avoiding, tolerating, or exploiting a particular disturbance regime. Management practices that mimic the natural disturbance regime, in terms of spatial scale, frequency, intensity, seasonality, effects on habitat structure, and other attributes, are more likely to maintain native biodiversity than are management practices that create conditions unlike those occurring in the natural landscape. For example, many eastern deciduous forest communities are characterized by frequent, small (10^1 to 10^3 m²) disturbances that create canopy gaps in which most tree regeneration and growth occur (Runkle 1985). It is reasonable to assume that selective cutting in these forests would better mimic the natural disturbance regime than would clear-cutting.

In another forest type, the longleaf pine (*Pinus palustris*) ecosystem of the southeastern coastal plain, even-aged management (clear-cutting) has been claimed to mimic the "uneven-aged mosaic of even-aged stands" that early investigators described for virgin longleaf pine forests (Lennartz 1989). A detailed study of an existing virgin longleaf pine population (Platt, Evans, and Rathbun 1988) indeed found an uneven-aged mosaic of even-aged patches, but with patch diameter averaging only about 25 meters. The grain of this natural mosaic differs considerably from the grain produced by 16-ha clear-cuts. Furthermore, although longleaf pine requires bare mineral soil in which to germinate, the small soil disturbances produced by gopher tortoises (*Gopherus polyphemus*) and pocket gophers (*Geomys pinetus*) create conditions vastly different from the mechanical site preparation that removes most of the herbaceous cover over an entire clear-cut (Means 1987). Fires expose mineral soil in these ecosystems but, unlike mechanical site preparation, do not eliminate the rich, native ground cover dominated by wiregrass (*Aristida stricta*).

Conservation strategies, even for single species, will be most effective when they address ecological phenomena at multiple spatial scales and levels of organization. Furbish's lousewort (*Pedicularis furbishiae*), a federally endangered endemic, inhabits banks of the St. John River in northern Maine and adjacent New Brunswick, sites that are subject to frequent, severe disturbances from ice scour and bank slumping. Although disturbances periodically eliminate individual lousewort populations along 10 m to 100 m or so of riverbank, the larger metapopulation requires such disturbance; within a few years after a disturbance, lousewort plants are shaded out by invading woody species. Either too much or too little disturbance would threaten Furbish's lousewort (see chapters 2 and 10). Events at a still-larger spatial scale pose a new danger to the species; heavy logging in the watershed appears to be contributing to increased variation in water levels and higher rates of population extinction (Menges 1988). Global climate change could further endanger the lousewort and other inhabitants of this unique riverbank community, as changes in temperature or precipitation alter hydrological and disturbance regimes. Thus, effective conservation of Furbish's lousewort may have to consider at least four spatial scales: (1) local habitats (within-population), (2) a set of local habitats (metapopulation), (3) the St. John River watershed, and (4) the Earth's atmosphere.

PROBLEMS IN TIME

Problems in time correspond to our limited temporal scale of concern in conservation strategy; in many cases, these problems again involve disturbance. In a general way, most human impacts on biodiversity represent one or both of two things: (1) a change in the environmental regime, often related to disturbance; and, (2) an increased rate of change. Changes in disturbance regimes threaten biodiversity when they introduce stresses or events either qualitatively or quantitatively different from the disturbances or stresses to which organisms have adapted over evolutionary time. Pollution, slash-and-burn agriculture (over large areas), and introduction of feral herbivores to islands are examples of changes in disturbance regimes that can have devastating effects on biodiversity. Disturbance regimes change naturally over time with changing climate. Hot, dry periods, for example, will be characterized by more frequent or intense fires in many ecosystem types. The problem is that we have increased the rate of environmental change far beyond what most species have experienced during their evolutionary histories, and beyond the rate of potential genetic adaptation for many taxa. Global warming over the next few decades may occur at least ten times faster than the warming at the close of the Pleistocene (Peters 1988).

Our present system of mostly small, isolated reserves is not well suited to maintaining viable populations in the long term, due to problems of demographic and genetic stochasticity, and to natural catastrophes (Soulé 1987). This is true even if we assume a stable climate. If we now recognize the reality of climate change, if only at the natural rates inferred from paleoecological studies, fragmentation-related problems are magnified due to the need for organisms to disperse long distances through landscapes riddled with urban areas, roads, agricultural fields, clear-cuts, and other artificial barriers. Furthermore, our "coarse filter" (community-level) approach to conservation, in those rare instances when we expand beyond individual species, has focused on species associations that we now know are ephemeral rather than on physical habitat types (as defined, say, by substrate, soil, and topography) that are much more persistent over time (Noss 1987; Hunter, Jacobson, and Webb 1988). In general, our temporal scale of concern has been limited; we have implicitly assumed a stable environment and are unprepared for change.

What would be the shape of a temporally expanded conservation strategy? To begin, it might be based less on short-term economic expediency and growth and more on long-term ecological and economic sustainability (Ehrlich and Ehrlich 1981). Second, we might adopt a more dynamic, nonequilibrium view of natural communities, accepting the fact that species combinations are constantly changing over time; such a view might concentrate on maintaining appropriate physical conditions and ecological processes with less emphasis on particular associations of species. Third, we would recognize the necessity of habitat connectivity and continuity for species migration across regions and continents and seek to maintain and restore broad habitat corridors at these scales (Hunter et al. 1988; Noss and Harris 1990). Finally, we might supplement our interest in sustaining viable populations of existing species with the notion of future evolutionary diversification. The end of evolution for large mammals in North America (Frankel and Soulé 1981) is a grim prospect indeed.

AN ETHICAL SEQUENCE

Perhaps the narrow spatiotemporal breadth of current conservation approaches is a reflection of the limited range of biological complexity in which we recognize value; that is, our science has been as narrow as our ethics. Although scientists are not wont to discuss values and often insist that their work is value-neutral, conservation biologists must at least acknowledge that values conflicts frequently underlie the problems they study (Naess 1986). Otherwise, they may attribute false or incomplete causes (mechanisms) and suggest impractical, technocratic solutions to conservation problems that are rooted in human values and attitudes.

Paralleling the hierarchy of biological organization to which advocates of biodiversity so commonly refer is another hierarchy (Figure 9.1), often called an ethical sequence (Leopold 1949; Nash 1989). To what levels of this ethical sequence should we assign value? Charles Darwin

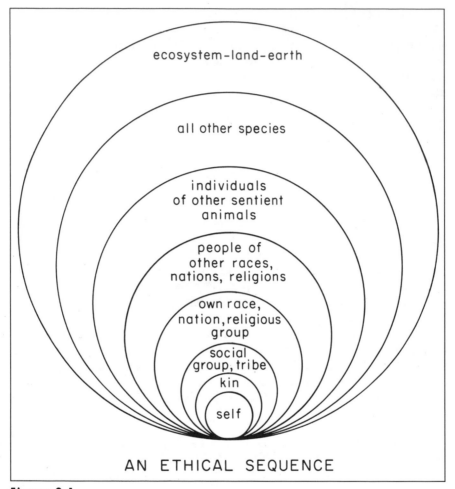

AN ETHICAL SEQUENCE

Figure 9.1
An Ethical Sequence (see Leopold 1949; Nash 1989). The sequence is portrayed as a nested hierarchy to emphasize that concern for higher levels is an extension, not a replacement, of traditional ethical concern for lower levels. The ecosystem-land-Earth is equivalent to a broader, ecological self. (A still broader concern, for the universe or cosmos, is philosophically attractive but of less relevance to conservation and could diffuse concern for Earth.)

(1871) pointed out that compassion for other living things is simply an extension of moral sentiments naturally selected to permit harmonious relationships among individuals in groups. Hence, people may expand beyond a concern for themselves, their family and social group, and even their own species to recognize value in other living things. This extension of ethics should not appear unreasonable; there is no objective reason for believing that humans are superior to any other species. Since Darwin, we know that we are kin to all life. If we recognize intrinsic value in humans (as do all major ethical traditions), then it is logical to recognize such value in other species. Naess (1989) calls ethical extension "Self-realization" because it seems to involve an expanded sense of self. Although transpecific altruism may make little sense in terms of individual fitness, it may be absolutely essential to the survival of the larger, ecological self.

Many people recognize value in *individuals* of "sentient" animals, that is, those most similar to humans in consciousness, particularly when they are also "cute and cuddly." Recognizing value in species and other groups, in snakes, plants, and invertebrates, and in higher levels of organization such as ecosystems, is much less common (Callicott 1986; Kellert 1986). There is an alarming correspondence between the public's bias toward charismatic megavertebrates and the implementation of the Endangered Species Act, where most listing and recovery effort has been devoted to such "higher" forms. When, if ever, will the public be ready for a Native Ecosystems Act (Noss 1991)?

The animal rights (or animal liberation) movement is reportedly the fastest-growing component of the "environmental" movement today but has many problems from a conservationist perspective (Callicott 1980). Animal liberationists often argue that the holistic viewpoint of ecologists such as Aldo Leopold is cruel and even fascist for putting the good of the community above the good of the individual. Suffering of individual higher vertebrates, not the health of the land or the status of biodiversity, is the primary concern of animal liberationists, who have explained to me that predation and natural disturbances such as fire are "evil" and will be eliminated as we evolve toward a more perfect world. Animal liberationists frequently defend the rights of introduced animals such as horses, burros, goats, and rabbits, usually at the expense of native species of plants and "lower" animals.

Clearly, an ethical underpinning for conservation biology must recognize fundamental value in higher-order elements of biodiversity if it is to support the attention to ecosystems, landscapes, overall biodiversity, and ecological processes that is imperative for effective conservation. Following Darwin, Leopold (1949) asserted that an extension of ethics to "the land" is "an evolutionary possibility and an ecological necessity." Leopold's famous line from *The Land Ethic*, "A thing is right when it tends to preserve the integrity, stability, and beauty of the biotic community. It is wrong when it tends otherwise," has become a rallying cry of holistic environmentalists. It is important to understand that concern for the ecosystem-land-Earth neither replaces nor overrides but rather supplements traditional ethics (Callicott 1987). A concern for all levels of organization in the ethical hierarchy, but guided by an appreciation of context provided by higher levels, might be an appropriate basis for a multiscale conservation strategy.

CONCLUSION

Our knowledge of which components of biodiversity are most critical to protect, for ecological and utilitarian reasons, is limited. Should we focus on species? If so, which species—only the rarest? Or maybe keystone species? Or should we focus on genes, communities, habitat structure, ecosystems, landscapes, the whole Earth? We know that ecological processes operate at multiple scales and levels of organization, yet research is only beginning to elucidate many of these phenomena. Which are most important? Important for what or whom? Our values remain anthropocentric.

In ecological science, no single scale of observation yields full understanding; exploration of a broad spectrum of spatial and temporal scales is recommended (Wiens et al. 1986). In conservation practice, the need to manage at multiple scales and levels of organization is becoming obvious, but our tendency has been to think small. In ethics, the defensibility of thinking small has vanished with increasing ecological understanding. Experience suggests that a narrow focus on local diversity, short time frames, and lower levels of biological organization (such as the individual self or individual animals) has unfortunate consequences for global biodiversity. Thus, a major lesson is to consider each phenomenon in a broader, ultimately global, context; that is, to "think big."

ACKNOWLEDGMENTS

I thank Steve Cline and Ann Hairston for comments on earlier drafts of this manuscript.

LITERATURE CITED

Allen, T.F.H., and T.B. Starr. 1982. *Hierarchy: Perspectives for ecological complexity.* Chicago: University of Chicago Press.

Bormann, F.H., and G.E. Likens. 1979. *Pattern and process in a forested ecosystem.* New York: Springer-Verlag.

Callicott, J.B. 1980. Animal liberation: A triangular affair. *Environ. Ethics* 2:311–38.

———. 1986. On the intrinsic value of nonhuman species. In *The preservation of species: The value of biological diversity*, ed. B.G. Norton, 138–72. Princeton, N.J.: Princeton University Press.

———. 1987. The conceptual foundations of the land ethic. In *Companion to a Sand County Almanac: Interpretive and critical essays*, ed. J.B. Callicott, 186–217. Madison: University of Wisconsin Press.

Cooper, W.S. 1913. The climax forest of Isle Royale, Lake Superior, and its development. *Bot. Gaz.* 55:1–44, 115–140, 189–235.

Darwin, C. 1871. *The descent of man and selection in relation to sex.* London: Murray.

Davis, M.B. 1981. Quaternary history and the stability of forest communities. In *Forest succession: Concepts and applications*, ed. D.C. West, H.H. Shugart, and D.B. Botkin, 132–53. New York: Springer-Verlag.

Ehrlich, P., and A. Ehrlich. 1981. *Extinction: The causes and consequences of the disappearance of species.* New York: Random House.

Frankel, O.H., and M.E. Soulé. 1981. *Conservation and evolution.* Cambridge: Cambridge University Press.

Hunter, M.L., G.L. Jacobson, and T. Webb. 1988. Paleoecology and the coarse-filter approach to maintaining biological diversity. *Cons. Biol.* 2:375–85.

Kellert, S.R. 1986. Social and perceptual factors in the preservation of animal species. In *The preservation of species: The value of biological diversity*, ed. B.G. Norton, 50–73. Princeton, N.J.: Princeton University Press.

Lennartz, M. 1989. Reply to Edward C. Fritz. *Nat. Areas J.* 9:4.

Leopold, A. 1949. *A Sand County almanac.* New York: Oxford University Press.

Means, D.B. 1987. *Impacts on diversity of the 1985 Land and Resource Management Plan for National Forests in Florida.* Report to The Wilderness Society. Washington, D.C.: The Wilderness Society.

Menges, E.S. 1988. Conservation biology of Furbish's lousewort. Final report to Region 5, U.S. Fish and Wildlife Service. Indianapolis: Holcomb Research Institute, Butler University.

Naess, A. 1986. Intrinsic value: Will the defenders of nature please rise? In *Conservation biology: The science of scarcity and diversity*, ed. M.E. Soulé, 504–15. Sunderland, Mass.: Sinauer.

———. 1989. *Ecology, community, and lifestyle: Outline of an ecosophy.* Translated and revised by D. Rothenberg. Cambridge: Cambridge University Press.

Nash, R.F. 1989. *The rights of nature: A history of environmental ethics.* Madison: University of Wisconsin Press.

Noss, R.F. 1983. A regional landscape approach to maintain diversity. *BioScience* 33:700–6.

———. 1987. From plant communities to landscapes in conservation inventories: A look at The Nature Conservancy (USA). *Biol. Cons.* 41:11–37.

———. 1991. From endangered species to biodiversity. In *Balancing on the brink: A retrospective on the Endangered Species Act*, ed. K.A. Kohm, 227–46. Washington, D.C.: Island Press.

Noss, R.F., and L.D. Harris. 1986. Nodes, networks, and MUMs: Preserving diversity at all scales. *Environ. Manage.* 10:299–309.

————. 1990. Habitat connectivity and the conservation of biological diversity: Florida as a case study. In *Proceedings of the 1989 Society of American Foresters Conference*, 24–27 September 1989, Spokane, Wash., 131–35. Bethesda, Md.: Society of American Foresters.

O'Neill, R.V., D.L. DeAngelis, J.B. Waide, and T.F.H. Allen. 1986. *A hierarchical concept of ecosystems*. Princeton, N.J.: Princeton University Press.

Peters, R.L. 1988. Effects of global warming on species and habitats: An overview. *Endang. Spec. Update* 5:1–8.

Pickett, S.T.A., and J.N. Thompson. 1978. Patch dynamics and the size of nature reserves. *Biol. Cons.* 13:27–37.

Pickett, S.T.A., and P.S. White, eds. 1985. *The ecology of natural disturbance and patch dynamics*. Orlando, Fla.: Academic Press.

Platt, W.J., G.W. Evans, and S.L. Rathbun. 1988. The population dynamics of a long-lived conifer. *Am. Nat.* 131:491–525.

Rabinowitz, D., S. Cairns, and T. Dillon. 1986. Seven forms of rarity and their frequency in the flora of the British Isles. In *Conservation biology: The science of scarcity and diversity*, ed. M.E. Soulé, 182–204. Sunderland, Mass.: Sinauer.

Ricklefs, R.E. 1987. Community diversity: Relative roles of local and regional processes. *Science* 235:167–71.

Runkle, J.R. 1985. Disturbance regimes in temperate forests. In *The ecology of natural disturbance and patch dynamics*, ed. S.T.A. Pickett and P. S. White, 17–33. Orlando, Fla.: Academic Press.

Samson, F.B., and F.L. Knopf. 1982. In search of a diversity ethic for wildlife management. *Trans. N. Am. Wildl. Nat. Res. Conf.* 47:421–31.

Scott, J.M., B. Csuti, K. Smith, J.E. Estes, and S. Caicco. 1991. Gap analysis of species richness and vegetation cover: An integrated conservation strategy for the preservation of biological diversity. In *Balancing on the brink: A retrospective on the Endangered Species Act*, ed. K.A. Kohm, 282–97. Washinton, D.C.: Island Press.

Shugart, H.H., and D.C. West. 1981. Long-term dynamics of forest ecosystems. *Am. Sci.* 69:647–52.

Soulé, M.E., ed. 1987. *Viable populations for conservation*. Cambridge: Cambridge University Press.

Wiens, J.A., J.F. Addicott, T.J. Case, and J. Diamond. 1986. Overview: The importance of spatial and temporal scale in ecological investigations. In *Community ecology*, ed. J. Diamond and T.J. Case, 145–53. New York: Harper & Row.

PART

III

THE BIOLOGY AND GENETICS OF POPULATIONS

• • •

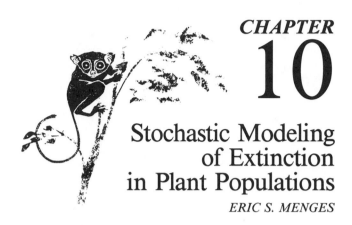

Stochastic Modeling
of Extinction
in Plant Populations

ERIC S. MENGES

ABSTRACT

Population viability analyses (predicting the future of small populations) have developed concepts relating largely to genetic threats, although environmental and demographic factors may be of greater immediate concern. Using twenty-eight published empirically derived projection matrices of various perennial herbs and trees, I model the behavior of plant populations by introducing temporal variation (stochasticity) in demographic parameters (mortality, growth, reproductive status, and reproductive output) into matrix projections of stage-structured populations. Stochastic modeling of population behavior allows estimation of extinction probabilities and minimum viable population.

Demographic stochasticity (DS) generates little variation in population dynamics. In contrast, moderate environmental stochasticity (ES) causes extinction risk for many populations with positive population growth under deterministic conditions. Increased ES causes increased extinction probabilities, decreased population sizes, decreased average time until extinction, and increased percentage of years with negative growth. Sensitivity to ES is greatest for populations with low finite rates of increase. Variation in mortality and growth among stages is far more important than variation in reproductive output. Simulation results generally agree with mathematical theory.

Using these analyses, I define a demographic version of minimum viable population (MVP) based on life history, degree of variation due to ES or DS, and acceptable levels of extinction probability over defined time intervals. Moderate ES may set higher MVPs than those necessary to counter short-term genetic effects of small population size. This technique could be extremely useful in conservation biology (e.g., managing endangered populations and preserve design) if data were available on temporal variation in demographic parameters.

INTRODUCTION

Effective management of populations and their habitats is required to control the risk of extinction (e.g., Schonewald-Cox et al. 1983; Soulé 1986, 1987). Individuals and organizations concerned with conserving biological diversity face challenging questions when financial resources or political realities limit the number and size of preserves. How does one decide whether a rare plant population is thriving or in danger of extinction? Is a given reserve large enough to support a viable population? Would additions safeguard a protected population? Is active intervention necessary to rescue a declining population (e.g., Goodman 1980)? Should a small, unprotected population be ignored because it is doomed to extinction? Understanding the relationships between population size and extinction probability is crucial to conservation biology (Shaffer 1981). Conservation biology needs to answer these questions and therefore requires techniques to predict or anticipate future population changes to determine the interventions necessary to assure species persistence, and to predict extinction probability from measurable population parameters. Although conservation biology has a vocabulary to begin this understanding, including a name for the subfield of "population viability analysis" (Soulé 1986), there are insufficient empirical data and accessible tools for data manipulation to routinely fashion predictions of population viability. The goal of this chapter is to suggest a tool that may be useful in guiding data collection and providing predictions. That tool is stochastic population modeling based on observed demographic parameters and their temporal variation.

The Conceptual Basis: Extinction Probability and Minimum Viable Population

The many causes of extinction can all be classified as systematic or stochastic. Among the latter are environmental stochasticity, demographic stochasticity, natural catastrophes, and genetic stochasticity (Shaffer 1981). All stochastic factors increase in importance at low population sizes, with environmental stochasticity and natural catastrophes being more important than demographic stochasticity for most population sizes (Shaffer 1987). A minimum viable population (MVP) is sufficiently large to minimize the probability of extinction from some or all forms of stochasticity over a long time period.

The conceptual basis of MVP has been developed largely within a population genetics framework (Frankel and Soulé 1981), although in specific cases estimates have been made from nongenetic criteria (see following). Low population sizes may lead to loss of genetic variation that may cause inbreeding depression (Franklin 1980; Frankel and Soulé 1981) and loss of evolutionary flexibility (Hamrick, Linhart, and Milton 1979; Levin 1984). In plants, many rare species or small and isolated populations develop low levels of within-population genetic variance (e.g., Ledig and Conkle 1983; Prentice 1984; Karron 1987a, 1987b; Waller, O'Malley, and Gawler 1988). Exceptions are not uncommon, however (e.g., Meagher, Antonovics, and Primack 1978; Loveless and Hamrick 1984).

Estimates of genetic MVP so far are general and based on approximations derived from *Drosophila* and cattle rather than for endangered species of interest. For example, an effective population size of 500 is an oft-cited MVP value necessary for an equilibrium between additive genetic variance gains through mutation and losses due to sampling error. This value is based on data of bristle characteristics in homogeneous lines of *Drosophila* (Franklin 1980) and obviously may be inappropriate for other organisms or situations.

Genetically based MVP (especially those based on assumptions from other organisms) may be irrelevant for many plant populations. Small neighborhood sizes (Levin 1981; Rai and Jain 1982; Meagher 1986); local genetic differentiation (e.g., Linhart 1974; Hamrick 1982; Silander 1985); and greater occurrence of and tolerance to inbreeding, apomixis, and vegetative reproduction (e.g., Antonovics 1968; Jain 1976; Park, Fowler, and Coles 1984) imply that deleterious recessives have been exposed to selection and lost. A direct linkage between low electrophoretic variation and inbreeding depression or loss of evolutionary flexibility has not been well established for plants and awaits coordinated ecological and genetic studies. Stronger evidence of such links comes from such animals as the cheetah (O'Brien et al. 1985).

Finally and most importantly, an MVP defined by genetic constraints may often be secondary to the demographic difficulties of small population size in varying environments. Extinctions of small populations may occur before deleterious genetic changes become evident, especially in plants.

Predicting Population Dynamics

Assessing extinction probabilities and minimum viable population sizes on nongenetic grounds requires an understanding of population dynamics and the factors that drive changes in population size. Predictions of future plant population behavior often use matrix-projection techniques to project a population with a given age structure forward through time, with transfers among ages defined by an invariant age-projection matrix (Lewis 1942; Leslie 1945; Lefkovitch 1965; Keyfitz 1968). Given the assumption that conditions do not change, matrix projection can calculate advantageous mathematical properties of the population, including equilibrium finite rate of increase (lambda, λ), equilibrium age structure, and reproductive values (Leslie 1945; Caswell 1982a, 1982b). Recent theoretical advances allow populations structured by stage and/or size to be analyzed similarly (Lefkovitch 1965; Goodman 1967; Hubbell and Werner 1979; Caswell 1982a, 1982b; Law 1983, others). In predicting an individual plant's demographic properties, stage/size generally is superior to age (Werner and Caswell 1977; Gross 1981; Lacey 1986) and has the biological advantage of explicitly considering important stages of a plant's life cycle (e.g., dormant seeds, clonal offspring) as well as processes associated with those stages (e.g., seed germination behavior, size-specific reproductive output).

The major drawback to deterministic forms of matrix projection is the assumption of unchanging conditions. This leads to unrealistic long-term predictions of exponential growth at stable stage structure, leading to either an infinite population size or certain extinction, depending on whether lambda (λ) is greater than or less than 1.

Of course, the demographic parameters used to formulate elements of projection matrices are rarely constant. Mortality, growth, reproductive status, and reproductive output vary over time in the same population (e.g., Klemow and Raynal 1981; Davy and Jefferies 1981; Bradshaw 1981; Waite 1984). Different populations located on similar sites also commonly differ in demographic parameters (Werner and Caswell 1977; Bierzychudek 1982; Pinero, Martinez-Ramas, and Sarukhan 1984). Therefore, both the predictions and the basis of deterministic matrix projection of plant population behavior are unrealistic. Through varying age structure, the effects of stochasticity render many equilibrium ecological analyses useless.

One advantage to stochastic simulation of matrix projection is the explicit recognition of extreme events, including extinction. The quantification of extinction probability is crucial to management of endangered and rare species. For a given life history, a given type and quantity of variation in demographic parameters, a working definition of extinction threshold, and a time frame of interest, the probability of extinction can be calculated from multiple stochastic simulations using the same input parameters. This is the basic approach of this paper. In addition, given a definition of acceptable extinction probabilities, a demographic minimum viable population can be calculated.

Models used to simulate population dynamics and to predict extinction probability (e.g., Lewontin and Cohen 1969; Richter-Dyn and Goel 1972; Leigh 1981; Strebel 1985; Wright and Hubbell 1983) have generally not been both stage-structured and stochastic (see discussion in Shaffer and Samson [1985]). Analytical solutions to particular types of stochastic population growth are available (e.g., Pollard 1966; Sykes 1969; Tuljapurkar and Orzack 1980; Cohen 1979a, 1979b; Braumann 1981; Goodman 1984) but have not used empirically derived data to predict population behavior. Compared to analytical solutions, computer simulation models are more flexible and can incorporate compensatory mechanisms and systematic pressures on a species or its habitat. Their formulation is also relatively simple compared to that of analytical models.

No explorations of MVP and extinction probability have been made using data from any plant species, although stochastic simulations have been used to investigate optimal life histories (Kachi and Hirose 1985). Bierzychudek (1982) used a particular type of stochastic stage-structured projection, randomly choosing between two projection matrices to simulate population growth of jack-in-the-pulpit. Menges (1989, 1990) used a similar approach with three matrices based on four years of data, also subdivided by site, to contrast population viability over various environments for Furbish's lousewort. Among animal species, stochastic age or stage-structured simulations have been applied to grizzly bears (Shaffer and Samson 1985; Knight and Eberhardt 1985), barnacles (Wethey 1985), and spruce budworm (Slade and Levenson 1984). The present paper explores the consequences of two types of variation, and of different levels of variation, on population behavior of twenty-eight populations of perennial herbs and trees. In particular, the effects of stochastic variation on extinction probability and minimum viable population are explored.

METHOD

Stochastic Modeling of Population Growth: General Protocol and Assumptions

I used a simulation model (POPPROJ) to project population growth of various species, generate variation in demographic parameters, and calculate various statistics. The program used standard matrix algebra to project the stage distribution at each time-step (year) by multiplying the previous stage distribution by a projection matrix.

Stochastic runs used projection matrices that varied with time. For this study, two types of variation were simulated. *Environmental sto-*

chasticity represents events that affect the entire population but impact different stages uniquely (Sykes 1969 model 3; Lewontin and Cohen 1969; May 1973; Cohen 1979a). To model environmental stochasticity, I allowed all matrix elements to vary independently each year. Once the year's matrix was set, it was applied to the entire population. The environment is defined operationally in terms of population responses and does not necessarily refer to external variation (see discussion for elaboration). *Demographic stochasticity* represents chance events affecting individuals in a very small population, even in a constant environment; for example, the fact that individuals either survive or die, and so do not themselves have survival rates (Pollard 1966; May 1973; Kieding 1975). To model demographic stochasticity, I used the same matrix each year but applied appropriate probabilities of transfer among stages to each individual in the population. Demographic stochasticity applies to transfers representing mortality, change in stage, and change in reproductive state but does not apply conveniently to reproductive output. In some runs, demographic stochasticity was combined with environmental stochasticity in reproductive output.

For environmental stochasticity, values for each nonzero matrix element were drawn from a normal distribution with mean as indicated by the average matrix element (in the average matrix) and variance input for that element. (Evidence on statistical distributions of demographic parameters over time is sparse; different distributions might have some effects on results.) Unreasonable values (e.g., reproductive output less than zero, survivorship less than zero or greater than one) were truncated to zero or one. This procedure assumed that an extremely bad year would have no greater effect on a population than a year bad enough to reduce survivorship or reproductive output to zero. An analogous statement applies to favorable years. Alternate protocol might also affect quantitative results. However, truncation actually *increased* the mean values of demographic components under extreme ES so that the modeled effects of ES on EP (extinction probability) are conservative. If the sum of nonreproductive elements in a column was greater than one, implying that overall survivorship was greater than one, the procedure was repeated for the whole column until reasonable values resulted. The program was flexible with regard to alternate reproductive states. Input parameters defined which elements in the matrix represented reproduction (e.g., seeds, clonal offshoots) and allowed individual matrix elements in those rows to vary upward beyond one.

For this study, environmental stochasticity was modeled in a simple and conservative manner. First, each matrix element was assumed to vary independently of all other elements. If the effects of environmental stochasticity were correlated among several transfers (e.g., a bad year increases mortality of several stages), different results would occur. If the correlations were positive, they would most likely cause more extreme results. A negative structure would be advantageous to populations (Caswell 1983). Second, zero autocorrelation among conditions was assumed; the effects of severe environments were not carried over from one year to another. Finally, no density-dependent regulation was assumed, because it often has little effect on field populations of plants (e.g., Shaw 1987). Inclusion of these features will probably increase probabilities of extinction, so the model runs are conservative with respect to the effects of stochasticity on extinction.

Projection Matrices

These explorations of stochastic effects on population growth were based on twenty-eight published projection matrices for plant species (Table 10.1). These species represent "biennials," perennial iteroparous herbs,

Table 10.1
Plant Projection Matrices Used in This Study

Reference	Life History[1]	Matrix Type	Species	No. of Matrices[2]	No. of Classes	Range of lambda[3]
Fiedler 1987	I,P	Stage	*Calochortus* spp.	8	4–5	0.96–1.39
Werner and Caswell 1977	S,B	Stage & age	*Dipsacus sylvestris*	16	7–8	0.0–2.60
Bierzychudek 1982	I,P	Stage	*Arisaema triphyllum*	4	7	0.85–1.32
Hartshorn 1975	I,T	Stage	*Pentaclethra macroloba*	1	15	1.002
Enright 1982	I,T	Stage	*Araucaria hunsteinii*	4	8–14	0.99–1.05
Enright and Ogden ·1979[4]	I,T	Stage	*A. cunninghamii*	3	4	0.39–1.02
Pinero et al. 1984	I,T	Stage	*Astrocaryum mexicanum*	4	7	0.99–1.03
Burns and Ogden 1985	I,T	Stage	*Avicennia marina*	1	7	1.22

[1]I = iteroparous; S = semeloparous; P = perennial herb; A = annual herb; B = biennial herb; T = tree.
[2]Representing different species, populations, and/or years.
[3]lambda = population growth rate at equilibrium, assuming deterministic population growth.
[4]Another species reported overlaps with Enright (1982).

and trees with a range of life histories. One particular feature of life histories was emphasized for comparisons: lambda (λ), the equilibrium finite rate of increase under deterministic conditions. In most cases, the projection matrix represents the results of a particular population in a given year. Because variation in demographic parameters has rarely been characterized over more than two years for the whole life cycle, the type and quantity of variation is explored as a sensitivity analysis.

The studies summarized in Table 10.1 represent the vast majority of appropriate published studies. A far greater number of studies either (1) do not completely characterize the life cycle, commonly omitting seed germination and dormancy or clonal production; or (2) report most or all components of a life cycle in many separate tables and figures. In the latter case, constructing a complete life table would require making key assumptions.

Simulation Procedure

Although the POPPROJ program is flexible in many ways, certain conventions were used to permit comparisons among species and types of variation. (Changes in this protocol are not thought to be important enough to qualitatively affect results.) Most simulations were begun with 200 individuals arranged in a stable stage distribution. When population size dipped to ten individuals or fewer, the population was termed extinct and the run terminated. All simulations were terminated after 100 years. Because demographic stochasticity was slow to run at large population sizes, runs were terminated when population size reached 5,000.

Each stochastic simulation produced different results. I generally performed 50 simulations using each set of input parameters. (Thus, statements of "no extinction probability" refer specifically to 0/50 extinctions in 100 years; the true EP may be nonzero but is certainly not large. Additional simulations (to 200) suggested that 50 simulations provided sufficient precision for extinction probabilities. Statistics for final population size and years to extinction apply only to nonextinct populations and extinct populations, respectively. These values are reported only if based on sample sizes of 5 (of 50) or more.

The degree of environmental stochasticity (ES) was varied by changing an arbitrary base level of variation. This base variation index is a variance/mean ratio of 1 percent for seed production and variance in mortality, growth, and clonal reproduction of 0.0001 (standard deviation 0.01). This base ES produced little change relative to the deterministic case. (In a few cases, lower variances were used for extremely small or large transitions where a variance of 0.0001 biased the mean of base runs.) To change the ES levels, variances were systematically increased from base levels by a multiplier (variation index). The ES levels were changed until little additional effect on population behavior was noted. In the results, I refer to specific ES levels as weak, moderate, strong, or extreme. However, these adjectives are only relative, because data are lacking on how demographic parameters typically vary.

RESULTS

Demographic versus Environmental Stochasticity

Moderate levels of environmental stochasticity caused extinction for many life histories. In contrast, demographic stochasticity generated very little variation and extremely small probabilities of extinction for most life histories (Table 10.2). However, populations with λ less than 0.99 that did not go extinct in 100 years under deterministic conditions were severely affected by DS.

As expected, introduction of stochasticity resulted in year-to-year variation in population sizes (Figure 10.1) and stage structure. Equilibrium rates of population growth and equilibrium stage structures were not reached in 100 years. However, DS resulted in minor fluctuations around deterministic, exponential growth. ES resulted in more marked fluctuations and a lowering of the average growth rate. Moderate levels of ES (less than 1,000, often less than 100) kept population size below 10,000 for populations with λ less than 1.05. Most populations with λ

Table 10.2
Overall Effects of Environmental (ES) versus Demographic (DS) Stochasticity on Extinction Probability (EP)

Type of Stochasticity and Relative Level	Level of ES[1]	Number of Populations			
		Little Affected	10%–49% EP	50%–90% EP	Greater than 90% EP
ES, weak	10	21	0	0	3
ES, moderately weak	25	17	3	1	3
ES, moderate	100	13	0	0	11
ES, strong	500	11	1	0	12
ES, very strong	1,000	9	2	1	12
ES, extreme	3,000	2	2	4	16
DS, lambda >1	—	17	3	1	0
DS, lambda <1	—	2	0	0	3

[1]Relative to base level: Variance in reproductive output is 1% of the mean; variance in other elements is 0.0001.

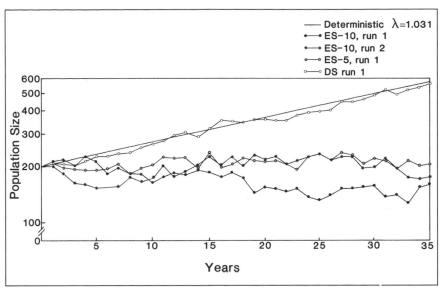

Figure 10.1
Environmental stochasticity causes variation in population size with time and also slows average population growth. Demographic stochasticity has minor effects relative to the deterministic case, and these effects are dampened as population size increases. This graph shows individual simulations with Astrocaryum mexicanum.

greater than 1.2 were not controllable by any level of ES, even those high enough to cause some extinctions. In between $\lambda = 1.05$ and $\lambda = 1.12$, fairly high levels of ES limited population size below 100,000.

Effects of Increasing Environmental Stochasticity

As environmental stochasticity (ES) increased, populations were increasingly likely to become extinct within 100 years. This relationship generally exhibited thresholds; a minimum level of variation was necessary to generate a nonzero extinction probability (EP) (Figure 10.2). Likewise, populations suffering greater than 80 percent EP were not particularly sensitive to further increases in ES. Between upper and lower thresholds, EP increased linearly, with more sensitive species showing faster increases in EP with increases in ES (Figure 10.2).

Sensitivity to environmental variation was a function of life history. Populations with highest deterministic finite rates of increase (lambda) were not subject to extinction unless environmentally induced variation in demographic parameters reached relatively high levels (Figure 10.2).

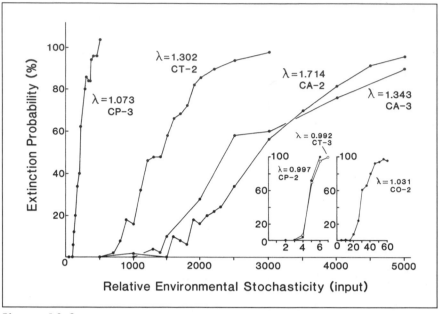

Figure 10.2
*Extinction probabilities increase with the level of environmental stochas-
ticity for four species of* Calochortus *studied by Fiedler (1987). CP = C.
pulchellus, CT = C. tiburonensis, CA = C. albus, CO = C. obispoensis;
2 = 1982–1983, 3 = 1983–1984. Species with the lowest deterministic
growth rates (lambda) are most sensitive to environmental stochasticity.
Note that the two insets at lower right have different X axes.*

In contrast, species with lambda less than or near 1 (stable population
size) were sensitive to variation two orders of magnitude less than populations with lambda greater than 1.2. The variation level necessary to
produce a given level of EP increased sharply with increasing deterministic lambda for most life histories. Populations with lambda slightly less
than 1 (declining populations) were sensitive to ES at levels less than 25.
Fairly stable populations were sensitive to ES from 50 to 215 ($1.0 > \lambda
> 1.1$ with one exception). Finally, rapidly expanding populations (λ
greater than 1.15) were relatively insensitive to ES; only ES at 1,400 or
greater was sufficient to induce 50 percent EP for these species. At a given
level of variation, EP decreased across populations with increasing
lambda. Even populations that were, on the average, declining were not
saved by greater environmental variation.

Other factors besides lambda affected population sensitivity to ES. I
selected eleven populations that were unusually sensitive or insensitive

for their lambda (as judged by scatter of ES vs. EP) and analyzed their life histories relative to those with similar lambda but different sensitivity to ES. Less sensitive populations had the same or a greater number of life history stages (4 of 6 populations). Because ES was applied independently to matrix elements, multiple stages buffered populations against catastrophic mortality or reproduction failure. To the extent that ES actually affects multiple stages and multiple transfer elements similarly (transition element intercorrelation), this buffering is a model artifact. Similarly, populations sensitive to ES for their lambda had fewer life history stages (3 of 5 populations), reproduction at earlier stages (5/5), and greater proportion of nonzero elements (3/5). These characteristics are found in short-lived species.

In summary, sensitivity to ES is greatest for populations with low λ (less than 1.15), and for short-lived species. However, two populations of *Arisaema triphyllum* with λ greater than 1.2 were completely insensitive to ES, showing no extinction until unreasonably high levels (greater than 10,000) of ES. This species reproduces both clonally and by seed, and the multiple reproduction modes buffered the species against reproductive failure under ES. (*Arisaema* populations with λ less than 1.0 were not particularly unusual, however.) Again, transition element intercorrelation, not modeled here, might reduce the difference between clonal and nonclonal species.

Increasing ES had other effects besides increasing EP (Table 10.3). For individual populations, increasing ES decreased the average time to extinction and decreased final population size. The negative effects of

Table 10.3

Effects of Increasing ES on Population Behavior

	Number of Populations			
	Increases	**Decreases**	**No Change or Variable**	**Not Applicable**
Years to extinction				
$\lambda > 1$	16	0	4	0
$\lambda < 1$	6	1	1	0
Final size				
$\lambda > 1$	0	10	3	8
$\lambda < 1$	2	4	1	1
Percent negative years				
$\lambda > 1$	20	0	0	0
$\lambda < 1$	1	3	5	0
Actual element mean relative to input				
$\lambda > 1$	16	2	2	0
$\lambda < 1$	3	2	2	1

increasing ES are at least partly due to increasing the percentage of negative years. All populations with λ greater than 1 suffered in this regard. The relationship of EP to percent negative years was consistent across life histories for populations with one mode of reproduction and λ greater than 1. Over 80 percent of these populations reached 50 percent EP when 55 percent to 64 percent of years had negative growth.

Environmental Variation in Life History Components

The threat posed to population survival by environmental variation appeared almost entirely due to variation in mortality, growth, and reproduction status and not to variation in reproductive output. In fact, the addition of reproductive variation to stochasticity in mortality and growth did not consistently increase or decrease extinction probability. Variation in reproductive output by itself was not sufficient to generate nonzero EP, except for extremely high levels of variation.

When Is Demographic Stochasticity Important?

Demographic stochasticity had significant effects on EP (for an initial population size of 200) for only a few populations. Extinction was more likely given DS for species with lambda near 1 (Figure 10.3). In these species, chance variation in individual fates can be crucial to the survival of small populations. However, not all such populations were affected by DS. Also, sensitivity to ES also was particularly great in populations sensitive to DS, which has its greatest effects on very small populations. The following section explores the effect of initial population size on EP, under DS and ES.

Initial Population Size and Minimum Viable Population

Stochasticity puts small populations at risk, but larger populations may be able to buffer ES or DS, at least for the short term. I followed previous MVP studies in defining minimum viable population (MVP) as the minimum initial population size necessary to reduce EP below 5 percent for 100 years. Initial size was then varied iteratively, with other simulation conventions unchanged (100 years, fifty simulations, extinction defined as ten or fewer individuals).

Patterns were similar for all species and projection matrices. For example, for *Astrocaryum mexicanum* at Site BB, DS implies an MVP

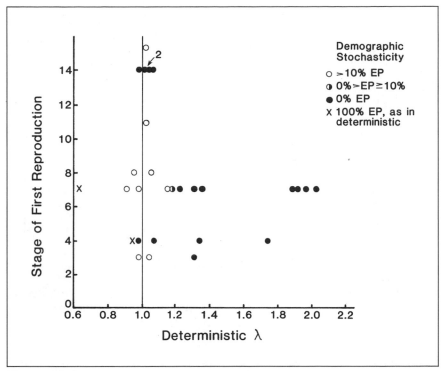

Figure 10.3
Demographic stochasticity increases extinction probability to modest levels only in populations with delayed reproduction and modest lambda.

equal to 50, while modest ES levels produce MVP greater than 100 (Figure 10.4). As expected, greater MVP is required to buffer larger environmental variation.

DISCUSSION

Stochasticity and Population Growth: Simulation Results and Mathematical Theory

A major finding of this research concerns the types of stochasticity likely to affect plant population viability. Environmental stochasticity (ES) clearly produces greater fluctuations and poses a greater threat of extinction than demographic stochasticity (DS). Mathematical theory agrees with these conclusions (e.g., Sykes 1969; Shaffer 1987; Goodman 1987) at least unless population size is very small, where DS can be important

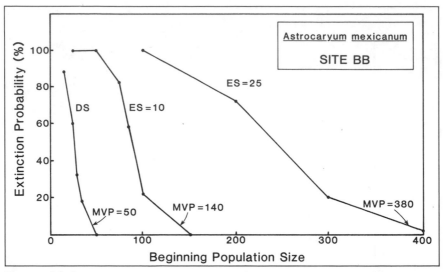

Figure 10.4
Minimum viable population (MVP) sizes necessary to limit extinction probabilities to less than 5 percent for Astrocaryum mexicanum *at site BB (Pinero et al. 1984), for DS and two levels of ES.*

(Kieding 1975; Shaffer 1987). Therefore, a minimum viable population (MVP) necessary to counter ES will be larger than that required to counter DS. For example, Shaffer and Samson (1985) considered both ES and DS in modeling EP and MVP of grizzly bears. Knight and Eberhardt (1985), considering only DS in a different model, found lower EP than did Shaffer and Samson (1985). Population fluctuations produced by ES tend to mimic fluctuations seen in long-term studies (e.g., Bradshaw and Doody 1978; Bradshaw 1981; Wells 1981), suggesting that ES could have a major effect on population dynamics for many plant species.

Increasing ES increases variation in population trends and growth rates and increases the extinction probability (EP) for almost all life histories. Again, these results generally are supported by analytical treatments (Lewontin and Cohen 1969; Roughgarden 1979, 390; Tuljapurker and Orzack 1980; Braumann 1981; Goodman 1984). Most models of population dynamics under ES (or ES relative to deterministic projection) create an overall tendency for slower population growth. However, Cohen (1979b) found that stochasticity could on the average have negative or positive effects if ES was autocorrelated through time. Unfortunately, knowledge of autocorrelation in ES is as difficult to ascertain from empirical data as are overall estimates of variability through time. Positive autocorrelation (meaning a particular environment affects several demographic parameters similarly) probably would increase the amplitude of population fluctuations and augment extinction risk.

Other effects of increasing ES found in this study include decreased time to extinction (see Goodman 1987), decreased population size, and a greater proportion of years with declining populations. Analytical results show an inverse relationship of sensitivity to ES and average lifetime of a population (Leigh 1981), agreeing with my analysis of outliers in a graph of EP versus ES.

The simulation results summarized here clearly show that populations with lowest growth rates are most sensitive to both DS and to low levels of environmentally induced variation in demographic parameters (ES). Few theoretical analyses have been concerned with comparison of life histories. Several treatments, however, have shown that as ES increases, the importance of mean population growth rate in controlling population dynamics and extinction is lessened (e.g., Ginzburg et al. 1982). This agrees with this study in that very high levels of ES produce extinction over a large range of life histories.

Populations with low lambda may have evolved in "stable" environments having few environmental fluctuations. If so, my finding of high sensitivity to ES in populations with low lambdas appears counterintuitive. The solution to this contradiction lies in the definition of ES used in this research: variation in demographic parameters. Therefore, an external environmental event (e.g., a drought) can be buffered to a greater or lesser extent by the population or its ecosystem. A population buffered against external environmental variation will show little variation in demographic parameters (Tuljapurkar and Orzak 1980), although its homeostatic reactions may require greater expenditures, such as increased reproductive effort to achieve the same reproductive output (Caswell 1983). A buffered population will experience little ES as defined here. The key issue in defining population dynamics and stability is how environmental variation is perceived by and dealt with by the organism (May 1973; Hastings and Caswell 1979; Caswell 1983). Thus, we adopt the operational view of Mason and Langenheim (1957, 339) that "in order to be classed as an environmental phenomenon a given phenomenon must be operationally significant to an organism." This view simplifies measuring ES by focusing its measurement on population responses.

The importance of ES in causing extinction is primarily due to variations in mortality and growth and not to reproductive output (see also Knight and Eberhardt 1985). The stochastic result has a deterministic analogue: Population growth rate (lambda, λ) is most sensitive to mortality rates, especially in early stages (Caswell and Werner 1978).

Population behavior is, to a great extent, controlled by "weakest links" in time or in life history. Crucial life history stages and dynamic rates control population growth under deterministic and stochastic scenarios. Also, critically unfavorable years may play a major role in causing extinctions and near-extinctions, because long-term population growth depends on geometric means of short-term growth rates (Lewontin and Cohen 1969). The detection, quantification, and mitigation of crucial processes within and between years will continue to influence work on endangered species. Modeling results suggest that increasing ES increases the proportion of such disastrous years. Populations with reduced variation in demographic parameters may have greater viability even if deterministic fitness measures such as lambda are relatively low (cf. Tuljapurkar and Orzack 1980). Reduction of variation in demographic parameters may be a future part of species management. At this time, however, we lack long-term data on environmental periodicity and its effects on population dynamics.

Importance and Implications of Demographic Approaches

Previously, I defended the emphasis on nongenetic sources of stochasticity. The MVP analyses described in this chapter suggest that ES may be a greater and more immediate threat to small plant populations than short-term genetic effects, because populations of 200 (greater than short-term genetic MVP of 50) are often under severe threats from moderate levels of ES. At the very least, demographic considerations should be included in population viability analysis (Wilcox 1986). Demographic analyses may, in fact, measure genetic effects on survivorship or fecundity. In other words, this emphasis does not deny the possibility of genetic effects and does not necessarily allow complete separation of genetic and nongenetic effects.

In plant species conservation, stochastic demographic modeling can be important in several practical ways. First of all, by focusing data collection on the dynamics of individuals (vs. measuring changes in population size) and on quantifying the entire life cycle, it develops a holistic view of the life cycles and maximizes the probability of discovering biologically relevant information. The importance of ES, especially in mortality, also underscores the need for long-term monitoring of the life history of rare plants, and for a quantification of variation in demographic parameters, possible within the context of many monitoring programs.

Demographic monitoring and modeling also can be used to guide management. In particular, experimental treatments can be contrasted for their effects on population viability, and adjusted to minimize EP, similarly to deterministic matrix projection analyses (e.g., Goodman

1980; Vaughan 1981; Silander 1983; Flipse and Veling 1984). Analyses that identify limiting stages or transitions in plant life histories (Caswell 1982a, 1982b, Goodman 1982) can also suggest mitigation strategies. Reduction of the effects of ES may be more important than augmenting population size in promoting population viability (Goodman 1987).

Finally, it should be possible to estimate MVP based on various forms of threats to populations. In this chapter, I have proposed a methodology for evaluating separately the effects of ES and DS, although they could be combined (also with genetic considerations), as done for the grizzly bear (Shaffer and Samson 1985; Wilcox 1986). Understanding MVP requirements will guide managers to optimize their uses of limited resources for preservation strategy and reserve size and design. Knowledge of MVP for particular species will also affect the design of preserve systems (e.g., White and Bratton 1980; Wilcox and Murphy 1985).

ACKNOWLEDGMENTS

I have benefited from discussions with and comments from a number of colleagues, including Tom Armentano, Marlin Bowles, Becky Dolan, Candace Galen, Peggy Fiedler, Kent Holsinger, Jeff Karron, Mary Palmer, Mark Shaffer, and Bruce Wilcox. Stacia Yoon assisted with computer runs, and Ginger Williams, Pat Bache, Jim Rogers, Christie Prinz, and Colleen Baker helped prepare the manuscript. Thanks also to the authors of the empirical studies on which this research is based.

LITERATURE CITED

Antonovics, J. 1968. Evolution in closely adjacent plant populations. 1. Evolution of self-fertility. *Heredity* 23:219–37.

Bierzychudek, P. 1982. The demography of jack-in-the-pulpit, a forest perennial that changes sex. *Ecol. Monogr.* 52:335–51.

Bradshaw, M.E. 1981. Monitoring grassland plants in Upper Teesdale, England. In *The biological aspects of rare plant conservation*, ed. H. Synge, 495–514. Chichester: John Wiley and Sons.

Bradshaw, M.E., and J.P. Doody. 1978. Plant population studies and their relevance to nature conservation. *Biol. Cons.* 14:223–42.

Braumann, C.A. 1981. Population adaptation to a "noisy" environment: Stochastic analogs of some deterministic models. In *Quantitative population dynamics*, ed. D.G. Chapman & V.F. Gallucci, 39–59. Farland, Md.: International Co-operative.

Burns, B.R., and J. Ogden. 1985. The demography of the temperate mangrove [*Avicennia marina* (Forsk.) Vierh.] at its southern limit in New Zealand. *Aust. J. Ecol.* 10:125–33.

Caswell, H. 1982a. Optimal life histories and the maximization of reproductive value: A general theorem for complex life cycles. *Ecology* 63:1218–22.

———. 1982b. Stable population structure and reproductive value for populations with complex life cycles. *Ecology* 63:1223–31.

———. 1983. Phenotypic plasticity in life history traits: Demographic effects and evolutionary consequences. *Am. Zool.* 23:35–46.

Caswell, H., and P. Werner. 1978. Transient behavior and life history analysis of teasel (*Dipsacus sylvestris* Huds.). *Ecology* 59:53–66.

Cohen, J.E. 1979a. Comparative statics and stochastic dynamics of age-structured populations. *Theor. Pop. Biol.* 16:159–71.

———. 1979b. Long-run growth rates of discrete multiplicative processes in Markovian environments. *J. Math. Anal. Appl.* 69:243–51.

Davy, A.J., and R.L. Jefferies. 1981. Approaches to the monitoring of rare plant populations. In *The biological aspects of rare plant conservation*, ed. H. Synge, 219–32. Chichester: John Wiley and Sons.

Enright, N.J. 1982. The ecology of *Araucaria* species in New Guinea. III. Population dynamics of sample stands. *Aust. J. Ecol.* 7:227–37.

Enright, N.J., and J. Ogden. 1979. Applications of transition matrix models in forest dynamics: *Araucaria* in Papau New Guinea and *Nothofagus* in New Zealand. *Aust. J. Ecol.* 4:3–23.

Fiedler, P.L. 1987. Life history and population dynamics of rare and common mariposa lilies (*Calochortus* Pursh: Liliaceae). *J. Ecol.* 75:977–95.

Flipse, E., and E.J.M. Veling. 1984. An application of the Leslie matrix model to the population dynamics of the hooded seal, *Cystophora cristata* Erxleben. *Ecol. Model.* 24:43–59.

Frankel, O.H., and M.E. Soulé. 1981. *Conservation and evolution.* Cambridge: Cambridge University Press.

Franklin, I.R. 1980. Evolutionary change in small populations. In *Conservation biology: An evolutionary-ecological perspective*, ed. M.E. Soulé and B.A. Wilcox, 134–50. Sunderland, Mass: Sinauer.

Ginzburg, L.R., L.B. Slobodkin, K. Johnson, and A.G. Bindman. 1982. Quasiextinction probabilities as a measure of impact on growth. *Risk Anal.* 2:171–81.

Goodman, D. 1967. On the reconciliation of mathematical theories of population growth. *J. Royal Statist. Soc., Series A* 130:541–53.

———. 1980. Demographic intervention for closely managed populations. In *Conservation biology: An evolutionary-ecological perspective*, ed. M.E. Soulé and B.A. Wilcox, 171–95. Sunderland, Mass: Sinauer.

———. 1982. Optimal life histories, optimal notation, and the value of reproductive value. *Am. Nat.* 119:802–23.

———. 1984. Risk spreading as an adaptive strategy in iteroparous life histories. *Theor. Popul. Biol.* 25:1–20.

———. 1984. The demography of chance extinction. In *Viable populations for conservation*, ed. M.E. Soulé, 11–34. Cambridge: Cambridge University Press.

Gross, K.L. 1981. Predictions of fate from rosette size in four "biennial" plant species: *Verbascum thapsus, Oenothera biennis, Daucus carota,* and *Tragopogon dubious. Oecologia* 48:209–13.

Hamrick, J.L. 1982. Plant population genetics and evolution. *Am. J. Bot.* 69:1685–93.

Hamrick, J.L., Y.B. Linhart, and J.B. Mitton. 1979. Relationships between life history characteristics and electrophoretically detectable genetic variation in plants. *Ann. Rev. Ecol. Syst.* 10:173–200.

Hartshorn, G.L. 1975. A matrix model of tree population dynamics. In *Tropical ecological systems: Trends in terrestrial and aquatic research*, ed. F.B. Golley and E. Medina, 45–51. New York: Springer-Verlag.

Hastings, A., and H. Caswell. 1979. Role of environmental variability in the evolution of life history strategies. *Proc. Nat. Acad. Sci.* 76:4700–03.

Hubbell, S.P., and P.A. Werner. 1979. On measuring the intrinsic rate of increase of populations with heterogeneous life histories. *Am. Nat.* 113:277–93.

Jain, S.K. 1976. The evolution of inbreeding in plants. *Ann. Rev. Ecol. Syst.* 7:469–95.

Kachi, N., and T. Hirose. 1985. Population dynamics of *Oenothera glazioviana* in a sand-dune system with special reference to the adaptive significance of size-dependent reproduction. *J. Ecol.* 73:887–901.

Karron, J.D. 1987a. The pollination ecology of co-occurring geographically restricted and widespread species of *Astragalus* (Fabaceae). *Biol. Cons.* 39:179–93.

———. 1987b. A comparison of levels of genetic polymorphism and self-compatibility in geographically restricted and widespread plant congeners. *Evol. Ecol.* 1:47–58.

Keiding, N. 1975. Extinction and exponential growth in random environments. *Theor. Popul. Biol.* 8:49–63.

Keyfitz, N. 1968. *Introduction to the mathematics of populations*. Reading, Mass: Addison-Wesley.

Klemow, K.M., and D.J. Raynal. 1981. Population ecology of *Melilotus alba* in a limestone quarry. *J. Ecol.* 69:33–44.

Knight, R.R., and L.L. Eberhardt. 1985. Population dynamics of Yellowstone grizzly bears. *Ecology* 66:323–34.

Lacey, E.P. 1986. Onset of reproduction in plants: Size- versus age-dependency. *Trends Ecol. Evol.* 1:72–75.

Law, R. 1983. A model for the dynamics of a plant population containing individuals classified by age and size. *Ecology* 64:224–30.

Ledig, F.T., and M.T. Conkle. 1983. Gene diversity and genetic structure in a narrow endemic, Torrey pine (*Pinus torreyana* Parry ex carr.). *Evolution* 7:79–86.

Lefkovitch, L.P. 1965. The study of population growth in organisms grouped by stages. *Biometrics* 21:1–18.

Leigh, E.G., Jr. 1981. The average lifetime of a population in a varying environment. *J. Theor. Biol.* 90:213–39.

Leslie, P.H. 1945. On the use of matrices in certain population mathematics. *Biometrika* 33:183–212.

Levin, D.A. 1981. Dispersal versus gene flow in plants. *Ann. Missouri Bot. Gard.* 68:233–53.

Levin, D.A. 1984. Genetic variation and divergence of a disjunct *Phlox. Evolution* 38:223–25.

Lewis, E.G. 1942. On the generation and growth of a population. *Sankhya* 6:93–96.

Lewontin, R.L., and D. Cohen. 1969. On population growth in a randomly varying environment. *Proc. Nat. Acad. Sci.* 62:1056–60.

Linhart, Y.B. 1974. Intra-population differentiation in annual plants. I. *Veronica peregrina* L. raised under non-competitive conditions. *Evolution* 28:232–43.

Loveless, M.P., and J.L. Hamrick. 1984. Ecological determinants of genetic structure in plant populations. *Ann. Rev. Ecol. Syst.* 15:65–95.

May, R.M. 1973. *Stability and complexity of model ecosystems.* Princeton, N.J.: Princeton University Press.

Mason, H.L., and J.H. Langenheim. 1957. Language analysis and the concept environment. *Ecology* 38:325–40.

Meagher, T.R. 1986. Analysis of paternity within a natural population of *Chamaelirium luteum.* I. Identification of most likely male parents. *Am. Nat.* 128:199–212.

Meagher, T.R., J. Antonovics, and R. Primack. 1978. Experimental ecological genetics in *Plantago.* III. Genetic variation and demography in relation to survival of *Plantago cordata,* a rare species. *Biol. Cons.* 14:243–57.

Menges, E. 1990. Population viability analysis for an endangered plant. *Cons. Biol.* 4:41–62.

O'Brien, S.J., M.L. Roelke, L. Marker, A. Newman, C.A. Winkler, D. Meltzer, L. Colly, J.F. Evermann, M. Bush, and D.A. Wildt. 1985. Genetic basis for species vulnerability in the cheetah. *Science* 227:1428–34.

Park, Y.S., D.P. Fowler, and J.F. Coles. 1984. Population studies of white spruce. II. Natural inbreeding and relatedness among neighboring trees. *Can. J. For. Res.* 14:909–13.

Pinero, D., M. Martinez-Ramos, and J. Sarukhan. 1984. A population model of *Astrocaryum mexicanum* and a sensitivity analysis of its finite rate of increase. *J. Ecol.* 72:977–91.

Pollard, J.H. 1966. On the use of the direct matrix product in analyzing certain stochastic population models. *Biometrika* 53:397–415.

Prentice, H.C. 1984. Enzyme polymorphism, morphometric variation and population structure in the restricted endemic, *Silene diclinis* (Caryophyllaceae). *Biol. J. Linn. Soc.* 22:125–43.

Rai, K.N., and S.K. Jain. 1982. Population biology of *Avena.* IX. Gene flow and neighborhood size in relation to microgeographic variation in *Avena barbata. Oecologia* 53:299–305.

Richter-Dyn, N., and N.S. Goel. 1972. On the extinction of colonizing species. *Theor. Popul. Biol.* 3:406–33.

Roughgarden, J. 1979. *Theory of population genetics and evolutionary ecology: An introduction.* New York: Macmillan.

Schonewald-Cox, C.M., S.M. Chambers, B. MacBryde, and W.L. Thomas. 1983. *Genetics and conservation.* Menlo Park, Calif.: Benjamin/Cummings.

Shaffer, M.L. 1981. Minimum population sizes for species conservation. *BioScience* 31:131–34.

————. 1987. Minimum viable populations: Coping with uncertainty. In *Viable populations for conservation*, ed. M.E. Soulé, 69–86. Cambridge: Cambridge University Press.

Shaffer, M.L., and F.B. Samson. 1985. Population size and extinction: A note on determining critical population sizes. *Am. Nat.* 125:144–52.

Shaw, R.G. 1987. Density-dependence in *Salvia lyrata*: Experimental alteration of densities of established plants. *J. Ecol.* 75:1049–63.

Silander, J.A. 1983. Demographic variation in the Australian desert cassia under grazing pressure. *Oecologia* 60:227–33.

————. 1985. The genetic basis of the ecological amplitude of *Spartina patens*. II. Variance and correlation analysis. *Evolution* 39:1034–52.

Slade, N.A., and H. Levenson. 1984. The effect of skewed distributions on vital statistics on growth of age-structured populations. *Theor. Popul. Biol.* 26:361–66.

Soulé, M.E., ed. 1986. *Conservation biology. The science of scarcity and diversity.* Sunderland, Mass.: Sinauer.

————. ed. 1987. *Viable populations for conservation.* Cambridge: Cambridge University Press.

Strebel, D.E. 1985. Environmental fluctuations and extinction in single species. *Theor. Popul. Biol.* 27:1–26.

Sykes, Z.M. 1969. Some stochastic versions of the matrix model for population dynamics. *Am. Stat. Assoc. J.* 64:111–30.

Tuljapurkar, S.D., and S.H. Orzack. 1980. Population dynamics in variable environments. I. Long-run growth rates and extinction. *Theor. Popul. Biol.* 18:314–42.

Vaughan, D.S. 1981. An age structure model of yellow perch in western Lake Erie. In *Quantitative population dynamics*, ed. D.G. Chapman and V.F. Gallucci, 189–216. Fairland, Md.: International Co-operative.

Waite, S. 1984. Changes in the demography of *Plantago coronopus* at two coastal sites. *J. Ecol.* 72:809–26.

Waller, D.M., D.M. O'Malley, and S.C. Gawler. 1988. Genetic variation in the extreme endemic *Pedicularis furbishiae*. *Cons. Biol.* 1:335–40.

Wells, T.C.E. 1981. Population ecology of terrestrial orchids. In *Biological aspects of rare plant conservation*, ed. H. Synge, 281–96. Chichester: John Wiley and Sons.

Werner, P.A., and H. Caswell. 1977. Population growth rates and age vs. size distribution models for teasel (*Dipsacus silvestris* Huds.). *Ecology* 58:1103–11.

Wethey, D.S. 1985. Castastrophe, extinction, and species diversity: A rocky intertidal example. *Ecology* 66:445–56.

White, P.S., and S.P. Bratton. 1980. After preservation: Philosophical and practical problems of change. *Biol. Cons.* 18:241–55.

Wilcox, B.A. 1986. Extinction models and conservation. *Trends Ecol. Evol.* 1:46–48.

Wilcox, B.A., and D.D. Murphy. 1985. Conservation strategy: the effects of fragmentation on extinction. *Am. Nat.* 125:879–87.

Wright, S.J., and S.P. Hubbell. 1983. Stochastic extinction and reserve size: A focal species approach. *Oikos* 41:466–77.

CHAPTER 11

The Effects of Inbreeding on Isolated Populations: Are Minimum Viable Population Sizes Predictable?

ROBERT C. LACY

ABSTRACT

Management of nature reserves, of multiple-use lands, and of captive breeding programs requires knowledge of the minimum population sizes below which the combined effects of random genetic changes and demographic variation would likely result in extinction. One prerequisite to estimating such minimum viable population sizes is the determination of the effects of inbreeding on fitness. Two hypotheses make distinct predictions about the relative tolerance of populations to inbreeding: If inbreeding depression results primarily from the expression of deleterious recessive alleles, then selection would have removed most such genes from populations with long histories of inbreeding, and those populations would be resistant to further inbreeding impacts. If inbreeding depression occurs because of a general selective advantage of heterozygosity throughout the genome, then previously inbred populations would have reduced fitness presently and would fare no better under future inbreeding than would large and heterogeneous populations. We tested the hypothesis that small, isolated populations of Peromyscus *mice would show less depression in fitness when inbred than would large, central populations. Remnant, insular populations had one-quarter to one-third the genic diversity of large, central populations. Although the populations varied greatly in the rate of loss of fitness (measured as infant viability) when experimentally inbred, the severity of inbreed-*

ing depression did not correlate with initial genic diversity of the stocks or, therefore, with the size and degree of insularity of the wild populations. Neither simple theory of inbreeding depression could account for the varied responses of the populations. It remains an important task for conservation biologists to discover phylogenetic, ecological, or genetic predictors of genetically minimum viable population sizes.

INTRODUCTION

Wildlife populations that were once large, continuous, and diverse have been reduced to small, fragmented isolates in remaining natural areas, nature preserves, or even zoos. Black rhinoceroses (*Diceros bicornis*) once numbered in the hundreds of thousands, occupying much of Africa south of the Sahara. Now a few thousand survive in a handful of parks and reserves, each supporting at most a few hundred animals. It is unlikely that any but the few largest of these support enough rhinos to constitute long-term viable populations. Tigers once roamed over much of Asia, from Indonesia north through China, Korea, and into Siberia and east to Iran and Turkey. Now isolated populations survive in parks in India, in remote areas of Indochina and Sumatra, and along the Sino-Soviet border. Today there are far more Siberian tigers (*Tigris tigris altaica*), the largest of the tiger subspecies, in zoos than in the wild. Smaller, less well-known species are also in decline: The Puerto Rican crested toad (*Peltophryne lemur*) once occupied most or all of the coastal lowlands of Puerto Rico, breeding in temporary ponds following heavy rains. The lowlands of Puerto Rico have been so intensively urbanized and converted to agriculture that the only breeding site remaining is the parking lot of a public beach. (Remarkably, the species is officially listed as only "threatened.") Since the arrival of Europeans, all six nonvolant native land mammals of Puerto Rico have gone extinct.

If the loss of natural habitats were not enough of a threat to the survival of species, the reduction and fragmentation of populations set in motion secondary processes that can cause further decline and eventual extinction. First, small populations face demographic risks. When mates are few and far between, many animals may not reproduce. When only a few animals reach reproductive maturity each year, all may be males or all females. The last five dusky seaside sparrows were males. With few animals in a population, disease, a severe storm, or just bad luck may cause the extinction of a local population. In the 1940s a hurricane eliminated one of the two populations of whooping cranes that remained at that time. The disappearance of local populations can quickly amount to global extinction if populations are too far apart to allow recolonization of vacated habitats.

On top of the demographic risks, small populations also face genetic risks. When numbers of breeding animals get very low, inbreeding—mating between relatives—becomes inevitable and common. It has been recognized for centuries that inbred animals often have a high rate of birth defects, slower growth, higher mortality, and lower fecundity, the phenomenon referred to as "inbreeding depression." Inbreeding depression has been well documented in domesticated stock (Darwin 1868; Falconer 1981; Wright 1977) and zoo populations (Ralls, Ballou, and Templeton 1988). Interestingly, the few studies of inbreeding in wild populations (e.g., great tits: van Noordwijk and Scharloo 1981; superb blue wren: Rowley, Russell, and Brooker 1986; baboons: Bulger and Hamilton 1988) have produced little evidence that inbreeding is costly under natural conditions.

An additional problem with inbreeding (or with any small, closed population, even in the absence of inbreeding) is that genetic variants are occasionally lost by chance, and the restoration of variation by mutation is much too rare an event to be significant in small populations (Lacy 1987). As the population becomes increasingly homogeneous it also becomes increasingly susceptible to disease (O'Brien and Evermann 1988), new predators, changing climate, or any environmental change. Without diversity, a change might be lethal to all individuals in the population, none having the adaptations necessary to survive the change. Accordingly, concern has been expressed that cheetahs (*Acinonyx jubatus*), which have been found to have very little genetic variation (O'Brien et al. 1983), may be especially vulnerable to disease, parasites, and extinction (O'Brien et al. 1985).

Unfortunately, the demographic and genetic consequences of the decimation and fragmentation of wildlife populations are synergistic. As an isolated population becomes small, difficulties in finding mates and other demographic risks may cause further decline of the population. Inbreeding and the loss of genetic diversity may lower fitness and lead to yet further decline, making demographic problems worse and in turn making genetic problems worse. Ultimately, the feedback between genetic and demographic decline can lead the population ever more rapidly toward extinction. This process has been aptly, though depressingly, termed an "extinction vortex" (Gilpin and Soulé 1986), and the size below which a population is likely to get drawn into the extinction vortex is the minimum viable population size, or MVP (Shaffer 1981).

How big is MVP? It is likely to be different for different species, and we do not know what MVP is for *any* species. Estimates of MVPs for natural populations (e.g., grizzly bears: Shaffer 1983; red-cockaded woodpeckers: Reed, Doerr, and Walters 1988; black-footed ferrets: Groves and Clark 1986) have been based on very few data, a lack of feedback among demographic and genetic processes, and the simplifying assumption that losses of genetic diversity will impact all species equally. If we cannot estimate accurately the MVP for species of concern, why not just keep all populations well above any reasonable estimate of MVP? Unfortunately, money, space, and human resources are inadequate. Species are going extinct that we could save in the wild or at least propagate in zoos, because no one provides the resources needed (Conway 1986). We cannot give to every species the intensive care we give to condors and ferrets. If we keep many more of some species than is necessary, we are unnecessarily dooming other species to extinction.

It is essential, therefore, that we estimate accurately the MVP for any species (of which there are many) requiring intensive management for its survival. Clearly, one aspect of estimating MVP is to determine how much inbreeding can be tolerated before the deleterious effects become of consequence. For which species can we allow first-cousin matings, or even brother-sister matings, and for which would even third-cousin matings be disastrous? The answer is critical to the establishment of nature preserves for wild populations, and for the management of zoo space and other captive breeding facilities for endangered species.

The common explanation for inbreeding depression rests on the presence of deleterious recessive genes in virtually all diploid organisms. The accumulated deleterious mutations within a population constitute that population's "genetic load" (Wallace 1970). One consequence of inbreeding is that it makes it much more likely that an individual is homozygous for a rare allele. Inbreeding is usually measured by the inbreeding coefficient F, defined as the probability that the two alleles at a genetic locus in the inbred individual descended from a single gene in an ancestor shared by the parents (Wright 1969). F is thus proportional to the increase in homozygosity, or loss of heterozygosity, in the inbred individual relative to noninbred individuals within the population. For example, in the extreme case of inbreeding, when a self-compatible hermaphrodite mates with itself, there would be a 50 percent chance that two identical copies of one or the other of the two alleles at any genetic locus would be transmitted to the inbred offspring ($F = .50$), and the inbred offspring would be expected to be heterozygous at half as many genetic loci as was its parent. With lesser inbreeding, matings between siblings, cousins, aunts and nephews, and so on, the mathematics change but the general trend does not: Inbred individuals have much higher probabilities of being homozygous for a gene present in a parent (including deleterious genes that may be very rare in the overall population) than do noninbred individuals, simply because genetically related parents share many genes. Ballou (1983) presents methods for calculating inbreeding coefficients from pedigrees.

The concept of "lethal equivalents," often used to quantify the genetic load of a population, is based on the assumption that inbreeding depression results from homozygosity of deleterious recessive genes. The number of lethal equivalents in a population is the mean number of lethal recessive genes per animal, if all deleterious effects are assumed to be clustered into fully lethal alleles rather than alleles conferring reduced but nonzero viability (Morton, Crow, and Muller 1956). Ralls et al. (1988) analyzed data on inbreeding depression in forty captive populations of mammals. They found a median of 3.14 lethal equivalents, but with considerable variation between species, ranging from 15.16 lethal equivalents in Wied's red-nosed rat to negative values (implying beneficial, or at least nonharmful, effects of inbreeding) in four species. Indeed, some animals and about one-third of plants regularly inbreed without apparent problems (Hamrick 1983; Selander 1983).

An alternative explanation for inbreeding depression is that heterozygosity per se is advantageous; that is, all genes are in a sense deleterious recessives, perhaps because of the greater flexibility afforded to heterozygotes (Lerner 1954; Mitton and Grant 1984). Evidence for such "general heterosis" can be seen in the failure of animal breeders to produce fully inbred strains of almost any animal other than house mice. In thousands of years of breeding livestock, it is remarkable that no one has succeeded in producing a healthy, fully inbred strain of any species of domesticated livestock. The potential economic value of pure-breeding livestock strains can be appreciated in the predominant use of hybrid crosses between inbred plant crop varieties in modern agriculture. Yet the vigor of the hybrid plants relative to all inbred strains attests to the inability of plant breeders to purge totally inbred lines of plants of undesirable genes. Evolutionary biologists have debated the preceding theories of inbreeding depression for decades, without resolution. A knowledge of whether deleterious recessives or general heterosis is the primary cause of inbreeding depression or, perhaps more meaningfully, knowledge of in which species each predominates is critical to our understanding of extinction processes and to the management of endangered populations.

To calculate an MVP for any species, we would need to know not only the relationship between fitness and inbreeding (perhaps measured by the number of lethal equivalents), but also how much loss of fitness a population could suffer before it was imperiled. Making a wild guess that a typical population could withstand a genetically imposed decrease of 10 percent in fitness, we can calculate from the data in Ralls et al. (1988) that the median mammal population could withstand a mean inbreeding coefficient of .067 (about the equivalent of a first-cousin mating), and their upper- and lower-quartile populations could withstand inbreeding levels of .037 and .234, respectively. Over 10 generations, the number of breeding individuals needed to keep cumulative inbreeding below these levels would be 72, 133, and 19 for the median, upper-, and lower-quartile populations, respectively. To maintain 90 percent of original fitness over 100 generations would require population sizes about tenfold greater. Wied's red-nosed rat would suffer a 10 percent decrease in fitness after inbreeding coefficients reached just .0069 (about the level resulting from the mating of individuals that have one great-grandparent in common) and would require 722 animals in the breeding population for 10 generations and 7,222 for 100 generations. Thus, there is great variation in the effects that inbreeding has on captive mammal populations, and consequently the MVPs (crudely estimated) for mammals vary over several orders of magnitude. MVPs would almost certainly vary even more if we could consider also the differing capacities of the populations to survive decreased fitness in their varied natural environments.

To estimate whether the MVP for a species is on the order of hundreds or thousands, we would need either to measure the severity of inbreeding depression in a long-term experiment (and, preferably, under varied conditions) or to find ecological, genetic, or taxonomic parameters that allow accurate predictions to be made without intensive study of each population. Ralls et al. (1988) did not find interorder differences among levels of inbreeding depression observed in populations of primates, rodents, and artiodactyls. If deleterious recessives are the primary cause of inbreeding depression, we could expect that populations having experienced episodes of inbreeding during population "bottlenecks" (population crashes and subsequent recovery) should be able to tolerate additional inbreeding in the future. Natural selection eliminates the genes that cause inbreeding depression (by the death of their bearers) when they are exposed in homozygotes, and populations with histories of inbreeding should become largely purged of deleterious recessive genes. Figure 11.1 shows (solid line) the accumulation of inbreeding (or loss of heterozygosity) that would result from repeated matings between full siblings. With such extreme inbreeding 25 percent of the heterozygosity is lost each generation. The figure also shows (dashed line) the expected reduction in the genetic load of recessive lethal genes that would occur with continued sib-matings. Evidence that recessive deleterious alleles do contribute to inbreeding depression can be seen in the high frequency of genetic defects and lowered fitness observed when mice are inbred in the lab, followed by the development of healthy inbred strains once those deleterious alleles (and many of the initial inbred lines) are removed by selection (Connor and Bellucci 1979; Lynch 1977; Strong 1978). Rao and Inbaraj (1980) found no effect of inbreeding on fetal growth and development in a population of humans with a long-term practice of inbreeding.

Populations that have long been large enough so that inbreeding is extremely rare may harbor large genetic loads, natural selection never

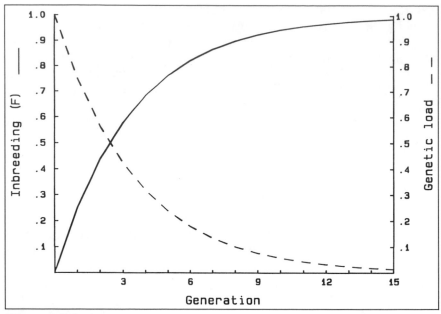

Figure 11.1
Cumulative inbreeding (solid line) and expected loss of genetic load (recessive lethal genes) (dashed line) resulting from fifteen generations of matings between full siblings.

having had the opportunity to purge the population of deleterious recessive alleles. Figure 11.2 illustrates the effects that repeated full-sib matings might be expected to have on a large, genetically diverse central population and a small, genetically depauperate insular population. In both, inbreeding initially causes a loss of fitness as deleterious recessive alleles are expressed in homozygotes. The selective deaths caused by these recessive alleles, however, remove the genetic load from the population, resulting in a restoration of fitness in the subsequent generations. In a central population with an initially large genetic load, the drop in fitness resulting from inbreeding is steeper than in more insular populations with initially lower genetic loads. Thus, black-footed ferrets, which now exist only in captivity, may be relatively insensitive to inbreeding. The last known wild population of ferrets had remained at twenty to fifty adults for about fifty years until it went extinct in 1986 and thus was subjected to at least low levels of inbreeding for about thirty-three generations (Lacy and Clark 1989). If the ferrets have been purged of most of their genetic load (i.e., they are toward the right end of the curve in Figure 11.2), perhaps further inbreeding will not hamper efforts at re-

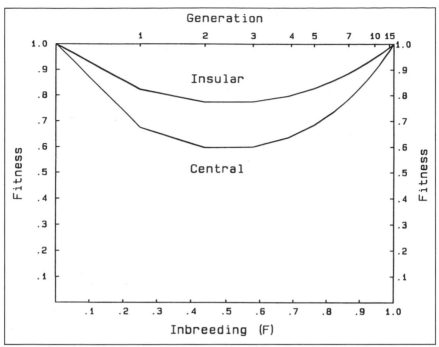

Figure 11.2
Expected relationship between cumulative inbreeding (F) and fitness in an insular population and in a central population, if inbreeding depression results primarily from deleterious recessive genes. At the top is given the generations of full-sib matings required to reach the levels of inbreeding on the bottom axis. Relative fitness on the Y axis is calibrated on an assumption that the central population has initially 3.14 lethal equivalents, and the insular population begins with half that genetic load.

covery of the species. On the other hand, California condors have been at very low numbers for few generations (generation time is probably ten times longer than in ferrets) and may still have considerable deleterious recessive genes that could cause problems for the captive breeding efforts if inbreeding becomes unavoidable in the small remnant population. The final blow to species such as the black rhino and elephant, which recently numbered in the hundreds of thousands, may be genetic defects arising in small, isolated reserves where inbreeding becomes common.

Considering now the alternative explanation of inbreeding depression, if heterozygosity is broadly advantageous, then those populations that have lost much variability during bottlenecks may be closer to a critically low level of heterozygosity and thus may be even more susceptible to further problems when inbred than are populations that still contain high levels of variation. Figure 11.3 shows the expected fate under inbreeding of a hypothetical central population with no history of inbreeding and considerable heterozygosity, and an insular population that has lost half of its heterozygosity during past bottlenecks. In this figure, I assume that the fitness of a population is directly proportional to extant heterozygosity and that natural selection is inefficient at retaining heterozygosity within the population during rapid inbreeding. (If heterozygote advantage is general, affecting genes throughout the genome, then there would be little variation in heterozygosity among the inbred progeny, and natural selection favoring the more heterozygous descendants would be ineffective at preventing the loss of heterozygosity and fitness

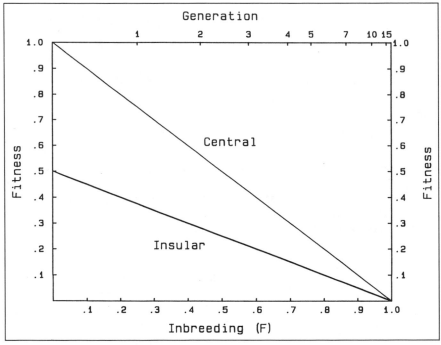

Figure 11.3
Expected relationship between cumulative inbreeding (F) and fitness in an insular population and in a central population, if inbreeding depression results primarily from heterosis. At the top is given the generations of full-sib matings required to reach the levels of inbreeding on the bottom axis. Relative fitness on the Y axis is calibrated on the assumption that the central population has twice the heterozygosity of the insular population and that fitness is directly proportional to heterozygosity.

from the population.) As such a population becomes increasingly inbred, its fitness would steadily and, over historical time, irreversibly decline if heterosis is the principal cause of inbreeding depression.

As important as an elucidation of the underlying mechanism of inbreeding depression is to evolutionary theory, to an understanding of the role of variability in natural populations, and to the management of small populations, the basic models of inbreeding depression have never been adequately tested. The hypotheses represented in Figures 11.2 and 11.3 make very different predictions about the effects of long-term population bottlenecks and the vulnerability of isolated populations to inbreeding depression. We do not yet know the extent to which populations can be purged of their genetic loads, nor whether smaller and more isolated populations are more or less sensitive to inbreeding depression.

AN EXPERIMENTAL TEST OF THE EFFECT OF BOTTLENECKS ON INBREEDING DEPRESSION

To confirm our understanding of inbreeding and, further, to begin to develop predictors of the severity of inbreeding depression in populations, in 1983 and 1984 Bruce Brewer collected mice of the genus *Peromyscus* from island populations that had undergone severe declines and had therefore probably experienced episodes of inbreeding, and also from large mainland populations of virtually infinite sizes. The mice were brought to Cornell University, where Bruce was a graduate student at the time, and a systematic study of the effects of inbreeding on each population was begun (Brewer 1988). Subsequently, Bruce and his mice went to the Chicago Zoological Park. As a curator of mammals at the zoo, Bruce focused his energies on animal management, and he turned the mouse inbreeding research project over to me and my research associates, Melissa Foster and Glen Alaks, in the Department of Conservation Biology. I will now summarize the results of the study, described in detail in Brewer et al. (1990).

The deer mice of the genus *Peromyscus* are abundant and widespread throughout North America. As with most species, the range consists of a mosaic of habitats that support local populations at varying densities and with varying degrees of isolation from neighboring populations. The populations sampled for this study, listed in Table 11.1, range from some centered in thousands of miles of continuous habitat to island-endemic subspecies. Eight populations of two species (*P. leucopus*, the white-footed mouse, and *P. polionotus*, the beach mouse or old-field mouse) were studied. One island population of beach mice, *P. polionotus leucoce-*

Table 11.1
Taxa, collecting sites, and degrees of insularity of the populations, mean heterozygosity (H) of the founders used to establish laboratory stocks, and the number of lethal equivalents determined from the regression of litter viability against inbreeding (I = island; P = peninsula; RM = range margin; C = central).

Taxon	Collecting Locality	Degree of Insularity	H	Lethal Equivalents
P. polionotus phasma	Anastasia State Park, Fla.	I	.040	.114
P. p. niveiventris	Canaveral National Seashore, Fla.	P	.088	.358
P. p. rhoadsi	Lake Placid, Fla.	RM	.087	−.156
P. p. subgriseus	Levy Co., Fla.	C	.126	.474
P. p. subgriseus	Ocala National Forest, Fla.	C	.075	−.638
P. p. leucocephalus	Santa Rosa Island, Fla.	I	.053	.602
P. leucopus tornillo	Southwest Texas	C	.106	.786
P. l. noveboracensis	Ithaca, N.Y.	C	.112	.628

phalus, was collected on Santa Rosa Island in the Florida section of the Gulf Islands National Seashore. The other island endemic, *P. p. phasma*, was trapped in Anastasia State Park, along the Atlantic coast of northern Florida. This subspecies has been eliminated from formerly suitable habitat on the mainland, is now limited to less than twenty-five acres of remaining habitat, and has been listed as endangered by the U.S. Fish and Wildlife Service (1987). *Peromyscus p. niveiventris* were collected from the Canaveral National Seashore further south on the Atlantic coast of Florida. Though still common in the National Seashore, Canaveral Air Force Station, Kennedy Space Center, and Merritt Island National Wildlife Refuge, this subspecies that once extended from Miami to Daytona is probably now restricted to the protected areas on Cape Canaveral and is listed as threatened. Beach mice inhabit frontal dunes and nearby grassy fields, habitats that put them in direct conflict with humans and human-associated rats, cats, and house mice. Direct habitat destruction (conversion to condominiums), predation by cats, and competition with house mice have severely reduced most beach mouse populations. Hurricanes occasionally decimate populations of beach mice occupying the frontal dunes, washing over much of the remnant habitats (Meyers 1983). Three of five Gulf Coast subspecies are listed as endangered (U.S. Fish and Wildlife Service 1987), including those inhabiting keys immediately to the east and immediately to the west of Santa Rosa Island, and the subspecies *P. p. decoloratus* that once occupied the beaches between Canaveral and Anastasia Island is probably now extinct (Humphrey and Barbour 1981).

Inland populations of *P. polionotus* inhabit open woodlands, prairies, and old agricultural fields on sandy soils. They are much darker than are beach mice and are commonly referred to as "old-field mice." At the southern end of the range of the species, *P. p. rhoadsi* were captured near Lake Placid, Florida. Two populations of the subspecies *P. p. subgriseus* were collected in the Ocala National Forest and along roadsides in Levy County, Florida. The old-field mouse is widespread in the extensive open forests and old fields of northern Florida and most of Georgia, South Carolina, and Alabama. White-footed mice, *P. leucopus noveboracensis*, were trapped in Ithaca, New York. This subspecies inhabits the northeastern quadrant of the U.S. and is one of the most abundant mammals throughout most of its range. *Peromyscus l. tornillo* were trapped in southwestern Texas. The subspecies is found in the extensive semi-arid lands of western Texas, New Mexico, and southeastern Colorado.

Genetic variability within each population was assessed by the technique of starch gel electrophoresis. Samples of blood or tissue are placed in a gel, a voltage is applied across the gel for a few hours to separate molecules based on charge and shape, and then the gel is removed and stained for specific proteins. Proteins show up as bands on the gel, and variants with different mobilities through the gel (about one-third of the molecular variants) are easily visualized, allowing assessment of genetically encoded variation within and between individuals. The wild-caught mice that were the founders of the lab stocks used in this experiment were analyzed by electrophoresis. Genetic variability in these founder stocks was quantified as the mean heterozygosity of each population across the thirty genes examined from proteins in liver, kidney, and muscle samples (Table 11.1).

The genetic variability in the founder stocks followed expectations based on presumed histories of the populations. Those populations that are large and abundant, are not known to have experienced population bottlenecks, and probably rarely inbreed in the wild were found to have high variability. Those that are island isolates, greatly reduced in numbers by habitat destruction and predation by cats, are subject to occasional decimation when hurricanes hit, and may be forced to inbreed during population crashes have only about one-third to one-half the heterozygosity observed in the more central populations. The heterozygosities given in Table 11.1 are close to those reported by other investigators that have studied these populations (Selander et al. 1971; Garten 1976) with one notable exception: The population from Anastasia Island has only about half the variability than had been found when it was sampled ten to fifteen years earlier. During the past decade this population was reduced in size as a result of habitat loss and apparently lost much genetic variability as a result. This loss of variability is comparable to the loss expected after about three generations of brother-sister (or parent-off-spring) inbreeding, though presumably the loss occurred during more prolonged but less severe inbreeding during population contraction.

After establishment of lab stocks, in each generation some mice were paired with relatives to produce inbred litters while others were paired with unrelated mice to produce outbred control litters. Prior to inbreeding, the control lab populations differed little in initial litter sizes. Survival of the offspring to weaning at twenty days of age varied among populations, being lowest for the beach mice from Anastasia Island (primarily because females frequently killed their litters, and occasionally their mates, just after giving birth), and highest in the two populations of *P. leucopus*. Whether the aggressiveness of the Anastasia subspecies is in some way a result of the rapid range contraction and loss of genetic variability it has experienced recently or is an adaptation to an aspect of their island habitat (perhaps competition from house mice) is unknown.

To assess the effects of inbreeding in each population, we used regression analysis to determine the relationship between litter survival and the inbreeding coefficient of the litter. We analyzed only those litters born to lab-reared descendants of the founder stocks, because the histories of the wild-caught mice were uncontrolled. Because the first litter conceived by each female was typically smaller and had higher mortality than subsequent litters, we statistically factored out this parity effect from the regression model. (Further statistical analyses examining the effects of inbreeding of the dam, inbreeding of the sire, and age of the dam, and examining inbreeding effects on initial litter size and litter growth are presented in Brewer et al. [1990]. The results of those more extensive analyses are well represented by the effect of inbreeding on litter viability, and the conclusions reached were no different than those that follow.)

The regression lines fit to the relation between litter viability and litter inbreeding for each of the eight populations are shown in Figure 11.4. The regressions clearly demonstrate the extremely varied responses of the populations to inbreeding. Whereas most populations showed inbreeding depression, as expected, two showed increased offspring survival with inbreeding. (Neither increase was statistically significant, however.) Although the populations varied considerably in their responses to inbreeding, they did not follow the pattern predicted by a deleterious recessive theory of inbreeding depression (Figure 11.2) or the predictions of the heterosis theory (Figure 11.3). One island population (Anastasia) had low viability before inbreeding but was minimally affected by inbreeding; the other island population (Santa Rosa) was among those showing the most inbreeding depression. The two mainland populations of *P. leucopus* suffered more litter mortality when inbred, but overall there was no trend for the more insular populations to be more or less

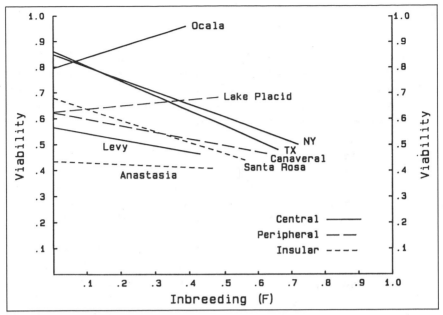

Figure 11.4
Relationships between cumulative inbreeding and the viability of litters in eight populations of Peromyscus *mice. Least-square regression lines are shown over the range of inbreeding coefficients produced in the breeding experiment. Effects of dam parity (first versus later litters) have been removed statistically, and Y intercepts are for multiparous dams.*

affected than the more central ones. We examined the data for any curvilinear responses to inbreeding (such as those in Figure 11.2) but found that second- and higher-order polynomial regressions did not fit the data better than did the simple linear relationships shown in Figure 11.4. Thus we did not find evidence that the populations were becoming purged of their genetic loads of recessive deleterious genes as they became more inbred. (Although it is not shown here, we also did not find that inbred litters of inbred parents fared any better than did inbred litters of non-inbred parents.) The number of lethal equivalents in each population can be estimated as twice the slope from the regression of the natural logarithm of litter survival against inbreeding (Morton et al. 1956; Ralls et al. 1988). That measure of genetic load, given in Table 11.1, shows the same lack of trends seen in Figure 11.4. It is perhaps notable that the numbers of lethal equivalents estimated for the *Peromyscus* populations we studied were all lower than most of the values given in the compilation by Ralls et al. (1988).

INTERPRETATION

The severity of the inbreeding depression was not predictable from knowledge of the ecology and genetics of populations of *Peromyscus* mice. This suggests that the effects of inbreeding are controlled by a small number of genes and that the presence or absence of those genes in a population is more or less a chance phenomenon. Perhaps some populations are unlucky and have a number of recessive, harmful genes that cause problems if the population declines rapidly; other populations may not harbor a significant genetic load. Perhaps general heterosis is the predominant factor in some populations, and deleterious recessives prevail in others. In any case, the response to inbreeding may well be determined by historical accident, not the frequency or recency of past population bottlenecks.

Unfortunately, we do not know whether the sampling of the gene pool that determined how the mouse populations were affected by inbreeding occurred as a founder effect in the establishment of the wild populations (so that the effects of inbreeding would be just as random among wild populations as they were among our lab stocks derived from the wild), or whether the between-stock diversity in inbreeding response occurred when we sampled the wild populations to obtain founders for our lab populations. Note, however, that the establishment of our lab stocks was probably not much different from the establishment of captive breeding stocks for endangered species protection, the reintroduction of animals to former habitat, or natural colonization. In each case a small number of founders—perhaps five to twenty—from one population become the nucleus of the new population.

CONCLUSION

What does all this mean for the conservation of endangered species? Our somewhat depressing findings suggest that we may not be able to predict with confidence the severity of inbreeding depression in populations in critical need of conservation. Within a genus, and even among subspecies within a species, the effects of inbreeding were chaotic. Thus, I have little confidence that we will soon be able to make informed guesses as to the minimum genetically viable population of any species without first conducting extensive studies of the genetics and the responses to inbreeding of the particular population of interest. Cheetahs, with almost no variability, may or may not be able to inbreed. Similarly, when the few black-footed ferrets in captivity are inbred (there are so few ferrets that soon all will be related), we won't know ahead of time whether the inbreeding will doom the species.

Certainly, more optimistic interpretations are possible. First, perhaps not enough mice were tested to detect trends in the data. That may be so, but if so then the trend is a very weak one, too weak to be a significant aid in the management of endangered species. Templeton and Read (1983) designed a captive breeding plan for Speke's gazelle, an endangered North African antelope, on the premise that past inbreeding would have removed the genetic load from the already inbred population. (As in the case of the black-footed ferret, however, they had virtually no choice but to inbreed the remnant population.) In the case of the beach mice, genetic load theory suggests that during past bottlenecks the island populations would have become relatively resistant to the effects of further inbreeding and could perhaps be managed without concern about inbreeding. For one of the two island populations (Santa Rosa) that decision would likely lead to disaster. Yet perhaps these mouse populations were atypical; either deleterious recessive genes or heterosis may explain inbreeding depression in most (other) species. That may be so, but unfortunately no one has yet demonstrated a consistent trend among any set of species that strongly supports either, or any, theory of inbreeding depression. Finally, although the aspects of ecology and genetics examined in this study were not good predictors of the severity of inbreeding depression, there probably are other biological predictors that would give us a means, a priori, to determine the fate of a population should it be subjected to inbreeding. That is also true, but so far no one has suggested what those better predictors might be. I hope that someone does develop a theory of inbreeding depression that is general enough to predict accurately the inbreeding effects for most populations. Because of the incredible diversity and complexity of the ecology, genetics, and evolution of species, however, I doubt that a simple theory will provide a powerful predictive tool for managing the increasingly many species that are threatened with extinction. I suspect that for some time yet we will have to treat each species as a special case to be examined for the causes of decline, for the threats to continued survival, and for the most effective course of action to achieve recovery of the species to safe numbers.

The "science" of conservation biology is quite new, though it draws much from the more developed fields of ecology, demography, genetics, and wildlife management. Presently, conservation biologists are busy applying knowledge from their diverse areas of training to problems that need immediate solutions. There is an understandable desire to demonstrate to the scientific and conservation communities that conservation biology is a real science that can provide the needed answers. Thus there is a tendency to claim that we know a lot more than we do, and to oversimplify the tasks before us. Yet I think we must admit that conservation biology is still a very imprecise science. We need to identify the most important questions and then design experiments to provide the answers. And of course we must apply that knowledge to the urgent task of conservation just as soon as we can.

ACKNOWLEDGMENTS

Many people, agencies, and funding sources contributed importantly to the mouse inbreeding study. Peter Brussard provided the original inspiration for the study and the continued prodding to see it through to completion. Bruce Brewer, Melissa Foster, and Glen Alaks collected far more of the data than I, and Bernie May, Risa Rosenberg, Anna Voeks, Suzanne Jones, David Featherston, and the keeper staff of the Brookfield Zoo Lion House (Brookfield, Ill.) all contributed to mouse collection, mouse maintenance, or lab analyses. The Florida Game and Fresh Water Fish Commission, Florida Department of Natural Resources, Gulf Island National Seashore, Southern Pacific Railroad, Canaveral National Seashore, and Ocala National Forest provided permits, permission to collect, and assistance in locating mice. Facilities were provided by the Section of Ecology and Systematics, Cornell University, and by the Chicago Zoological Park. Funding was provided by the Chicago Zoological Society and grants from Sigma Xi; Wildlife Preservation Trust International; the Mellon Foundation; and the Institute of Museum Services, an agency of the National Foundation for the Arts and Humanities.

LITERATURE CITED

Ballou, J. 1983. Calculating inbreeding coefficients from pedigrees. In *Genetics and conservation: A reference for managing wild animal and plant populations*, ed. C.M. Schonewald-Cox, S.M. Chambers, B. MacBryde, and W.L. Thomas, 509–20. Menlo Park, Calif.: Benjamin/Cummings.

Brewer, B.A. 1988. An investigation of protein electrophoresis as a predictor of inbreeding depression in captive populations of *Peromyscus*. In *Dissertation Abstracts International* 49:5131. Cornell University, Ithaca, New York.

Brewer, B.A., R.C. Lacy, M.L. Foster, and G. Alaks. 1990. Inbreeding depression in insular and central populations of *Peromyscus* mice. *J. Heredity* 81:257–66.

Bulger, J., and W.J. Hamilton III. 1988. Inbreeding and reproductive success in a natural chacma baboon, *Papio cynocephalus ursinus*, population. *Anim. Behav.* 36:574–78.

Connor, J.L., and M.J. Bellucci. 1979. Natural selection resisting inbreeding depression in captive wild housemice (*Mus musculus*). *Evolution* 33:929–40.

Conway, W.G. 1986. The practical difficulties and financial implications of endangered species breeding programs. *Int. Zoo Yearbook* 24/25:210–19.

Darwin, C. 1868. *The variation of animals and plants under domestication.* London: John Murray.

Falconer, D.S. 1981. *Introduction to quantitative genetics.* New York: Longman.

Garten, C.T., Jr. 1976. Relationships between aggressive behavior and genic heterozygosity in the oldfield mouse, *Peromyscus polionotus. Evolution* 30:59–72.

Gilpin, M.E., and M.E. Soulé. 1986. Minimum viable populations: Processes of species extinction. In *Conservation biology: The science of scarcity and diversity*, ed. M.E. Soulé, 19–34. Sunderland, Mass.: Sinauer.

Groves, C.R., and T.W. Clark. 1986. Determining minimum population size for recovery of the black-footed ferret. *Great Basin Nat. Memoirs* 8:150–159.

Hamrick, J.L. 1983. The distribution of genetic variation within and among natural populations of plants. In *Genetics and conservation: A reference for managing wild animal and plant populations*, ed. C.M. Schonewald-Cox, S.M. Chambers, B. MacBryde, and L. Thomas, 335–48. Menlo Park, Calif.: Benjamin/Cummings.

Humphrey, S.R., and D.B. Barbour. 1981. Status and habitat of three subspecies of *Peromyscus polionotus* in Florida. *J. Mamm.* 62:840–44.

Lacy, R.C. 1987. Loss of genetic diversity from managed populations: Interacting effects of drift, mutation, immigration, selection, and population subdivision. *Cons. Biol.* 1:143–58.

Lacy, R.C., and T.W. Clark. 1989. Genetic variability in black-footed ferret populations: Past, present, and future. In *Conservation biology and the black-footed ferret*, ed. U.S. Seal, E.T. Thorne, M.A. Bogan, and S.H. Anderson, 83–103. New Haven: Yale University Press.

Lerner, I.M. 1954. *Genetic homeostasis.* New York: J. Wiley and Sons.

Lynch, C.B. 1977. Inbreeding effects upon animals derived from a wild population of *Mus musculus. Evolution* 31:526–37.

Meyers, J. M. 1983. *Status, microhabitat, and management recommendations for Peromyscus polionotus on Gulf Coast beaches Report.* Atlanta: U.S. Fish and Wildlife Service.

Mitton, J.B., and M.C. Grant. 1984. Associations among protein heterozygosity, growth rate, and developmental homeostasis. *Ann. Rev. Ecol. Syst.* 15:479–99.

Morton, N.E., J.F. Crow, and H.J. Muller. 1956. An estimate of the mutational damage in man from data on consanguineous marriages. *Proc. Nat. Acad. Sci. U.S.A.* 42:855–63.

O'Brien, S.J., and J.F. Evermann. 1988. Interactive influence of infectious disease and genetic diversity in natural populations. *Trends Ecol. Evol.* 3:254–59.

O'Brien, S.J., M.E. Roelke, L. Marker, A. Newman, C.A. Winkler, D. Meltzer, L. Colly, J.F. Evermann, M. Bush, and D.E. Wildt. 1985. Genetic basis for species vulnerability in the cheetah. *Science* 227:1428–34.

O'Brien, S.J., D.E. Wildt, D. Goldman, C.R. Merril, and M. Bush. 1983. The cheetah is depauperate in genetic variation. *Science* 221:459–62.

Ralls, K., J.D. Ballou, and A. Templeton. 1988. Estimates of lethal equivalents and the cost of inbreeding in mammals. *Cons. Biol.* 2:185–93.

Rao, P.S.S., and S.G. Inbaraj. 1980. Inbreeding effects on fetal growth and development. *J. Med. Genet.* 17:27–33.

Reed, J.M., P.D. Doerr, and J.R. Walters. 1988. Minimum viable population size of the red-cockaded woodpecker. *J. Wildl. Manage.* 52:385–91.

Rowley, I., E. Russell, and M. Brooker. 1986. Inbreeding: Benefits may outweigh costs. *Anim. Behav.* 34:939–41.

Selander, R.K. 1983. Evolutionary consequences of inbreeding. In *Genetics and conservation: A reference for managing wild animal and plant populations*, ed. C.M. Schonewald-Cox, S.M. Chambers, B. MacBryde, and L. Thomas, 201–15. Menlo Park, Calif.: Benjamin/Cummings.

Selander, R.K., M.H. Smith, S.Y. Yang, W.E. Johnson, and J.B. Gentry. 1971. Biochemical polymorphism and systematics in the Genus *Peromyscus*. I. Variation in the old-field mouse (*Peromyscus polionotus*). Studies in Genetics. VI. *Univ. Texas Publ.* 7103:49–90.

Shaffer, M.L. 1981. Minimum population sizes for species conservation. *BioScience* 31:131–34.

———. 1983. Determining minimum viable population sizes for the grizzly bear. *Int. Conf. Bear Res. Manage.* 5:133–39.

Strong, L.C. 1978. Inbred mice in science. In *Origins of inbred mice*, ed. H.C. Morse, 45–67. New York: Academic Press.

Templeton, A.R., and B. Read. 1983. The elimination of inbreeding depression in a captive herd of Speke's gazelle. In *Genetics and conservation: A reference for managing wild animal and plant populations*, ed. C.M. Schonewald-Cox, S.M. Chambers, B. MacBryde, and L. Thomas, 241–61. Menlo Park, Calif.:Benjamin/Cummings.

U.S. Fish and Wildlife Service. 1987. *Recovery plan for the Choctawhatchee, Perdido Key and Alabama Beach Mouse.* Atlanta: U.S. Fish and Wildlife Service.

van Noordwijk, A.J., and W. Scharloo. 1981. Inbreeding in an island population of the great tit. *Evolution* 35:674–88.

Wallace, B. 1970. *Genetic load: Its biological and conceptual aspects.* Englewood Cliffs, N.J.: Prentice-Hall.

Wright, S. 1969. *Evolution and the genetics of populations. Vol. 2. The theory of gene frequencies.* Chicago: University of Chicago Press.

———. 1977. *Evolution and the genetics of populations. Vol. 3. Experimental results and evolutionary deductions.* Chicago: University of Chicago Press.

Conservation of Asian Primates: Aspects of Genetics and Behavioral Ecology that Predict Vulnerability

MARY PEARL

ABSTRACT

*Primate conservationists monitor the numbers of primate popula-
tions around the world as they decline at the hands of human activity.
However, it is important to make realistic assessments of the survival
prospects of different endangered primate species. Considerations
should be made of behavioral, genetic, and ecological traits that
might create greater or lesser vulnerability to extinction given the
same human impacts. A brief consideration of two rare Asian monkeys, the lion-tailed macaque and the proboscis monkey, illustrates
the need to focus on the unique complex of traits of each species to
devise realistic conservation strategies. General unifying models of
vulnerability to extinction are relatively unhelpful.*

INTRODUCTION

Over recent decades, primate conservationists have monitored the decline of primate numbers at the hands of human activity worldwide. Less emphasis has been placed on understanding what factors intrinsic to particular primate taxa predispose them to greater risk of extinction. Such an understanding is needed to predict with more precision the populations of greatest vulnerability. Typically, reports of the conservation status of primates are compilations or contain summaries of the most recent and accurate censuses available (e.g., Wolfheim 1983; Mittermeier and Cheney 1987; Richard and Sussman 1987; Oates, Gartlan, and Struhsaker 1982). Such broad descriptions and equally broad prescriptions for reserve establishment are key to initial conservation action. In the context of conservation biology, however, it is important to devise means for predicting the relative population viability of different primate taxa in various habitats. Some primatologists have recognized this and have attempted to correlate ecological and/or demographic parameters with endangerment or population decline (Happel, Nass, and Marsh 1987; Marsh, Johns, and Ayers 1987; Johns and Skorupa 1987; Dobson and Lyles 1989; Robinson and Redford 1989). However, conclusions remain elusive due to the multiplicity of environmental parameters and genetic and behavioral traits relative to the sophistication of mathematical treatments. Placing priorities for conservation action must follow an understanding of the characteristics of relatively successful and unsuccessful populations of primates. For Asian primates, it is particularly critical to address the issue of vulnerable taxa because the region's dense human population, coupled with exploding economic activity, means that primates will be crowded into ever-smaller habitat where intrusive methods of management such as translocations and diet supplementation will be necessary (Western 1989; Conway 1989).

Some obstacles exist in the development of a means to predict vulnerability. First is the problem of definition of terms. Both *rarity* and *vulnerability* are words that have been used in conjunction with primates whose populations are in danger of decline. Rarity is a characteristic that depends on the interaction of the habitable sites available to a species, population, or form, and traits intrinsic to it. It is a general term covering several patterns of relative abundance and distribution. Understanding the causes of rarity in a particular species is key to assessing its vulnerability because rare populations are particularly liable to extinction (Nitecki 1984). Vulnerability is the intersection of rarity with anthropogenic events; various taxa have particular idiosyncrasies that respond negatively to certain human activities.

Second, there is the problem of choosing the unit of concern. Species are typically chosen but are in part artifacts of taxonomy (e.g., there is more genetic difference between a white-eyed vireo and a yellow-eyed vireo than between chimps and humans; Diamond 1986). Some primate conservationists have recognized this problem by placing conservation priority on species with fewer closely related taxa (Oates 1986; Eudey 1987).

Finally, there is uncertainty about the correct context for assessing vulnerability if the relative priority of conservation is to determine the likelihood of extinction of species or of whole ecosystems. The most common recommendation for reserve design is to optimize diversity by maximizing chances for invasion and minimizing emigration and loss through the creation of large protected areas with a minimal circumference (e.g., Diamond and May 1976). Yet as Harper (1981, 201) states,

> Conservationists may often be more concerned to guard a particular species against extinction rather than to conserve diversity per se. There is no reason to suppose that any conservation policy designed to maintain diversity will be appropriate for maintaining the presence of a particular, specified rare species. Indeed, the optimal shape, size and condition of a reserve appropriate for a particular rare species may, in every case, be biologically specific and peculiar to its life-cycle pattern, its mode of dispersal, its reaction to predators and all of those other peculiarities that distinguish the biology of individual species. Certainly, the rules for maintaining community diversity may be quite different from those appropriate for safeguarding particular species and these latter rules may be so species-specific as to defy generalization.

In the practice of conservation, managers are on the one hand mandated to protect vulnerable species (e.g., the Endangered Species Act of 1983), while solutions from international agencies (e.g., United States Agency for International Development 1987) offer funds for the protection of biological diversity. Clearly, identifying which communities are vulnerable, either because they are assembled in such a way that they are sensitive to the loss of certain species or because species introduced by humans can have a great negative impact (Pimm 1987; Western et al. 1989), is a priority for the practice of good conservation. For the present, however, it is species that we have more developed expertise in managing.

The following discussion will focus on species, namely the Old World monkeys, or catarrhines, of Asia as the units of conservation concern

(Figure 12.1) rather than communities. Species can be important because they are "keystones" whose loss is likely to trigger the greatest number of secondary extinctions, because they are directly or indirectly economically important, or because they are charismatic to humans and rally support among people for protection of the ecosystems they represent. Certainly, many primates are among the last category of importance. Far too little is known about keystone biological species, and special efforts must be devoted to identifying them because the fate of so many other species is tied to them. The important role of animal species to the viability of plant species is one that has been studied broadly, because half or more of some rain forests' canopy trees use mammals and birds for seed dispersal (van der Pijl 1972). To date, however, there are no published reports of the role of Asian primates in seed dispersal. The following brief discussion will therefore draw on African and neotropical

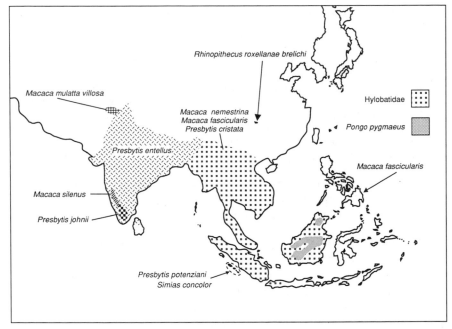

Figure 12.1
Map of Asia showing the location of primates mentioned in this chapter. The following major islands have a distinctive array of primates species in addition to those already indicated on the map: Sumatra—Macaca fascicularis, M. nemestrina, and Presbytis cristata; Java—M. fascicularis and P. cristata; and Borneo—M. nemestrina, M. fascicularis, P. cristata, and Nasalis larvatus.

examples. Mangabeys (*Cercocebus* spp.) in the Kibale forest of Uganda are "quality seed dispersers" in that they, like other primates with cheek pouches, spit out seeds one at a time, so that resultant seedlings do not face competition (Struhsaker, pers. com.). However, many forest primates do eat seeds, and passage through the large intestine appears to destroy some kinds of seeds (Hladik and Hladik 1969) so that primates are sometimes seen as a threat to forests. Yet if one monitors tree germination and growth, the number of seedlings germinated exceeds by a thousandfold the number of adult trees. At least for the Kibale forest, highest mortality occurs in the sapling stage so that primates cannot be seen as a cause of significant tree mortality (Struhsaker, pers. com.). Furthermore, in the neotropics at least, the majority of seeds dispersed by primates and deposited on the forest floor in their dung are either killed by seed predators or moved by secondary dispersers (Chapman 1989). At an earlier stage in the reproductive cycle of trees, fruit and flower destruction by frugivorous primates reduces the number of seeds drastically. Ayres (1986), who works in the flooded forests of Amazonia, estimates that an average-sized troop of forty-five to fifty bald uakaris (*Cacajao calvus*) will reduce the fruit crop of one species of tree (*Echweilera albiflora*) in his study area by 60 percent. Yet this tree is the second-most abundant tree in the *várzea* forests. Ayres and Prance (1989) conclude that reduction of the flower and fruit crop causes the tree to divert more resources to the remaining fruits, thus increasing seed size. Larger seeds in turn may lead to a higher probability of successful germination.

VULNERABILITY

Although the division is somewhat arbitrary, it can be useful to divide consideration of vulnerability of Asian catarrhines into two broad categories that include (1) characteristics intrinsic to the species and (2) human-engendered causes of vulnerability. Similar human impacts on a habitat can affect species differently; therefore a species' vulnerability to extinction is partly a function of its own biology. The most important factor in a species' prognosis is population size: All else being equal, rare species are more likely to become extinct than common ones (Western et al. 1989). Rarity can be temporal, spatial, or both. If spatial, rarity refers to a low population density throughout a geographical range, or a higher-density but narrowly restricted geographical range (see chapter 2). Rabinowitz (1981) proposed a typology of rare species based on three characteristics: geographic range, habitat specificity, and local population size. Causes of rarity tied to a species' biology include factors intrinsic to genetics, behavior, and ecology.

GENETIC FACTORS

How genetic variability in wild primate populations relates to population size and rarity has not been studied. Only a few dozen endangered species, much less endangered primates, have been examined genetically. Of work done to date, no genetic studies of natural populations have produced any evidence of inbreeding among cercopithecines monkeys, where sex-biased dispersal prevents inbreeding both in any one social group and in the population as a whole. For species that ordinarily outcross in nature, the loss of heterozygosity during inbreeding generally is associated with inbreeding depression, a decline in viability, growth, and fertility (Vrijenhoek 1989). Small populations frequently lose genetic variability and become inbred. For example, the tigers in the Chitawan National Park in India have an average inbreeding coefficient of 33 percent (Soulé 1989).

No research has been done to investigate the genetic consequences of the rather different social organization found among colobines, the other major group besides cercopithecines within the catarrhines (Old World monkeys). Whereas *Macaca*, the cercopithecine found in Asia, is characterized by multimale, multifemale groups with dispersion by males, among colobines one also sees harems, fission/fusion groups, and dispersion by both sexes. In both colobines and cercopithecines, genetic differentiation that ultimately leads to speciation may be due to a number of behavioral factors, including nonrandom migration, group fission, and nonrandom mating (Melnick and Pearl 1987). Small populations and fragmentation of habitat can disturb migration and fission.

Attempts to remedy genetic problems of inbreeding by artificial means (e.g., through artificial introduction of genetic material from distant populations) can backfire because of the possibility of outbreeding depression. This occurs when fecundity and/or viability declines following intraspecific hybridization because genes in a local population are primarily adapted to the genetic environment defined by other genes. Although the phenomenon has not been described for Old World monkeys, the neotropical owl monkey (*Aotus trivergatus*) has several chromosomal races (Ma 1981; Galbreath 1983; Pieczarka and Nagamachi 1988), and reproductive success is enhanced when individuals of similarly numbered chromosomes mate.

Beyond outbreeding depression is the phenomenon of interspecific hybridization. Though common in baboons (*Papio* spp.; Phillips-Conroy and Jolly 1986), hybridization does not occur naturally among macaques (Yoshikubo 1985; but see report of a possible *M. nemestrina* – *M. fascicularis* cross of unknown origin in Kavanagh and Payne 1987) and has not been reported for wild colobines.

BEHAVIORAL AND ECOLOGICAL FACTORS

Behavioral Plasticity

If dispersal of individuals is key to the genetic viability of local populations, then the means of dispersal is tied to behavioral and ecological factors. For species that are territorial, such as gibbons, colobines, and macaques, individuals move into new habitats by joining an established group that has cultural knowledge of optimal foraging sites, safe sleeping sites, scarce minerals, and so on. As Milton (1986, 248) said of the woolly spider monkeys (*Brachyteles arachnoides*) at Barreiro Rico, Brazil,

> [They] do not eat leaves from most of the more common tree species in their environment. How do they know to avoid these leaves and seek out leaves from rarer species? We simply do not know the answer. How do woolly spider monkeys know where these rare trees are? How do animals know when to visit these trees? There is a tremendous amount of learning of some type involved in this pattern of leaf eating, regardless of whatever innate components are involved.

The same could be said for colobines in Asia's dipterocarp forests, because that family of trees (Dipterocarpaceae) provides almost no food for primates (Chivers 1986). The equatorial rain forests of the Mentawi Islands of Indonesia, for example, consist of up to 59 percent dipterocarps. On the Mentawis' Siberut Island, only 2 percent of the trees at one study site were from the legume family (Whitten 1982), important in most primate diets. Resultant scattered food sources give rise to the very low biomass of primates found in the Mentawis as well as to the social groups atypically small and atypically structured for colobines (Bennett 1988). The point is that reintroduction schemes may be possible only when there exist core groups of animals *in situ* and on whose culture introduced animals can rely. Behavioral studies to determine home range sizes, group size and composition, and methods of group transfer (e.g., van Schaik and van Noordwijk 1985) are necessary for reintroduction strategies to be devised. Behavioral factors may limit the speed and effectiveness with which newly transplanted animals colonize vacant habitats. For cercopithecines that live in multimale groups, such as the long-tailed macaques (*Macaca fascicularis*) studied by van Schaik and van Noordwijk or rhesus monkeys (*Macaca mulatta*) whose behavior I studied in Pakistan (Pearl 1982), the fact that males can employ several behavioral strategies for joining new groups means that introductions of new males from captivity or distant areas may be feasible. For the simakobu (*Simias concolor*) and the Mentawi Island langur (*Presbytis potenziani*), the monogamous social structure presents a greater problem for artificial augmentation of the population.

It is possible that behavior patterns developed in a different kind of habitat can be preadaptive to survival in a new one. Kavanagh (1980) describes the behavioral flexibility of vervet monkeys (*Cercopithecus aethiops*). Savanna-dwelling vervets invaded a forest habitat, bringing with them some behaviors that evolved in the savanna yet were suited to survival in forests better than those of the original forest-dwelling vervet population. Compared to their forest-dwelling conspecifics, savanna vervets had less-predictable home ranges, a less-detectable call repertoire, stronger male vigilance, and the custom of hiding from dogs rather than confronting them. Forest vervets were then observed to adopt these different traits from the savanna monkeys. This flexibility refutes the assumption common in the 1960s and 1970s that all aspects of animal biology is determined by species-specific biological stratification and ecological niche (Dunbar 1988). A key finding of the 1980s was that there can be tremendous variation around species-typical means for most primates' ecological and social strategies. Dunbar laments the irony that just as we are coming to appreciate monkeys' tremendous responsiveness to changing environmental and demographic conditions, our human activities are exceeding the monkeys' capacity to respond.

Stress

Stress, a response when the potential for behavioral flexibility has passed, can be a factor in failure to reproduce, reduced foraging efficiency, suppressed immune responses, and therefore a resultant population decline. Heart rate studies on Rocky Mountain sheep (*Ovis canádensis*; Hutchins and Geist 1987) revealed that even brief disturbances from such human activities as logging and aircraft sounds had long-lasting effects on the animals. Chivers (1986) notes that although gibbons (*Hylobates* spp.) maintain their territories after logging, the stress associated with it affects breeding, resulting in a latent decline in population size that can extend twenty years. MacKinnon (1974) noted a decline in birth rate in a population of orangutans (*Pongo pygmaeus*) in a logged area and speculated that stress and overcrowding was the proximate cause. One antidote to some kinds of stress (e.g., airplane noise) can be habituation. Other sources of stress cannot be eliminated so easily. Hutchins and Geist point out that consumptive (i.e., hunting) and nonconsumptive uses of wildlife are incompatible, because when humans are seen as a threat, their presence causes stress. Thus schemes of "sustainable harvest" of primates may be intrinsically unsustainable.

Ecological Factors

Social organization is inextricably linked to ecology. Some behavioral-ecological parameters that might predict vulnerability include degree of seasonality of food resources and climate, degree of selectivity in food choice, and degree of tolerance for changes in substrate. Many mountain-dwelling ungulates (Hutchins and Geist 1987) and primates (*M. mulatta villosa*, Pearl, pers. obs.; *Rhinopithecus roxellanae brelichi*, Xie Jiahua, pers. com.) perform seasonal altitudinal migrations. For conservation objectives, it is vital that patches of seasonally used habitats, and unobstructed access between them, be conserved (Hutchins and Geist 1987). Migrations of many species, including orangutans, occur in the dipterocarp forests of southeast Asia because of the phenomenon of mast fruiting, or highly synchronized seed crops (Lisa Curran, pers. com.; John Mitani, pers. com.) from region to region. Such movements can skew animal censuses such that very low animal biomass is reported for these forests when they are devoid of fruits.

Diet

There is a broad range of primate diets. Some primates are *generalists* that feed on a wide range of food parts or food species (or both), whereas others are *specialists* that eat few parts and/or species. Animals in either category that make rapid and dramatic changes in their diet can be called *opportunists* (Richard 1985). The buffy saki (*Pithecia albicans*), for example, more of a generalist, feeds on a wide range of fruits, arils, seeds, leaves, and probably insects (Johns and Skorupa 1987). In contrast, the black saki (*Chiropotes satanas*) is more of a specialist and has a much more exclusively frugivorous diet. The black saki's most important fruit trees are felled in logging schemes, and as a result this species is very rare in logged forests, whereas the buffy saki survives well after such disturbance (Johns and Skorupa 1987). This is because the buffy saki not only has a broader diet, but also possibly because it can make rapid changes in the relative composition of various items in its diet.

Primates also can be more or less specialized in terms of their substrate use as well as their food resources. Animals that habitually use the ground in addition to trees can cross open terrain with far less stress than an exclusively arboreal species. Thus, if a road is cut through a forest, for example, any home range on the other side of the road is effectively cut off for the species more vulnerable in this regard. Therefore, to coexist with humans, it is better for primates *not* to be exclusively arboreal. Diminished habitat, however, regardless of substrate, in turn contributes to reduction in population numbers.

ECOLOGICAL CORRELATES OF ENDANGERMENT

Several researchers have attempted to match factors such as arboreality, home range size, group size, diet, habitat type, geographical range, diurnality versus nocturnality, body size, demography, and geological age of taxon to the degree of endangerment of a species (Happel et al. 1987; Johns and Skorupa 1987). Happel and her co-workers found that geographical ranges of under 100 km² were more frequent among primate species listed as endangered by the International Union for the Conservation of Nature and Natural Resources [World Conservation Union] Red Data Book (IUCN 1982). In other words, few threatened species have very large geographical ranges, and many threatened species have very small geographical ranges. Gestation length, in turn, was found to be a predictor of geographic range as well as of population density. Rare primates have either longer- or shorter-than-average gestation lengths, with short gestation correlated positively with high density. Other factors were not significant predictors of rarity (Happel et al. 1987).

In a review of studies of primate survival in disturbed habitats, Johns and Skorupa (1987) noted that large-bodied frugivores were most vulnerable but that the data formed an insufficient basis for predicting the outcome of primate species subjected to habitat disturbance. The low predictive power of ecological parameters on survival in such situations, the authors suggest, is perhaps due to the multiplicity of factors involved to the extent that no one factor is dominant. Robinson and Redford (1989) have also investigated body size in relation to vulnerability to local extinction. They conclude, after examining body size, diet, and population variation in 103 neotropical mammal species, that diet may predict extinction indirectly through its effects on body mass. Large animals occur at lower densities and have greater variation in density with location compared to small-bodied mammals. Thus, large primates may be relatively more vulnerable to localized extinction because human disturbance may differentially affect their densities. In this as well as other studies that aim to interpret the effects of ecological parameters on primate vulnerability, it is nearly impossible to either measure or exclude the human factor. For example, the fact that large-bodied animals are vulnerable compared with smaller counterparts is impossible to test wherever there is or has been hunting pressure, which selects for large prey.

EXTRINSIC FACTORS

A number of specific human activities have an impact on primate numbers (Table 12.1). These can be termed *extrinsic* causes of vulnerability

Table 12.1

Human activities that have an impact on primate numbers. Note that primary effects can lead to secondary impacts. For example, logging may fragment habitat, which in turn may reduce opportunities for migration and disturb gene flow.

Activity	Effects
Logging	Removal of substrate Reduction of diet Fragmentation of habitat Reduction in gene flow Inbreeding No access to isolated limiting resources (e.g., scarce minerals) Change in forest/open land ratio Changes in relative populations of different species Stress Diminished reproduction
Hunting/ removal	Reduction in density Increased demographic fluctuations Stress
Nontimber resource collection	Reduction in food resources Removal of substrate for prey species (e.g., insects)
Silvicultural treatments	Poisoning of noncommercial trees Poisoning of animals who consume products from these trees
Coastal zone development	Removal of mangrove Elimination of proboscis monkeys
Forest clearance	Removal of substrate, food Elimination of forest species
Shifting cultivation	Creation of open areas Increase of open-space dwellers Increase in crop raiders Reduction of forest species

as opposed to the genetic, behavioral, and ecological factors discussed previously. Various forms of habitat destruction are the foremost extrinsic source that produces vulnerability, ranging from outright destruction for an incompatible human use of the land, to isolation and insularization, to minimization of habitat to some threshold of long-term nonviability for primate inhabitants. Other human activities are directed squarely at removing primates as agricultural pests, as food for human or pet consumption, as bait, or, in the event of live capture, as pets or laboratory research subjects. Six categories of forest disturbance (Marsh et al. 1987) have had a negative impact in Asia: (1) removal of selected plant products by traditional means, (2) clearance or damage of the under-story, (3) changes in water regimes (such as irrigation or dams), (4) shifting cultivation, (5) selective commercial logging, and (6) large-scale forest clearance.

The first of these categories presents a double-edged sword for conservation practitioners. Although nontimber products provide a justification for forest protection, if the markets for them are successfully developed, the result could be overexploitation or clear-cutting for areas of commercial cultivation. Understory clearing, often for fodder for domestic animals or for plantations of shade-loving crops, destroys seedlings, and with them, the possibility of the regeneration of canopy trees. Changes in water regimes also present a long-term hazard to forest survival (or short-term, if a dam inundates former habitat!).

The fourth category of forest disturbance, shifting cultivation, increases the proportion of secondary forest at the expense of primary forest. In Asia, this change favors cercopithecines at the expense of colobines (Marsh et al. 1987). In peninsular Malaysia, the primate biomass in primary forest is dominated by colobines (*Presbytis* spp.), whereas secondary forest, riverine swamp, domestic orchards, padi fields, and so forth commonly host cercopithecines, particularly long-tailed macaques (Marsh et al. 1987). However, in most parts of primate habitat in Asia, habitat disturbance commonly takes the form of commercial logging as well as shifting cultivation. For open-country monkey species like most macaques (*Macaca* spp.) and the Hanuman langur (*Presbytis entellus*), *any* kind of disturbance that results in landscapes such as the meadowlike communities created by clear-cutting, grazing, and crop planting mimics the forest edge or savanna to which they originally were adapted. Goldstein (1984), citing preferential feeding on ground herbs that occur in disturbed sites, has argued that the rhesus monkey (*Macaca mulatta*) is a "weed" species—one that has spread with the disappearance of the forests of Asia during the Pleistocene glaciations and that continues to prosper in the face of anthropogenic deforestation.

The effects of logging on primate survival are tied to the type of logging practices and the requirements of the animals. One key factor in survival is whether or not timber sought by loggers is also a food resource for the monkeys in question. At one site in Malaysia, despite the fact that 45 percent to 50 percent of the forest was destroyed, primates survived (Johns and Skorupa 1987). In contrast, in a forest in Uganda with the same degree of damage but largely to trees that produced food for one or more resident species of primate, significant population declines resulted among five of the seven resident species (Johns and Skorupa 1987). Other discussions of logging and primate recovery (Marsh and Wilson 1981; Johns 1985; Marsh et al. 1987) also describe considerable differences between various species of gibbon, langur, and macaque and their pattern of recovery. Caldecott (1981) found that islands of tropical rain forest surrounded by monocultures of food value sometimes have a higher density of some primates (e.g., silvered leaf monkeys, *Presbytis cristata*; pig-tailed macaques, *Macaca nemestrina*).

Hunting of primates is not as great a problem in Asia as in Africa or South America because three of the major religions there (Islam, Buddhism, and Hinduism) have proscriptions against eating primates. Nonetheless, hunting by aboriginal groups is common in Indonesia, and Chinese hunt all species of monkeys in China and in other Asian countries in which they live (Mittermeier 1987; Bennett, pers. com.). In Asia, the major direct threat to primates is in their commercial trade, in that the majority of commerce for biomedical use in monkeys is in macaques, an Asian genus.

RESISTANCE TO REMEDIES: THE CASE OF TRANSLOCATION

Given the high numbers of primates of conservation concern relative to resources for their protection, vulnerable populations must be evaluated for their potential to profit from conservation action. One such action is translocation—that is, the intentional release of animals into the wild to establish, re-establish, or augment an existing population. Translocations are coming into increasing use in the face of dwindling natural populations of primates and other animals (Kleiman 1989). A survey of recent translocations (Griffith et al. 1989) reveals that the transfers least likely to succeed are those of captive-born animals of endangered species. This is bad news for those who see zoos as "arks" for holding animals over several generations until their habitat can be restored and secured. Other poor prospects for success in translocation are animals introduced into suboptimal habitat at the periphery of or outside their historical ranges. Successful translocations occur significantly more often where the released animals are drawn from an increasing rather than a stable or dwindling source population. Also, herbivores are more successful in establishing themselves than carnivores and omnivores (Griffith et al. 1989).

It is important to monitor translocations very closely so that we can learn from successes and failures. Yet most releases of primates have been unintentional, unmonitored, or both. Among approximately 175 species of primates, well-monitored reintroduction programs have occurred for very few. Three well-known examples are those for orangutans (*Pongo pygmaeus*; Aveling and Mitchell 1982), chimpanzees (*Pan troglodytes*; Brewer 1978), and golden lion tamarins (*Leontopithecus rosalia*;

Dietz 1985) (see Konstant and Mittermeier 1982 for a survey of a number of introductions, including haphazard and unintentional ones, of several neotropical species). All reintroduced animals spent a long time in captivity or were captive-born—the least likely candidates for success. Because of their high capacity for learning, however, great apes may fare well after reintroductions following very costly and lengthy training programs, the cost and duration of which in themselves comprise a major obstacle to implementation. With or without training sessions (Dietz et al. 1988; Benjamin Beck, pers. com.), the translocated golden lion tamarins had poor survival success. Of fifteen individuals introduced, only three remain alive today, and all receive an artificially supplemented diet (Beck, pers. com.). Between 1959 and 1961, fifteen to twenty black-headed spider monkeys (*Ateles geoffroyi*) were reintroduced to Barro Colorado Island. All but four (one male, three females) died. Twenty-five years later, a census of this highly protected reserve reveals only fifteen black-headed spider monkeys, including three of the original introduced animals (Milton 1986). For primates with complex diets, a successful reestablishment is a long-term prospect.

Even given animals with flexible behavior and drawn from an expanding wild population, without high habitat quality translocations of any kind are unlikely to be successful. Yet some findings suggest that prospects for viable populations depend on the specialization of a primate's diet. In suboptimal habitat, it has been found empirically that folivorous primates are more likely to be successful than frugivores. The biomass of frugivorous primates in a disturbed forest at Kibale, Uganda, averaged 59 percent lower than in undisturbed forest, whereas folivorous primates were down 39 percent (Johns and Skorupa 1987).

One can conclude from this discussion that although translocations are unlikely to be very successful for endangered primates, there are more and less likely candidates for such programs. Folivores and insectivores can find adequate diets in at least one kind of suboptimal habitat more easily than frugivores. As mentioned earlier, in addition to diet, behavioral traits related to type of group transfer or tolerance of potentially stressful stimuli are also factors that would affect chances for successful reintroduction. However, even ideal candidates for translocation may spell trouble if there are conspecifics or other species that might suffer from disease, competition, or genetic pollution (Konstant and Mittermeier 1982; Aveling and Mitchell 1982).

Beyond nonhuman primate-centered obstacles, there are human-engendered blocks to conservation remedies. On a large scale, political chaos and warfare have blocked conservation efforts. Closer to the animals themselves, corrupt protection officials, insufficient secure habitat, or public hostility pose threats to endangered species. Each of these factors currently plays a role in the decline of species in various places throughout southeast Asia.

CASE STUDIES

Now in place are conservation activities on behalf of a variety of endangered Asian primate species. In light of the foregoing general discussion of potential sources of vulnerability to extinction, I will now consider the long-term opportunities for more intensive intervention relative to the obstacles to the survival of two of the rarest Asian primate species: a cercopithecine, the lion-tailed macaque (*Macaca silenus*) and a colobine, the proboscis monkey (*Nasalis larvatus*). Both are large, attractive animals, potential "charismatic keystone species" for conservation that historically have enjoyed local tolerance and live in regions not currently beset by warfare.

*The Lion-Tailed Macaque (*Macaca silenus*)

Lion-tailed macaques are unusual among the macaques in that they are obligate rain-forest dwellers. They are nearly completely arboreal (Mangalraj Johnson 1985), unlike their congenerics, which are adapted to forest clearings and edges. *Macaca silenus* is confined to rain forests covering about 4,683 km² (5.8% of total forest cover) along the Western Ghat mountain ranges in the Indian states of Kerala, Karnataka, and Tamil Nadu. Most of this rain forest is degraded and widely scattered amidst clearings for plantations, human settlement, and hydroelectric schemes (Kumar 1985). Like other macaques, *M. silenus* lives in multimale groups characterized by female philopatry and male migration. Therefore, their long-term genetic viability depends on the ability of males to move from one group to another. This exchange has grown increasingly difficult as habitat continues to fragment and overall numbers decrease. Again, like many macaques, *M. silenus's* diet consists of a wide variety of fruits and arthropods (Mangalraj Johnson 1985). As might be predicted from the example of frugivores versus folivores of suboptimal habitat at Kibale, Uganda (Johns and Skorupa 1987), macaque numbers have continued to dwindle, whereas the sympatric Nilgiri langur (*Presbytis johnii*) has been growing in number, spreading into adjoining dry deciduous forests and even teak plantations (Mangalraj Johnson 1985). Theoretically, *M. silenus* numbers might grow because of arthropods that might abound in regeneration forest, but logging and cardamom-growing practice result in the removal of brushy plants and lianas that harbor prey insects.

Kumar (1985) makes the sobering statement that the pattern of declining numbers of *M. silenus* matches that of southeast Asian forest primates in every way but temporally. The reason for their precarious status is that habitat destruction in India began earliest (perhaps with the Neolithic, culminating in the disastrous wholesale clear-cutting for plantation and logging that began with the English colonial era), and the current situation of fragmented bits of degraded habitat is what awaits the future of southeast Asian primates. Land use in lion-tailed macaque habitat involves clearing all but economically useless trees for cardamom plantation, or clearing land completely for agricultural or dam-building schemes. In no case is collection of nontimber forest products (except for brush removal) an important land use. In other words, no economic incentive exists for rain-forest maintenance. Any keystone role lion-tails once played in forest regeneration is no longer of economic interest to local people.

Regarding genetic factors, preliminary research (Jolly and King 1985) reveals that lion-tails are not different from other macaques in their relatively high genetic variability and in a lack of any indication of inbreeding, which is characteristic of other macaques in multimale societies (Melnick, Pearl, and Richard 1984). Their social organization, as shown by their living in aggregated groups with promiscuous mating, appears relatively favorable for prospects for survival. This prediction is based on mathematical models of primate demographics and long-term viability (Dobson and Lyles 1989) that suggest the maintenance of such groups at smaller population densities than of species with more solitary and monogamous habitats. Lion-tailed macaques eat a very broad range of plants, insects, and other invertebrates (Mangalraj Johnson 1985), even if they are not as flexible in diet as their sympatric colobines. Their social behavior is flexible: The demographic composition of groups is variable (Mangalraj Johnson 1985), and the nature of intergroup encounters is also highly plastic (Kumar and Kurup 1985).

Lion-tailed macaques have been bred in captivity since 1839 (Conway 1985); there were 475 individuals in zoos and private collections around the world as of January 1989 (Laurence Gledhill, pers. com.). Some have been released into semicaptivity at the New York Zoological Society's endangered-species breeding station, and some are free-ranging on St. Catherine's Island off the coast of Georgia.

What should our response be to the current threatened status of lion-tailed macaques? Their chief threat is their low numbers and fragmented distribution. Thus the most promising remedy, theoretically, is to augment numbers and create more habitat. Augmenting numbers would appear to be feasible technically, given their social organization and the fact of varied modes of male transfer, as well as the stock of genetically varied, captive individuals that could be used in such a program. Stress from such human activities as brush removal or airplane overflights, as two examples, might be limiting population growth considering the high mortality among infants in plantation/forest habitat compared to those in all-forested habitat (Mangalraj Johnson 1985). A priority for research on captive animals, especially for the released group on St. Catherine's Island, would be an understanding of the effect of stress on their behavior.

Another critical subject for research on the free-ranging St. Catherine's Island lion-tailed macaques is their diet. Citing the unforeseen flexibility in feeding ecology of whooping cranes fostered by sandhills, Conway (1985) has suggested that it may be possible for captive-bred lion-tailed macaques to be introduced successfully into secondary forests outside the original range of wild lion-tails. Before going to the expense and through the moral implications of managing monkeys "at liberty" as opposed to in their natural setting, it would be useful to measure any lessening of dependence on provisioning in a captive-born group. The purpose of such a study would be to determine whether it would be possible for lion-tailed macaques to use a mosaic of different habitats. The record of introductions of captive animals outside historical habitats is poor, and *M. silenus* has not expanded its range beyond rain forests.

The foregoing discussions assume that the original, pristine rain-forest habitat in which lion-tailed macaques evolved will continue to succumb to human activity but that a remnant of contiguous forest, probably the Ashambu Hills and Silent Valley of Kerala State, will be preserved. It is far more economical to preserve animals in nature than in captivity. There are not enough spaces in zoos to maintain viable populations of even a small fraction of our endangered species (Conway 1986), even if we agreed that an ark of animals with no place to go was an acceptable goal. Thus a sensible conservation program for lion-tailed macaques would be to extend the reserve boundaries as well as to provide effective protection for existing reserves. But beyond this broad prescription, detailed research on population augmentation, including the introduction of preconditioned animals into uncharacteristic habitat and ongoing monitoring of the genetic and demographic structure of the base population and the world captive reservoir, will be necessary. If it is not possible to extend usable habitat beyond existing rain forest, or to augment artificially the existing wild population, conservation dollars should be targeted elsewhere.

The Proboscis Monkey (Nasalis larvatus)

Proboscis monkeys, like lion-tailed macaques, have a highly restricted range. They are confined to mangrove, riverine, and peat swamp forests off the coastal lowlands on Borneo and two other, much smaller islands of the northeast coast of Borneo (Bennett 1991). Both species therefore have very patchy distributions, although in the case of proboscis monkeys, it is naturally so rather than caused by the anthropogenic fragmentation of rain forest as in southern India. Nonetheless, the number of forest patches is declining, and because the distance between these patches has increased at the hands of human activity, the possibility of migration between patches has been severely reduced. Extensive surveys of proboscis monkeys has been accomplished only in Sarawak (Bennett 1986), and their status in Kalimantan and Sabah, which comprise about 75 percent of their range, is relatively unknown.

Unlike the lion-tailed macaque, the proboscis monkey is a highly selective feeder (Bennett and Sebastian 1988; Yeager 1989) that consumes only selected young leaves, seeds, and nonsweet fruits. Like all colobines, it has a sacculated stomach and fermentation digestion system. Its gut is enormous, over twice the size of other colobines (Chivers and Hladik 1980), and it also has a relatively large body size (Bennett in press). As a result, it requires a very large home range to find enough appropriate food each day.

The social organization of proboscis monkeys has been studied in fresh-water peat swamp habitat (Yeager 1990) and in mangrove forest (Bennett and Sebastian 1988). In both areas, the monkeys are found in either harem groups numbering from three to twenty-three members, consisting of an adult male, several adult females, and infants, juveniles, and female adolescents associated with the adult females; or in all-male aggregations of adults, adolescents, and juveniles. Whereas in peat swamp forest the composition of the harem groups is stable, Bennett (1991) reports that in her study area in mangrove forest, not only do males leave their natal groups, but at least some females also move between groups relatively often. All-male aggregations change composition almost daily.

One-male groups and all-male bands form a common pattern of social organization for colobines (Yeager 1990). Female transfer, however, is uncommon (Moore 1984). Also unlike most colobines, proboscis monkey groups frequently coalesce during the day and, more frequently, in their sleeping trees on riverbanks at night (Bennett 1991). We can conclude that proboscis monkeys have an extremely flexible variation on the colobine norm for social organization, perhaps in response to the relatively difficult demands of locating sufficient quantities of food for their extremely specialized diet in depauperate and seasonally variable habitat.

Colobines in general are notoriously difficult to keep in captivity, and the world captive population of proboscis monkeys as of 1986 consisted of twenty-six animals in six zoos. Of these, sixteen were born in captivity (*International Zoo Yearbook,* vol. 26). In other words, there are no more than ten founders for this population. In addition, no genetic work has been performed on these zoo animals nor on their wild counterparts. Therefore, though it might be possible from the standpoint of behavior to augment artificially a wild group, given that female transfer occurs naturally, the reservoir of animals and knowledge of their genetic parameters is inadequate.

Threats to proboscis monkeys on Borneo stem from the fact that much human development activity is centered along rivers, river mouths, and coasts (Bennett 1991). Selective logging in itself does not harm proboscis monkeys because most food trees are not commercially important. However, the common practice of silvicultural treatment to destroy noncommercial tree species, which follows much selective logging, is extremely detrimental. Another major threat to proboscis habitat is the commercial exploitation of mangroves to produce wood chips (Bennett 1991). In fact, it has been said that plans exist for the conversion of the entire mangrove fringe of Borneo for commercial use (Robert Goodland, pers. com.).

Before any conservation strategy is considered, accurate censuses must be carried out. Estimates calculated in the late 1980s of proboscis monkeys in Kalimantan have ranged from a high of 260,000 with 25,600 within reserves to 36,100 with 4,800 in reserves (cited in Bennett 1991). There are several very large reserves (e.g., Samunsam Wildlife Sanctuary in Sarawak or Gunung Palung in Kalimantan), but no reserve in any part of Borneo has been proven to encompass the entire range of any population of proboscis monkeys. Thus, following surveys, reserve boundaries must be renegotiated to include more habitat useful to proboscis monkeys. Because stocking from captive populations is impossible, relocations of monkeys from doomed habitats to enlarged reserves should be considered. This will require genetic assay of both local populations and field experimentation while numbers of monkeys remain high enough for such manipulation to be defensible. Despite the much higher numbers of proboscis monkeys than of lion-tailed macaques, their prospects for survival seem no brighter given their narrow niche, inadequate reserves, and poor response to human manipulation.

CONCLUSION

Despite the seemingly poor prognosis for lion-tailed macaques and proboscis monkeys, in neither case have we come to the end of the road in terms of conservation efforts. As conservation practitioners, people at the New York Zoological Society are engaged in both field and captive studies of both species. For example, in the field, status surveys are being conducted in Karnataka, India, and in Sarawak and Sabah, Borneo. Our nutrition department is conducting analyses of wild proboscis diet and is using the results to better manage the Zoo's captive breeding population (Ellen Dierenfeld, pers. com.).

With no time to lose, studies on endangered primates must become more tightly focused on issues proximate to *pragmatic* conservation action. Conservation research on Asian primates also must be based on analyses that consider not only the current status, but characteristics of the primate and its habitat that have led to its vulnerability. Some promising avenues of research that have been discussed in this chapter include (1) cytogenetic analyses to determine taxa of conservation concern and to identify genetic obstacles to successful population growth; (2) ecological studies on the role of primates in forest regeneration to justify the protection of primates within valuable forests; (3) behavioral studies to determine home range sizes, group size and composition, and methods of group transfer in less-known species—such research is a necessary prelude to translocation or reintroduction; (4) studies of the relative response of a variety of species to the same stressful stimulus to identify the relative risk of different primates in various land-use schemes; and (5) description and analysis of the genetic and demographic structure of local populations and captive reservoirs to monitor a species' health in the face of human encroachment and various conservation measures. It is my hope that readers will embark on work along these lines, because it can be both intellectually challenging and morally fulfilling.

ACKNOWLEDGMENTS

I appreciate the help of Liz Bennett and Carey Yeager in sharing results of their studies of proboscis monkeys with me, as well as their views on rarity in primates, and I also thank P. Fiedler, R. Sukumar, and M. Ayres for helpful comments.

LITERATURE CITED

Aveling, R., and A. Mitchell. 1982. Is rehabilitating orangutans worthwhile? *Oryx* 16:263–70.

Ayres, J. 1986. *Uakaris and the Amazonian flooded forest.* Ph.D. dissertation, University of Cambridge, U.K.

Ayres, J., and G. Prance. 1989. On the distribution of pithecine monkeys and Lecythidaceae trees in Amazonia. Unpublished ms. Wildlife Conservation Intl., N.Y., and Royal Botanical Gardens, Kew.

Bennett, E. 1986. *Proboscis monkeys in Sarawak: Their ecology, conservation, status and management.* New York: World Wildlife Fund Malaysia, Kuala Lumpur/New York Zoological Society.

———. 1988. Odd nosed colobines: Their ecology and conservation status. Unpublished ms. Wildlife Conservation Intl. (N.Y.) and World Wildlife Foundation—Malaysia Kuala Lumpur.

———. In press. *Proboscis monkey, Nasalis larvatus.* Monograph for series *Illustrated Monographs of Living Primates,* ed. J. Kaiser and M. van Roosmalen. Netherlands: Instituut voor Ontwikkelingsopdrachten.

Bennett, E., and A. Sebastian. 1988. Social organization and ecology of proboscis monkeys (*Nasalis larvatus*) in mixed coastal forest in Sarawak. *Int. J. Primatol.* 9:233–55.

Brewer, S. 1978. *The chimps of Mt. Asserik.* New York: Alfred Knopf.

Caldecott, J. 1981. Findings on the behavioural ecology of the pig-tailed macaque. *Malays. Appl. Biol.* 10:213–20.

Chapman, C. 1989. Primate seed dispersal: The fate of dispersed seeds. *Biotropica* 21:148–54.

Chivers, D.J. 1986. Southeast Asian Primates. In *Primates: The road to self-sustaining populations,* ed. K. Benirschke, 127–52. New York: Springer-Verlag.

Chivers, D.J., and C.M. Hladik. 1980. Morphology of the gastrointestinal tract in primates: Comparisons with other mammals in relation to diet. *J. Morphol.* 166:337–86.

Conway, W.G. 1985. Saving the lion-tailed macaque. In *The lion-tailed macaque: Status and conservation,* ed. P. Heltne, 1–12. New York: Alan R. Liss.

———. 1986. The practical difficulties and financial implications of endangered species breeding programs. *Int. Zoo Yearbook* 24/25:210–19.

———. 1989. The prospects for sustaining species and their evolution. In *Conservation for the 21st century,* ed. D. Western and M. Pearl, 199–209. New York: Oxford University Press.

Diamond, J. 1986. Foreword. In *Primates: The road to self-sustaining populations,* ed. K. Benirschke. New York: Springer-Verlag.

Diamond, J., and R. May. 1976. Island biogeography and the design of natural reserves. In *Theoretical ecology,* ed. R. May, 163–86. Oxford: Blackwell Scientific.

Dietz, J., M. Castro, B. Beck, and D. Kleiman. 1988. The effects of training on the behavior of golden lion tamarins reintroduced into natural habitats. *Int. J. Primatol.* 8:425.

Dietz, L. 1985. Captive-born golden lion tamarins released into the wild: A report from the field. *Primate Conserv.* 6:21–27.

Dobson, A., and M. Lyles. 1989. The population dynamics and conservation of primate populations. *Cons. Biol.* 3:362–80.

Dunbar, R.I.M. 1988. *Primate social systems.* Ithaca, N.Y.: Cornell University Press.

Eudey, A.A. 1987. *Action plan for Asian primate conservation: 1987–91.* Gland, Switzerland: World Conservation Union/Species Survival Commission Primate Specialist Group.

Galbreath, G.J. 1983. Karyotypic evolution in *Aotus. Amer. J. Primatol.* 4:245–51.

Goldstein, S.J. 1984. Feeding ecology of rhesus monkeys, *Macaca mulatta*, in northwestern Pakistan. Ph.D. dissertation, Yale University, New Haven.

Griffith, B., J.M. Scott, J.W. Carpenter, and C. Reed. 1989. Translocation as a species conservation tool: Status and strategy. *Science* 245:477–80.

Happel, R.E., J.F. Noss, and C.W. Marsh. 1987. Distribution, abundance, and endangerment of primates. In *Primate conservation in the tropical rain forest*, ed. C. Marsh and R.A. Mittermeier, 63–82. New York: Alan R. Liss.

Harper, J.L. 1981. The meanings of rarity. In *The biological aspects of rare plant conservation*, ed. H. Synge, 89–96. New York: John Wiley & Sons.

Hladik, A., and C.M. Hladik. 1969. Rapports trophiques entre végétation et Primates dans la forêt de Barro Colorado (Panama). *Terre et Vie* 23:25–117.

Hutchins, M., and V. Geist, 1987. Behavioral considerations in the management of mountain-dwelling ungulates. *Mountain Res. Devel.* 7:135–44.

Johns, A.D. 1985. Primates and forest exploitation at Tefe, Brazilian Amazonia. *Primate Conserv.* 6:27–29.

Johns, A.D., and J.P. Skorupa. 1987. Responses of rain forest primates to habitat disturbance: A review. *Int. J. Primatol.* 8:157–91.

Jolly, C., and A. King. 1985. Serogenetic analysis and the conservation of lion-tailed macaques. In *The lion-tailed macaque: Status and conservation*, ed. P. Heltne, 195–219. New York: Alan R. Liss.

Kavanagh, M. 1980. Invasion of the forest by an African savannah monkey: Behavioral adaptations. *Behaviour* 73:238–60.

Kavanagh, M., and J. Payne. 1987. Hybrid macaque in Sabah. *Primate Conserv.* 8:57.

Kleiman, D. 1989. Reintroduction of captive mammals for conservation. *BioScience* 39:152–61.

Konstant, W.R., and R.A. Mittermeier. 1982. Introduction, reintroduction and translocation of Neotropical primates: Past experience and future possibilities. *Int. Zoo Yearbook* 22:69–77.

Kumar, A. 1985. Patterns of extinction in India, Sri Lanka, and elsewhere in southeast Asia: Implications for lion-tailed macaque wildlife management and the Indian conservation system. In *The lion-tailed macaque: Status and conservation*, ed. P. Heltne, 65–89. New York: Alan R. Liss.

Kumar, A., and Kurup. 1985. Inter-troop interactions in the lion-tailed macaque, *Macaca silenus*. In *The lion-tailed macaque: Status and conservation*, ed. P. Heltne, 91–107. New York: Alan R. Liss.

Ma, N.S.F. 1981. Chromosome evolution in the owl monkey: *Aotus. Am. J. Primatol.* 14:255–63.

MacKinnon, J.R. 1974. The behavior and ecology of wild orangutans (*Pongo pygmaeus*). *Anim. Behav.* 22:3–74.

Mangalraj Johnson, T.J. 1985. Lion-tailed macaque behavior in the wild. In *The lion-tailed macaque: Status and conservation*, ed. P. Heltne, 41–63. New York: Alan R. Liss.

Marsh, C.W., A.D. Johns, and J.M. Ayres. 1987. Effects of habitat disturbance on rainforest primates. In *Primate conservation in the tropical rainforest*, ed. C.W. Marsh and R.A. Mittermeier, 83–107. New York: Alan R. Liss.

Marsh, C.W., and W.L. Wilson. 1981. *A survey of primates in peninsular Malaysian forests*. Final report. Universiti Kebangsaan, Malaysia.

Melnick, D., and M. Pearl. 1987. Cercopithecines that live in multimale groups. In *Primate societies*, ed. B. Smuts, D. Cheney, R. Seyfarth, R. Wrangham, and T. Struhsaker, 121–34. Chicago: University of Chicago Press.

Melnick, D., M. Pearl, and A. Richard. 1984. Male migration and inbreeding avoidance in wild rhesus monkeys. *Am. J. Primatol.* 7:229–43.

Milton, K. 1986. Ecological background and conservation priorities for woolly spider monkeys (*Brachyteles arachnoides*). In *Primates: The road to self-sustaining populations*, ed. K. Benirschke, 241–50. New York: Springer-Verlag.

Mittermeier, R.A. 1987. Effects of hunting on rain forest primates. In *Primate conservation in the tropical rain forest*, ed. C.W. Marsh and R.A. Mittenmeier, 109–46. New York: Alan R. Liss.

Mittermeier, R.A., and D.L. Cheney. 1987. Conservation of primates and their habitats. In *Primate societies*, ed. B. Smuts, D. Cheney, R. Seyfarth, R. Wrangham, and T. Struhsaker, 477–90. Chicago: University of Chicago Press.

Moore, J. 1984. Female transfer in primates. *Int. J. Primatol.* 5:537–89.

Nitecki, M.H., ed. 1984. *Extinctions*. Chicago: University of Chicago Press.

Oates, J.F. 1986. *Action plan for African primate conservation*. Gland, Switzerland: World Conservation Union/Species Survival Commission Primate Specialist Group.

Oates, J.F., S. Gartlan, and T. Struhsaker. 1982. A framework for planning African rain-forest primate conservation. *Int. Primatol. Soc. News* 1:4–9.

Pearl, M. 1982. Networks of social relations among Himalayan rhesus monkeys (*Macaca mulatta*). Ph.D. dissertation, Yale University, New Haven.

Phillips-Conroy, J.E., and C.J. Jolly. 1986. Changes in the structure of the baboon papio hybrid zone in the Awash National Park, Ethiopia. *Am. J. Phys. Anthropol.* 71:337–50.

Pieczarka, J.C., and C.Y. Nagamachi. 1988. Cytogenetic studies of *Aotus* from eastern Amazonia: Y/autosome rearrangement. *Am. J. Phys. Anthropol.* 14:255–63.

Pimm, S. 1987. Determining the effects of introduced species. *Trends Ecol. Evol.* 2:106–8.

Rabinowitz, D. 1981. Seven forms of rarity. In *The biological aspects of rare plant conservation*, ed. H. Synge, 205–17. New York: John Wiley and Sons.

Richard, A. 1985. *Primates in nature*. New York: W.H. Freeman.

Richard, A., and R. Sussman. 1987. Framework for primate conservation in Madagascar. *Am. J. Primatol.* 12:309–14.

Robinson, J., and K. Redford. 1989. Body size, diet, and population variation in neotropical forest mammal species: Predictors of local extinction? *Adv. Neotrop. Mammal.* 1989:567–94.

Soulé, M. 1989. Conservation biology in the 21st century: Summary and outlook. In *Conservation for the 21st century*, ed. D. Western and M. Pearl, 297–303. New York: Oxford University Press.

Templeton, A. 1986. Coadaptation and outbreeding depression. In *Conservation biology*, ed. M. Soulé, 105–116. Sunderland, Mass.: Sinauer.

USAID. 1987. Summary of AID biological diversity activities under the FY 1987 legislated earmark. United States Agency of International Development report 7/16/87.

van der Pijl, L. 1972. *Principles of dispersal in higher plants.* New York: Springer-Verlag.

van Schaik, C., and M.A. van Noordwijk. 1985. Male migration and rank acquisition in wild long-tailed macaques (*Macaca fascicularis*). *Anim. Behav.* 33:849–61.

Vrijenhoek, R. 1989. Population genetics and conservation. In *Conservation for the 21st century,* ed. D. Western and M. Pearl, 89–98. New York: Oxford University Press.

Western, D. 1989. Population, resources and environment in the 21st Century. In *Conservation for the 21st century,* ed. D. Western and M. Pearl, 11–25. New York: Oxford University Press.

Western, D., M. Pearl, S. Pimm, B. Walker, I. Atkinson, and D. Woodruff. 1989. An agenda for conservation action. In *Conservation for the 21st century,* ed. D. Western and M. Pearl, 303–24. New York: Oxford University Press.

Whitten, A.J. 1982. The ecology of singing in Kloss gibbons (*Hylobates klossi*), Siberut Island, Indonesia. *Int. J. Primatol.* 3:33–51.

Wolfheim, J. 1983. *Primates of the world: Distribution, abundance, and conservation.* Seattle: University of Washington Press.

Yeager, C. 1989. Feeding ecology of the proboscis monkey (*Nasalis Larvatus*) *Int. J. Primatol.* 10:497–530.

———. 1990. Proboscis monkey (*Nasalis larvatus*) Social organization: Group structure. *Am. J. Primatol.* 20:95–106.

Yoshikubo, S. 1985. Species discrimination and concept formation by rhesus monkeys *Macaca mulatta. Primates* 26:285–99.

CHAPTER
13

Genetic and Demographic Considerations in the Sampling and Reintroduction of Rare Plants

Essay by EDWARD O. GUERRANT, JR.

INTRODUCTION

We are witness to what paleontologists of the distant future may record as one of the few truly great mass extinctions of the last half billion years. Simultaneously, there is an increasing homogenization of regional biotas resulting from human-mediated long-distance dispersal, especially of weedy colonizing species. Thus these same future paleontologists could justifiably name our time the "Homogecene" epoch. Habitat destruction and fragmentation caused by human land use, combined with probable global warming and atmospheric ozone depletion, stress even further the ability of our remaining natural areas to support their native flora and fauna. We must act now if we are to bequeath to our distant descendants a world whose biodiversity is only minimally more degraded than the one we inherited.

Although species are becoming extinct at an alarming rate, hope springs eternal, and concrete steps are beginning to be taken to reduce the toll. A wide variety of individuals, private organizations, and governmental agencies worldwide are addressing themselves to different facets of the threat to the biodiversity we still enjoy.

This chapter reflects on one such conservation effort to forestall the extinction of individual rare species in the United States. It may correctly be argued that a species-level approach is almost necessarily an exercise in "crisis control" and that a habitat- or ecosystem-level approach is to be preferred. Nevertheless, given that our strongest legal measure available to protect biodiversity is the federal Endangered *Species* Act and that our understanding of the ecological dynamics of communities and ecosystems relative to population-level phenomena is limited, we can attain certain conservation goals by focusing on individual species and populations. Species-level efforts are not, however, a substitute for or an alternative to more comprehensive habitat- and ecosystem-based efforts: rather, they are complementary. So, too, must we pay more attention to ecological and evolutionary processes rather than lavishing the bulk of our intellectual and monetary resources on attempts to perpetuate as static entities aspects of landscapes that are inherently dynamic (see chapter 4).

To forestall the apparently imminent extinction of our most vulnerable plants is the common goal of the approximately twenty botanic gardens and arboreta throughout the United States that together compose a consortium known as the Center for Plant Conservation (CPC). This effort has two basic components:

1. To obtain and safely store off-site, or *ex situ,* genetically representative samples of seeds (or other propagules) without significantly compromising short-term survival prospects of the sampled populations. (Assuming this is completed successfully, the next task is as follows)
2. To be able to reestablish from these samples self-sustaining populations genetically comparable to those from which they were derived

An understanding of population genetics and plant demography is basic to the success of this effort, and, armed with empirical knowledge of the taxa and situation in question, we may well extend the longevity of populations and species. A large seed collection will not necessarily ensure that genetic diversity will be preserved, yet with proper management a sizable portion of the genetic diversity found in an ancestral population can be retrieved from an extremely small founding population (Templeton 1991).

It is important to emphasize that these *ex situ* species-based efforts are no substitute for on-site, or *in situ,* habitat or community-based conservation efforts. Rather, they offer a complementary tool to enhance the probability of success of these more comprehensive and absolutely irreplaceable *in situ* efforts. *Ex situ* samples serve as an insurance policy or hedge against population extinction and catastrophic loss of habitat. New populations can rise phoenix-like out of the ashes of extinction only if collections exist off-site. It is also critical to note that *ex situ* collections are not an end in themselves. Their ultimate value will be derived from how they are used and their effect, if any, on the long-term prospects for survival of rare plant species.

OBTAINING GENETICALLY REPRESENTATIVE COLLECTIONS

A symposium of population biologists and geneticists was held in 1989 by the CPC in part to address the question of how to obtain a genetically representative sample for conservation collections of endangered plants (Falk and Holsinger 1991). Recommendations that emerged are summarized in Table 13.1. Note that they are arranged hierarchically, focusing on four levels of biological organization. From most to least inclusive, they are (1) from which species should samples be taken; (2) how many

Table 13.1
Summary of recommended guidelines for genetic sampling for conservation collections of endangered plants (Center for Plant Conservation, 1991. Reprinted with permission.)

	Decisions			
	Which species to collect?	**How many populations sampled per species?**	**How many individuals sampled per population?**	**How many propagules taken per individual?**
Recommended range	–	1–5	10–50	1–20
Target level of biological organization	Species	Ecotype and population	Individual	Allele
Key consideration	Probability of loss of unique gene pool Potential for restoration or recovery	Degree of genetic difference among populations Population history	Log of population size Genetic mobility within population	Survivability of propagules Long-term use of collection
Factors affecting sampling decisions				
Collect more	High degree of endangerment Experiencing rapid decline Few protected sites Biological management required Recently or anthropogenically reduced Feasibility of successful maintenance in cultivation or storage Possibility of reintroduction or restoration	High diversity/limited gene flow among populations Imminent destruction of populations High observed ecotypic or site variation Isolated populations Potential for biological management and recovery Recent or anthropogenic rarity Self-fertilization	High diversity among individuals within each population Observed microsite variation Mixed-mating or outcrossing Fragmented historical populations Small breeding neighborhood size Low survivability of propagules Extremely large populations	Low survivability of propagules Planned use for reintroduction or restoration program

Economic potential	Herbaceous annual or short-lived perennial	Boreal distribution	
	Early- to mid-successional stage	Gymnosperm or monocot	
	Gravity-, explosively, or animal-dispersed seed	Woody perennial	
	Dicot or monocot	Late-successional stage	
	Temperate–tropical distribution	Animal- or wind-dispersed seed	
	Wind-dispersed seed	Temperate–tropical distribution	
	Outcrossing wind-pollinated	Early- to mid-successional stage	
	Late-successional stage	Dicot	
	Observed similarity among populations	Herbaceous annual or short-lived	
High integrity of communities	Long-lived woody perennial	Self-fertilizing	
	Gymnosperm	Explosively or gravity-dispersed seed	
Many protected sites	Boreal–temperate distribution	High survivability of propagules	
Natural rarity	Protected populations or naturally rare	Large breeding neighborhood size	Low annual reproductive output (indicates multiyear collecting strategy)
Stable condition	Closely clustered populations		
Low degree of endangerment	Low diversity/extensive gene flow among populations	Low diversity among individuals within each population	High survivability of propagules

Collect less/fewer

Note: Four basic practical decisions are addressed: which species to conserve, and the number of populations, individuals, and propagules to be sampled for each species. For each decision, a recommended range is shown, along with an indication of the relevant level of biological organization and a summary of key considerations. Factors that influence the sampling decision are listed in detail, with the most significant factors shown in italics. Factors suggesting larger or more extensive collections are shown at the top of each column, while those suggesting smaller or less extensive collections are given at the bottom.

(and which) populations per species should be sampled; (3) how many individuals per population should be sampled; and, finally, (4) how many seeds or other propagules should be taken from each individual sample.

For the sake of brevity, my discussion is limited to seed collections and will ignore other forms of plant propagules that must in some cases be taken. For example, seeds of many if not most of the plants with which two affiliated Florida gardens must deal are termed *recalcitrant,* meaning that they cannot usefully be stored off-site because they do not remain viable for any significant length of time. Consequently the gardens must work with cuttings or use other more capital- and labor-intensive means of propagation (Wallace 1990). Sampling decisions are, in principle, the same in these situations, but practical considerations make seed collection for *ex situ* conservation the method of choice where possible.

Five basic sampling decisions must be made—one at each of the four levels of biological organization and these are followed by a fifth concerning the sampled population's ability to sustain the determined level of collection (CPC 1991). Recommendations concerning each of these decisions, along with the genetic and demographic considerations, are discussed briefly in turn.

FIVE SAMPLING DECISIONS: GENETIC AND DEMOGRAPHIC CONSIDERATIONS

---

1. Which Species Should Be Collected?

The degree of endangerment of a species is considered the single best indicator of priority, but this is, of course, a highly subjective criterion. Clearly, if only a few sites are known and not protected, or if evidence exists that population numbers are declining rapidly, they can be considered highly endangered. Perhaps a more subtle criterion that would suggest a species is endangered is related to whether it was always rare or if its rarity is a result of human-mediated habitat destruction.

Species that have always been rare, at least those with historically small or sparse populations, may have ecological or genetic adaptations that facilitate persistence that remnant fragments of previously widespread species lack (Barrett and Kohn 1991; Menges 1991a, 1991b; see chapter 2 of this book). In a long-term study of prairie grasses Rabinowitz et al. (1989) demonstrated that the reproductive output of common species closely tracked environmental fluctuations in rainfall over nine years, whereas reproductive output of sparse species was less variable and less likely to suffer complete reproductive failure.

Though all endangered plants are in some sense rare, not all rare plants are necessarily endangered (Rabinowitz 1981; Holsinger and Gottlieb 1991). For example, in the state of Oregon, *Lomatium bradshawii* is known from over twenty sites and *L. erythrocarpum* from only two; yet the former is on both the state and federal lists of endangered plant species, whereas the latter is not (Oregon Dept. of Agriculture 1989; U.S. Fish and Wildlife Service 1991). The difference relates to their history and current circumstance. *Lomatium bradshawii* is a plant of the Willamette Valley where the vast majority of its prairie habitat has been destroyed. Habitat remnants are found in highly fragmented patches that typically have been invaded by aggressive weedy exotics. On the other hand, *L. erythrocarpum*, which apparently has always been rare, is native to the relatively undisturbed subalpine habitat in the Wallowa-Whitman National Forest.

2. How Many Populations Should Be Sampled per Species?

The CPC recommends sampling from one to five populations per species. In practical terms, the decision is often moot, because almost three-quarters of the roughly 2,300 plant species in the United States being tracked by the CPC are known from five or fewer populations (CPC 1991). Similarly, roughly 80 percent (38 of 47) of the endangered plant species surveyed in southwestern Australia are found in five or fewer populations (Brown and Briggs 1991).

The most important considerations in the decision of how many populations to sample are the degree of genetic difference among populations and the history of site disturbance. In general, more populations should be sampled from species with high diversity or when gene flow among populations is low than from more genetically uniform species or those with high rates of gene flow among populations. Factors that suggest high among-population diversity are noticeable ecotypic differentiation, isolation of populations either naturally or from recent habitat fragmentation, and certain breeding systems. For example, although in absolute terms inbreeding species typically have less total genetic variation than do outbreeding species, more of what little variation they do have is distributed among populations (Hamrick 1989). Thus, to acquire any given proportion of total variation, a higher proportion of an inbreeder's populations will probably have to be sampled.

The suggested range of populations of one to five to be sampled per species may seem surprisingly low. It is based in part on the extensive survey of electrophoretically detectable genetic variation conducted by

Hamrick and his colleagues (Hamrick and Godt 1989; Hamrick et al. 1991), and on the analyses of Brown and Briggs (1991). The former researchers surveyed 653 published studies of 449 species in 165 genera, finding that on average, approximately 50 percent of the loci were polymorphic within species. Of those polymorphic loci, approximately 78 percent of the allozyme diversity occurs within populations. Therefore, the vast majority of the genetic information in a species can be found within a "typical" population. The remaining 22 percent of variation that is distributed among populations can, however, be very important with respect to local adaptation and the ability of species or populations to respond to selection.

Of the eight characteristics Hamrick and his colleagues (Hamrick and Godt 1989; Hamrick et al. 1991) considered as potential explanatory variables, geographical range accounted for most of the little variation they could explain. Other significant variables included life form and breeding system. Specifically, they found that in general (1) widespread species have more variation than narrowly distributed endemic species, (2) long-lived perennials are more variable than short-lived perennials and annuals, and (3) outcrossing species are genetically more variable than selfers or those with mixed mating systems. It is important to note, however, that only about a quarter of the total variation among species was accounted for by these eight variables.

Though in an intuitive sense, the more populations represented the better, CPC recommendations incorporate a practical trade-off between the benefits and probability of acquiring significant new genetic variants with increased sampling, and the costs associated with collection of a particular species in relation to the needs of other species. In our world of great need and limited resources, it seems reasonable to keep low the number of sampled populations per species, at least until we have basic coverage of all threatened and endangered species.

3. How Many Individuals Should Be Sampled per Population?

As mentioned earlier, the bulk of a species' genetic information is found in any "typical" population. Consequently, decisions about how this and the next question are answered will determine to a great extent the conservation value of the samples collected. The purpose of sampling is to capture the significant fraction of a species' total genetic information found within populations. The recommended range, ten to fifty individuals per population (CPC 1991), is again perhaps smaller than what intuition might suggest.

The underlying reason for this apparently low number is that the allelic content of a sample is proportional to the logarithm of both the population size and the sample size. Consequently, a strong law of diminishing returns is in operation. Monomorphic alleles and those in relatively high frequency are captured in early samples, and the chance of obtaining additional rare alleles declines rapidly with increasing sample size. In a statistical sense, the first ten individuals sampled are as important as the next ninety (Brown and Briggs 1991).

Within the range of ten to fifty individuals recommended, key factors to consider in the design of a sampling scheme are the population size and degree of genetic communication among individuals within the population. Conservation biologists should collect seeds from more individuals if there is high diversity among individuals within the population. High diversity among individuals may be indicated by high microsite variation within populations if the plants are self-incompatible or if there is reason to believe the genetic neighborhood size is small. Conservation biologists should collect toward the high end of the recommended range of individuals if the chance of raising successful offspring is low. Although relatively more populations of inbreeders than outbreeders should be sampled, fewer individuals need be sampled in populations of inbreeders than outbreeders. This is because inbreeders are less variable within populations than are outbreeders.

4. How Many Propagules Should Be Collected from Each Individual?

The sample size recommended for this decision has a proportionally wider range than the others: one to twenty propagules. This is in part because of large differences in our ability to successfully grow plants from seeds or other propagules. Experience among CPC gardens suggests that perhaps an average probability of a seed reaching sexual maturity is about 10 percent, but much lower values are not uncommon. For example, Wallace (1990) discusses an example in Florida in which 2,000 seeds of the annual plant *Warea amplexifolia* were sown in an appropriate habitat. Six hundred seedlings emerged, of which only sixteen survived to reproduce—a net yield in plants per seed of less than 1 percent.

Genetic variation and continuity are best maintained when each founder individual contributes equally to a new population (Templeton 1990, 1991). In practice we do this relatively easily for female parents, simply by collecting equal numbers of seeds from each plant sampled, or perhaps more realistically by keeping track of seeds from individual plants. But it may be next to impossible to equalize the contribution of different males in the population, because paternal fitness can vary considerably among males (Schoen and Stewart 1986).

5. Under What Circumstances Is a Multiyear Collection Plan Indicated?

The foregoing four decision criteria appear reasonable, but the question remains as to what effect, if any, the sampling itself has on the short-term prospects for survival of the sampled populations. This is more strictly a demographic question, bringing to mind an old saying: "The cure was a success, but the patient died." If collecting seeds for conservation purposes significantly endangers the population sampled, then indeed the cure may well be worse than the disease, because sampling a rare plant population could give us a false sense of security.

Although there are intuitive guides as to how much seed collection some plant species can withstand relative to others, we do not yet have a good quantitative handle on the actual proportions that any species can tolerate. Clearly, populations of an annual plant that has no dormant seed bank, a case that may be approached if not realized by *Howellia aquatilis* (Lesica 1990), are more susceptible to damage by having seeds removed by collectors than are those of a long-lived polycarpic perennial (many of which often suffer years of complete reproductive failure). More theoretical work clearly is needed in this arena if conservation biologists are to be a force that benefits rather than degrades the plant species we seek to conserve. In the absence of specific direction on this issue, it is comforting to know that Menges (see chapter 10) found that "The threat posed to population survival by environmental variation appeared almost entirely due to variation in mortality, growth, and reproductive status and not to variation in reproductive output."

REESTABLISHING SELF-SUSTAINING POPULATIONS FROM COLLECTIONS

It is one thing to obtain a genetically representative seed sample of a population and quite another to be able to reestablish successfully a comparable population from such a collection. We are approaching an adequate theoretical understanding of the former, but the latter is a much more formidable challenge. The transition from population genetic and demographic theory to conservation practice is not an easy one. Theory informs us of the relevant variables and their relationships, yet if we know anything for certain, it is that the phenomena with which we are dealing are fundamentally probabilistic.

Theory: Population Viability and Vulnerability

Attempts to reestablish self-sustaining populations must necessarily consider factors that influence the extinction probability of populations. Recent theory pertaining to this question that incorporates genetic and demographic parameters has been organized around two related concepts, that of the *Minimum Viable Population* (MVP) (Shaffer 1981), and its derivative, *Population Vulnerability Analysis* (PVA) (Gilpin and Soulé 1986). Shaffer (1981, 132) declared that "a minimum viable population for any given species in any given habitat is the smallest isolated population having a 99% chance of remaining extant for 1000 years despite the foreseeable effects of demographic, environmental, and genetic stochasticity, and natural catastrophes." Though the actual values used in any particular case are subject to discussion, the basic idea remains. Gilpin and Soulé (1986) introduced the term population vulnerability analysis to describe the process whereby the MVP is estimated.

Although the population is probably rightly considered the basic unit of interest, it is perilous to neglect the more inclusive if less well understood *metapopulation* (Gilpin 1987), or population of populations. The work of Menges and his co-workers on Furbish's lousewort (*Pedicularis furbishiae*) (see Menges 1990) is both a splendid example of a PVA and a sobering reminder of the central role metapopulation dynamics may play in the persistence of many plant species. Empirical demographic studies of *P. furbishiae* have demonstrated that many individual populations are growing rather rapidly (i.e., they have finite rates of population increase considerably greater than 1.0), suggesting a secure future. However, a conservation strategy that simply protected these populations would not be wise. The combination of a high probability of catastrophic extinction of individual populations due to ice scouring or bank slumping along its riverine habitat, and the observation that *P. furbishiae* does not fare well in the face of competition with later successional shrubby vegetation, translates to ephemeral populations. Persistence of the Furbish's lousewort along the St. John River thus is dependent on metapopulation dynamics, that is, a positive balance between local colonization events and extinction rates.

It is important to note that MVP and PVA do not provide precise predictions of the future. Rather, they are very general models that serve to help organize conceptually the myriad factors that influence the possible fates of populations and their relative probability.

Extinction can result from either deterministic or stochastic causes (Shaffer 1981; Gilpin and Soulé 1986). Deterministic extinction results when something essential to the population is removed, or when something lethal is introduced (Gilpin and Soulé 1986). Deterministic extinction could be the result of systematic habitat destruction or conversion to human uses of all localities of a rare plant. There are a variety of stochastic causes of extinction.

Natural catastrophes mentioned earlier for the Furbish's lousewort are at one extreme in the range of *environmental stochasticity* that can lead to extinction. A less extreme but still dramatic example of environmental stochasticity that may have contributed to the apparent extinction of *Stephanomeria malheurensis* is annual variation in precipitation (Parenti and Guerrant 1990). *Stephanomeria malheurensis* is known from a single population estimated to comprise a maximum of 750 plants in any one year. However, *S. malheurensis* has experienced nearly 100-fold declines in population sizes from one year to the next, closely correlating with threefold reductions in precipitation (Gottlieb 1979, 1991). This annual plant does not have the apparently adaptive seed dormancy characteristics or germination requirements of its sympatric ancestral species, *S. exigua* ssp. *coronaria* (Gottlieb 1973; Brauner 1988). Thus, with an uncertain soil seed bank and some fatal fall germination (Gottlieb 1979; Parenti, pers. com.), it is perhaps not surprising that the species was thought to have become extinct by 1985.

Demographic stochasticity and *genetic stochasticity* also can lead to the extinction of populations. Both are statistical phenomena that become increasingly more important as population size decreases. Although stochastic variation in demographic parameters may by itself be important only in very small populations of fifty or fewer individuals (see chapter 10; Menges 1991b), Lande (1988, 1455) argues that both "theory and empirical examples suggest that demography is usually of more immediate importance than population genetics in determining the minimum viable sizes of wild populations."

Demographic stochasticity involves chance variation in all components contributing to individual survival and reproductive success. Genetic effects of small population sizes are well understood theoretically and so basic that they often appear in introductory courses in ecology and evolution. Small populations are subject to random loss of alleles by genetic drift. Inbreeding may also increase as population size decreases, with an accompanying increase in homozygosity and decrease in fitness.

Gilpin and Soulé (1986) and Gilpin (1987) have organized these extinction causes along with demographic and genetic principles into a heuristic model of PVA (Figure 13.1). It allows the relationship among various causes of extinctions to be viewed in such a way that probable

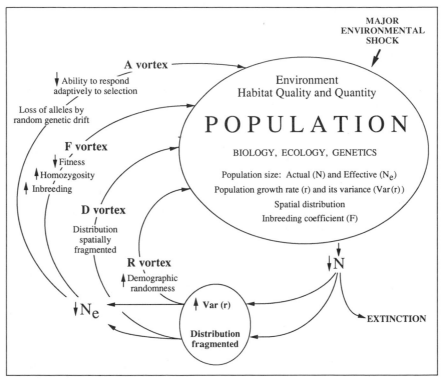

Figure 13.1

A population vulnerability analysis model showing how, following a major shock, various factors can interact to form positive feedback loops that may erode the population further and lead eventually to extinction (adapted from several figures in Gilpin and Soulé 1986, and Gilpin 1987). The large ellipse represents the population in question, including its environment, genotype, phenotype, and population structure and fitness. The factors listed are illustrative, and not exhaustive. The figure depicts a major shock that results in a reduced population size, increased variance in population growth rate and/or fragmentation of the population into relatively isolated patches. Thus compromised, the population becomes more vulnerable, and may enter one or more of the four extinction vortices. The term extinction vortex *illustrates well the runaway property of positive feedback systems, where an initial input becomes magnified.*

outcomes of population persistence or extinction may be analyzed. They draw attention to several positive feedback loops among parameters that can arise when populations suffer environmental shocks. The runaway property of positive feedback systems in general is reflected in the term *extinction vortex,* coined by Gilpin and Soulé (1986). The four extinction vortices they describe are not completely independent of one another, and real-world situations may incorporate parts of several.

A chance reduction in the population size (N) might increase the variance in the population growth rate ($Var\ (r)$), which in turn reduces population size, and so on. Gilpin and Soulé (1986) term this the R vortex. Alternatively, or additionally, a reduced N and an increased $Var\ (r)$ may lead to fragmentation of a once-continuous population into a number of smaller isolated populations. Each of these population fragments is more likely to succumb to extinction than is a single population of the same number of individuals. This is the D vortex, and it can lead to effective population sizes (N_e) vastly lower than the census population size (N). The two other extinction vortices, termed F and A, result from decreases in N_e. The F vortex refers to the inbreeding coefficient symbol (F); as N_e decreases, inbreeding is expected to increase. Consequently, as the value of the inbreeding coefficient increases, the level of homozygosity in a population will increase. As homozygosity increases, deleterious recessive alleles will reveal themselves, and fitness will decrease overall. This increase in inbreeding depression is relatively independent of the environment and further exacerbates the deteriorating population size, feeding back into the F extinction vortex. Finally, loss of evolutionary adaptability (A vortex) is another negative consequence of a decreased N_e. If an affected population is not consumed by the other extinction vortices, loss of alleles due to genetic drift caused by a low N_e will leave a population less able to respond adaptively to environmental change and so leave the species more vulnerable to extinction by the A vortex.

It is important to realize that from a genetic point of view, the size of a population may not be what it seems. Effective population size (N_e), or genetic neighborhood size, as opposed to census size (N), which is counted directly, is not necessarily an easy property to ascertain. For example, Harris and Allendorf (1989) devote an entire article to comparing different estimators of N_e in grizzly bears. Schmidt (1980) calculated the theoretical N_e in populations of three *Senecio* species if they were pollinated solely by either bumblebees or butterflies. The relatively few long-distance flights between plants taken by butterflies, relative to the short interplant visits of bees, resulted in the hypothetical N_e values ranging from on the order of the low tens if pollinated solely by bees to several thousands if pollinated exclusively by butterflies. Barrett and Kohn (1991) draw attention to the lack of available evidence of N_e for plant populations and point out that the few data that do exist show that the N_e may be considerably smaller than N.

Practice—Two Examples

I am aware of no cases where the genetic profile of an existing population has explicitly been investigated electrophoretically and where seed samples were taken and stored off-site, with the goal of establishing another self-sustaining population from the sample. Some of the best information we have in this regard comes serendipitously from on-going experimental recovery efforts of two federal- and state-listed endangered plants. Attempts to establish self-sustaining populations of *Stephanomeria malheurensis* and *Amsinckia grandiflora* are based exclusively on seed that had been maintained by academic botanists in off-site cold-storage for up to twenty-five years. They will be used to illustrate various points, but because both have an annual life cycle, they are not necessarily representative of vascular plants in general. I will briefly describe each separately and then discuss implications of these studies for reintroduction efforts in general.

Stephanomeria malheurensis. For a plant known only from a single locality and that apparently became extinct less than twenty years after it was first discovered, *Stephanomeria malheurensis* is extremely well studied scientifically (Gottlieb 1977, 1978, 1983). Recovery efforts by the U.S. Bureau of Land Management in conjunction with the U.S. Fish and Wildlife Service were possible only because Dr. L. Gottlieb, who originally described the species, had carefully maintained a seed stock at the University of California at Davis. It is from these *ex situ* samples that this self-pollinating species owes its continued existence (Parenti and Guerrant 1990).

Recovery efforts being made at the original site are more difficult than they might otherwise have been for at least two reasons. The original habitat has been significantly altered by the successful invasion of cheatgrass (*Bromus tectorum*) following a fire in 1972. Previously the area between shrubs or grass clumps had been mostly bare soil. This very aggressive weed, introduced from Europe, is perhaps the most abundant grass in the colder, drier parts of the Pacific Northwest. Larger-scale variation in environmental conditions may also complicate this situation (see chapter 9). Short-term climatic changes associated with the high rainfall years of 1981 to 1985, in which the levels of nearby Malheur and Harney lakes rose so dramatically that they formed a single body of water inundating farms and roads, may have contributed to the decline of *S. malheurensis* (Parenti, pers. com.; Brauner 1988; Wicklow-Howard 1991) and its sympatric progenitor *S. exigua* ssp. *coronaria*. From an estimated population of 40,000 in the early to mid-1970s (Gottlieb 1979), no more than 12 individuals of *S. exigua* ssp. *coronaria* were found annually between 1981 and 1987—the interval in which the lakes rose—and *S. malheurensis* apparently became extinct (Brauner 1988).

An electrophoretic investigation of thirteen loci showed that the highly self-pollinating *S. malheurensis* has, with one possible exception, a subset of the alleles found in its outcrossed ancestor, *S. exigua* ssp. *coronaria* (Gottlieb 1973). *Stephanomeria malheurensis* is monomorphic for eleven of the thirteen loci examined and has fewer alleles than *S. exigua* ssp. *coronaria* in the two polymorphic loci. Given its self-pollinating breeding system *S. malheurensis* is not likely to have changed much genetically the few years seed was increased by off-site propagation. Clearly, artificial selection during off-site propagation could have affected its genetic constitution, but genetic drift and inbreeding depression are probably not large factors.

Given this extreme environmental disruption, early recovery efforts are necessarily experimental. The basic recovery design involved planting seedlings and artificially watering them until the plants were well established in four separate rodent-proof wire-mesh exclosures. Each exclosure was dominated by one of four species common to the area: three natives, big sagebrush (*Artemisia tridentata* var. *wyomingensis*), Great Basin wild rye (*Elymus cinereus*), green rabbitbrush (*Chrysothamnus vicidiflorus*), and the exotic cheatgrass. All of the cheatgrass was removed from the native plant plots. Half of the cheatgrass plots were hand-weeded of half of their cheatgrass cover, and the other half were left undisturbed. Survivorship and reproduction were generally high the first year; 1,000 plants yielded 40,000 seeds. However, *S. malheurensis* individuals in both cheatgrass plots were significantly smaller, slower to bolt and flower, and less fecund than plants established in the other plots (Brauner 1988). Since then, significant events include the catastrophic loss in 1988 of *S. malheurensis* plants in the big sagebrush plot, due to herbivory by small mammals (Parenti and Guerrant 1990). In 1990, no seedlings or older plants were seen. The year was cold and dry during the period when seeds normally germinate, but several hundred plants appeared again in 1991, a year in which there were suitable germination conditions at the appropriate time of year (Parenti, pers. com.).

Amsinckia grandiflora. Native to central California, *Amsinckia grandiflora* is, like the previous example, a very narrowly distributed endemic. Until 1991, when a relatively large new population was discovered about 2 km away, it had been known to survive in only one general area that supported what some call one (Ornduff 1976) and others call two populations (Pavlik 1991). Several other populations are known historically, but none have been relocated and all are thought extirpated.

The original population(s) had been doing poorly throughout the last decade, having dwindled from several thousand in 1966 (Ornduff 1976) to ninety-two individuals (up from 23 in 1980, and down from 355 in 1988) and twenty-nine plants each in 1991 (Pavlik 1990, 1991). Consequently, the U.S. Fish and Wildlife Service has established a recovery plan that calls for the establishment of three new self-sustaining populations within the historical range to enhance existing populations (Pavlik 1991).

Unlike the highly inbred *Stephanomeria malheurensis*, *A. grandiflora* is distylous and therefore presumably more highly outcrossed (Ornduff 1976). Although no electrophoretic studies of natural populations of *A. grandiflora* were performed when the populations were large, one of two seed sources used in the reintroduction effort is thought to consist mostly of seed that had been gathered by Ornduff directly from those large natural populations and had not been grown since. The other source, which stems from the same original collections by Ornduff, is several generations removed, having been propagated off-site several times. Small samples from these two sources were compared electrophoretically. Seeds from the original population contain more genetic variation than those that had been propagated off-site: a higher percentage of polymorphic loci, more alleles per locus, and a higher mean heterozygosity (Pavlik 1990).

Between 1989 and 1990, an appropriate reintroduction site was located within the historical range of this species. Of 3,460 nutlets that were planted, approximately 50 percent (1,774) germinated, and about 32 percent (1,101) survived to reproduce, yielding approximately 35,000 new nutlets (Pavlik 1990), an order of magnitude increase. Demographic monitoring of a thoughtful experimental design enabled Pavlik to assess the effects of burning, hand-clipping, and application of a grass-specific herbicide on fates of these seeds. As in the case of *Stephanomeria*, competition from exotic annual grasses (mostly *Avena fatua*, *Bromus diandrus*, *B. mollis*, and *B. rubens*) had a significant negative effect on survivorship to reproduction, plant size, and reproductive output in *A. grandiflora*. Competition from annual grasses was most effectively reduced by application of a grass-specific herbicide. Hand-clipping actually increased competition late in the season, and burning was less effective than herbicide treatment because of annual grass recruitment subsequent to the burn.

Based on this initial success, establishment of three new populations was attempted between 1990 and 1991. The original experimental population was monitored as well. Based on the number of nutlets sown and rates of seedling emergence, survival, reproduction, and nutlet production, some of the new populations are expected to increase next year and others (including the extant native site) to decrease. Unlike the previous year, artificial treatment to reduce grass cover did not enhance *A. grandiflora* fitness parameters.

The difference in the effect of grass competition between the two years appears to be associated with rainfall patterns. Grasses appear to be more of a problem for *A. grandiflora* in springs following years with near- or above-average late-autumn rainfall (e.g., 1989) than they are in years of below-average late-autumn rainfall (e.g., 1990). Once again, variation in environmental parameters, in this case manifest in precipitation-dependent differences in competitive abilities of exotic grasses, appears to play a significant role in the fate of introduced and natural populations.

Common Threads and Divergent Paths

First and foremost, both of these restoration efforts took place with seed that had been stored off-site. In the case of *Stephanomeria*, these seeds may have been the only link between persistence and extinction. The work of the Center for Plant Conservation in developing genetically representative seed collections and storing them off-site is particularly timely, given the current rates of species and population loss.

Genetic diversity of samples stored off-site can, however, erode rapidly or be distorted, especially if they must be propagated off-site to counteract declining seed survivorship. It is sobering to consider that what little genetic variation existing in the more pristine *Amsinckia* nutlet sample was lowered even further in only a few generations of off-site propagation. Where there had been four polymorphic loci, one propagated sample was reduced to two, and another was reduced to only one polymorphic locus. To what degree this was due to selection versus random genetic drift cannot be assessed rigorously. However, different selective pressures on- and off-site cannot be discounted a priori. The problem of genetic change in captivity must be considered and surmounted if we are to return viable populations to the wild.

Templeton (1990, 1991) argues that the best way to minimize the impact of genetic drift in a closed population is to subdivide the population into breeding units between which there is minimal gene flow. Such a strategy might be appropriate in *ex situ* collections where it is necessary to propagate the plants to increase numbers or maintain seed viability. Each subpopulation would, of course, lose variation faster than would a panmictic population composed of all captive individuals. However, the population as a whole will retain more diversity when subdivided because the probability of all subpopulations losing the same allele at a polymorphic locus is less than the probability of the larger panmictic population becoming fixed for one allele. When the time comes to use plants that have been maintained in such a breeding regime, genetic variation could be restored by the mixing of the previously separated subpopulations.

Problems that might arise from such a plan include inbreeding depression and the possibility that different populations could become differentially adapted to their various captive environments. To minimize the chance of outbreeding depression resulting from distinct lines being crossed, Templeton (1990, 1991) recommends releasing only healthy individuals of F_2 or later generations. In this way, reintroduced populations do not have to suffer the consequences of adaptations to captivity that may have evolved in the inbred subpopulations.

Extreme variations in population sizes between years, due apparently to environmental variation, have been documented for both *Stephanomeria malheurensis* and *Amsinckia grandiflora*, both before and after reestablishment attempts. In the resurrected populations, demographic performance spans the range from complete failure to a potential order of magnitude increase. This variation apparently is attributable to environmental variation. Such variation may be the rule rather than the exception, and in these and future reintroduction efforts, attempts should be made to minimize its impact. A metapopulation strategy whereby multiple populations are established should perhaps be considered standard practice.

There appear to be at least two key reasons why reintroduction efforts should involve attempts to establish *multiple* populations, even if the original range of the species was limited to a single population. First is that the probability of catastrophic failure due to unforeseen environmental factors, such as generalist herbivory, is sufficiently high that long-term survival prospects are enhanced if risk is spread over as many discrete populations as can be established. Second, even under the best of circumstances reintroduced populations often will be started with a genetic subset of the original species. The degrading effects on genetic variation due to small population size can in part be mitigated if several different populations are started. Although each population might be smaller than one large one that could be started with limited seed available, more variation might be preserved if several populations are established. There is, of course, a complex trade-off to be evaluated with respect to the rate of loss of genetic variation due to different reintroduction strategies. The trade-off depends in part on the total number of seeds available for the project and the probability of environmental variation causing population size swings.

These two studies illustrate differential use of what some might consider to be "heroic" measures. In a discussion of a case where cuttings were used to establish a new population of *Pediocactus knowltonii* (Olwell et al. 1987), Pavlik (1991) argues strongly that if recovery efforts are to result in self-sustaining population and not short-lived rare-plant gardens, then recovery efforts must begin with the planting of seeds. He argues that by creating an "instant population" we obtain no information on a major demographic hurdle—that is, from seed to established juvenile. Though it is certainly true that for the population to be self-sustaining, future generations of sexually reproducing plants must eventually be derived from seeds produced on-site, I see no compelling reason why the initial generation must be derived from seed germinated at the site. It seems to me that the choice of how to establish the "starter" generation should be based on how best to get the maximum number of reproducing adults from the available material in the initiating generation. If that involves planting cuttings or germinated seedlings, as was the case in *Stephanomeria*, so be it. Long-term success, of course, necessitates (at least in sexual species) that new generations of plants be derived from seed. The decision to use seedlings in the case of *Stephanomeria* may simply have been a means to return to the soil the maximum number of seeds from a limited number available initially.

Both of these reintroduction efforts involved attempts to establish a native species in areas heavily infested with introduced annual grasses. Pavlik achieved very good results one year with the use of grass-specific herbicides. No herbicides have been used at the *Stephanomeria* site. Brauner (1988) discounts the use of fire as an effective tool to remove cheatgrass and suggests that herbicide use probably represents the only option that could be sustained at reasonable expense. Brauner further suggests that herbicide use is not currently an option because of an injunction against any herbicide use on land managed by the U.S. Bureau of Land Management. Such a decision might then represent an unfortunate consequence of a generally laudable land-management policy.

LOOKING TO THE FUTURE

The task of obtaining for *ex situ* storage genetically representative seed samples from our nation's most rare and endangered plants is well under way, but there still is much to do. Given the new CPC guidelines, satisfactory completion of this task appears largely to be a matter of time and resources. However, we have barely begun the task of understanding how to reestablish comparable self-sustaining populations from these samples. This is a task of immediate importance in an applied sense, and also one that holds great promise for the advancement of ecological and evolutionary theory.

The foregoing discussion provides a springboard from which a number of possible options emerge. This concluding section is not intended necessarily to advocate any particular course of action, but rather to stimulate discussion of various possible uses of seed collections to reestablish self-sustaining populations of rare species in native habitats, and also perhaps to establish new ones.

Insofar as the extinction probabilities of separate populations are independent (i.e., barring a regional environmental perturbation or widespread catastrophe), the chance of species extinction declines with each additional healthy population. If there is a 50 percent probability of a single population lasting 100 years, there is a 75 percent probability that at least one of two independent populations will last 100 years, and almost a 95 percent probability that at least one of four will survive a century. Species limited to a single population are at great risk. Consequently, it may be wise when reintroducing extirpated populations to establish multiple populations (i.e., a metapopulation) as a matter of course. Clearly, the ultimate chance of reducing the extinction possibility of plant species known from only one or a few populations would seemingly be enhanced if additional new populations are started (and given legal protection) *before* the native ones are extirpated.

Such an aggressively preemptive strategy may involve the establishment of more populations than existed originally. This would be the case with *Stephanomeria malheurensis* because only one population has ever been found. Despite the reluctance some biologists and government officials might have to such activities, just such a precedent has been set by the U.S. Fish and Wildlife Service (USFWS). To reduce the risk of extinction to *Pediocactus knowltonii*, a federally listed endangered species known to occur naturally at only a single site, the USFWS took cuttings from the known population and successfully established a second at a new site (Olwell et al. 1987). This seems a simple thing to do; why can it not be done more often?

Another possible strategy when establishing new populations might be to mix seed sources to increase the genetic diversity in the sample (Barrett and Kohn 1991; Holsinger and Gottlieb 1991). This certainly is not a universal panacea, because there is considerable risk of making things worse than if single-source seed supplies are used. Undesirable consequences include outbreeding depression, breaking up coadapted gene complexes, and otherwise squandering limited valuable seed. More theoretical and empirical work might allow us to identify circumstances in which such a strategy might be effective.

One final possibility to consider is how we might provide endangered species with the best opportunity possible to respond adaptively to conditions in our rapidly changing world. Sewell Wright's shifting balance theory describes a situation in which rapid adaptive evolution is particularly favored (Wright 1982 and references therein). Briefly, it involves a population, or rather a metapopulation, of relatively small, semi-isolated demes with low interdeme migration rates. By providing for some appropriate rate and pattern of migration among the multiple populations established, adaptation to changing conditions would be facilitated. But to adopt such a strategy would be to use genetic and demographic theory in an explicit attempt to change genetically species of naturally occurring rare plants. Although this obviously involves ethical issues about humans consciously altering the genetic makeup of current taxa, is not the ultimate goal of conservation to allow taxa not only to persist but also to undergo adaptive evolutionary change?

ACKNOWLEDGMENTS

I would like to extend my sincere appreciation to Dr. Peggy Fiedler for the opportunity to write this chapter, and to her, Dr. Bruce Pavlik, Katie Robinson, and Margie Gardner for providing critical comment on the manuscript. Dr. Bruce Pavlik generously provided me with unpublished reports of his work on *Amsinckia grandiflora*; Don Falk and the Center for Plant Conservation provided several chapters and the appendix of their book before publication; and Dr. Peggy Fiedler supplied me with various chapters of this book. Dr. Michael Flower turned a rough sketch into Figure 13.1 with a Macintosh computer. To all of these people I am grateful for these and many other kindnesses.

LITERATURE CITED

Barrett, S.C.H., and J.R. Kohn. 1991. Genetic and evolutionary consequences of small population size in plants: Implications for conservation. In *Genetics and conservation of rare plants,* ed. D.A. Falk and K.E. Holsinger, 3–30. New York: Oxford University Press.

Brauner, S. 1988. Malheur wirelettuce (*Stephanomeria malheurensis*) biology and interactions with cheatgrass: 1987 study results and recommendations for a recovery plan. Report to Burns District, Oregon, Bureau of Land Management. Submitted 20 June 1988.

Brown, A.H.D., and J.D. Briggs. 1991. Sampling strategies for genetic variation in *ex situ* collections of endangered plant species. In *Genetics and conservation of rare plants,* ed. D.A. Falk and K.E. Holsinger, 99–119. New York: Oxford University Press.

Center for Plant Conservation. 1991. Appendix. Genetic sampling guidelines for conservation collections of endangered plants. In *Genetics and conservation of rare plants*, ed. D.A. Falk and K.E. Holsinger, 225–38. New York: Oxford University Press.

Falk, D.E., and K.E. Holsinger, eds. 1991. *Genetics and conservation of rare plants.* New York: Oxford University Press.

Gilpin, M.E. 1987. Spatial structure and population vulnerability. In *Viable populations for conservation,* ed. M.E. Soulé, 125–39. Cambridge: Cambridge University Press.

Gilpin, M.E., and M.E. Soulé. 1986. Minimum viable populations: Processes of species extinction. In *Conservation biology: The science of scarcity and diversity,* ed. M.E. Soulé, 19–34. Sunderland, Mass.: Sinauer.

Gottlieb, L.D. 1973. Genetic differentiation, sympatric speciation and the origin of a diploid species of *Stephanomeria. Am. J. Bot.* 60:545–53.

————. 1977. Phenotypic variation in *Stephanomeria exigua* ssp. *coronaria* (Compositae) and its recent derivative species "malheurensis." *Am. J. Bot.* 64:873–80.

————. 1978. Allocation, growth rates and gas exchange in seedlings of *Stephanomeria exigua* ssp. *coronaria* and its recent derivative *S. malheurensis. Am. J. Bot.* 65:970–77.

————. 1979. The origin of phenotype in a recently evolved species. In *Topics in plant population biology*, ed. O.T Solbrig, S. Jain, G.B. Johnson, and P.H. Raven, 264–86. New York: Columbia University Press.

————. 1983. Interference between individuals in pure and mixed cultures of *Stephanomeria malheurensis* and its progenitor. *Am. J. Bot.* 70:276–84.

————. 1991. The Malheur wire-lettuce: A rare, recently evolved Oregon species. *Kalmiopsis. J. Native Plant Soc. Oregon* 1:9–13.

Hamrick, J.L. 1989. Isozymes and the analysis of genetic structure in plant populations. In *Isozymes in plant biology*, ed. D.E. Soltis and P.S. Soltis, 87–105. Portland: Dioscorides Press.

Hamrick, J.L., and M.J.W. Godt. 1989. Allozyme diversity in plant species. In *Plant population genetics, breeding, and genetic resources*, ed. A.D.H Brown, M.T. Clegg, A.L. Kahler, and B.S. Weir, 43–63. Sunderland, Mass.: Sinauer.

Hamrick, J.L., M.J.W. Godt, D.A. Murawski, and M.D. Loveless. 1991. Correlations between species traits and allozyme diversity: Implications for conservation biology. In *Genetics and conservation of rare plants*, ed. D.A. Falk and K.E. Holsinger, 75–86. New York: Oxford University Press.

Harris, R.B., and F.W. Allendorf. 1989. Genetically effective population size of large mammals: An assessment of estimators. *Cons. Biol.* 3:181–91.

Holsinger, K.E., and L.D. Gottlieb. 1991. Conservation of rare and endangered plants: Principles and prospects. In *Genetics and conservation of rare plants*, ed. D.A. Falk and K.E. Holsinger, 195–208. New York: Oxford University Press.

Lande, R. 1988. Genetics and demography in biological conservation. *Science* 241:1455–60.

Lesica, P. 1990. Habitat requirements, germination behavior and seed bank dynamics of *Howellia aquatilis* in the Swan Valley, Montana. Report submitted to Flathead National Forest, Montana.

Menges, E.S. 1990. Population viability analysis for an endangered plant. *Cons. Biol.* 4:52–62.

————. 1991a. Seed germination percentage increases with population size in a fragmented prairie species. *Cons. Biol.* 5:158–64.

————. 1991b. The application of minimum viable population theory to plants. In *Genetics and conservation of rare plants*, ed. D.E. Falk and K.E. Holsinger, 45–61. New York: Oxford University Press.

Olwell, P., A. Cully, P. Knight, and S. Brack. 1987. *Pediocactus knowltonii* recovery efforts. In *Conservation and management of rare and endangered plants*, ed. T.S. Elias, 519–22. Sacramento: California Native Plant Society.

Oregon Department of Agriculture, compiler. 1989. *State list of endangered and threatened plant species*. OAR-603-73-070. Updated 27 October 1989.

Ornduff, R. 1976. The reproductive system of *Amsinckia grandiflora*, a distylous species. *Syst. Bot.* 1:57–66.

Parenti, R.L., and E.O. Guerrant, Jr. 1990. Down but not out: Reintroduction of the extirpated Malheur wirelettuce, *Stephanomeria malheurensis*. *Endang. Spec. Update* 8:62–63.

Pavlik, B.M. 1990. Reintroduction of *Amsinckia grandiflora* to Stewartville. Report prepared for Endangered Plant Program, California Department of Fish and Game. Sacramento, Calif.

———. 1991. Reintroduction of *Amsinckia grandiflora* to three sites across its historic range. Report prepared for Endangered Plant Program, California Department of Fish and Game. Sacramento, Calif.

Rabinowitz, D. 1981. Seven forms of rarity. In *The biological aspects of rare plant conservation*, ed. H. Synge, 205–17. Chichester: John Wiley & Sons.

Rabinowitz, D., J.K. Rapp, S. Cairns, and M. Mayer. 1989. The persistence of rare prairie grasses in Missouri: Environmental variation buffered by reproductive output of sparse species. *Am. Nat.* 134:525–44.

Schoen, D.J., and S.C. Stewart. 1986. Variation in male reproductive investment and male reproductive success in white spruce. *Evolution* 40:1109–20.

Schmidt, J. 1980. Pollination foraging behavior and gene dispersal in *Senecio* (Compositae). *Evolution* 34:934–43.

Shaffer, M.L. 1981. Minimum population size for species conservation. *Bio-Science* 31:131–34.

Soulé, M.E., ed. 1987. *Viable populations for conservation*. Cambridge: Cambridge University Press.

Templeton, A.R. 1990. The role of genetics in captive breeding and reintroduction for species conservation. *Endang. Spec. Update* 8:14–17.

———. 1991. Off-site breeding of animals and implications for plant conservation strategies. In *Genetics and conservation of rare plants*, ed. D.A. Falk and K.E. Holsinger, 182–94. New York: Oxford University Press.

U.S. Fish and Wildlife Service, Department of Interior. 1991. *Endangered and threatened wildlife and plants*. 50 C.F.R. 17.11 and 17.12. 15 July 1991.

Wallace, S.R. 1990. Central Florida Scrub: Trying to save the pieces. *Endang. Spec. Update* 8:59–61.

Wicklow-Howard, M. 1991. Malheur wirelettuce research study annual report 1990. Report submitted to Burns District, Oregon. Bureau of Land Management.

Wright, S. 1982. Character change, speciation, and the higher taxa. *Evolution* 36:427–43.

PART

IIII

THE PRACTICE OF
NATURE CONSERVATION, PRESERVATION,
AND MANAGEMENT

CHAPTER
14

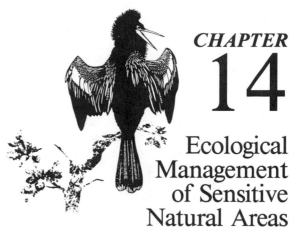

Ecological Management of Sensitive Natural Areas

C. RONALD CARROLL

ABSTRACT

Most of the world's remaining "wildlands" are relatively small areas that are located in highly intrusive economic landscapes. The populations, species interactions, and system properties enclosed in these areas will require proactive management efforts to prevent continual degradation. In this chapter, I present a "taxonomy" of natural areas and ecological processes that are especially sensitive and will require active management. These are (1) rare species, habitats, and resources; (2) small and fragmented habitats; (3) sites with vulnerable locations; (4) areas with low resilience to perturbations; and (5) keystone resources. I conclude with a discussion of three management issues that are of particular concern to conservation and that are characterized by differing levels of uncertainty and time scales. These are (1) the management of populations that are strongly influenced by human activity, (2) the changing economic landscape, and (3) the possible consequences of climate change. Throughout the chapter I attempt to highlight areas where major research gaps exist.

INTRODUCTION

The land and coastal regions of the world form an anthropogenic landscape. Embedded in a few places here and there in this landscape are the remnants of wildlands that now contain virtually all land-based natural biodiversity and that provide our only glimpses of ecosystem structure and function as they used to be. Most of these wildlands are too small and fragmented to survive as representative natural habitat without active ecological management. Indeed, for some taxa there may already be no natural areas left that are large enough to sustain them without active management intervention on their behalf. The dilemma faced by the applied ecologist—the ecological manager of wildlands—is how to provide these natural areas with the protection from outside forces of degradation without creating too much artificiality; that is, how to avoid converting the biotic communities within natural areas into museum pieces that have no evolutionary viability or into arbitrary welfare collections of plants and animals. With apologies to Hutchinson (1965), the manager must be both the minimalist director of the evolutionary play and caretaker of the ecological theater. Managers first must do what is necessary to ensure the survival of the species, but they cannot neglect efforts to approximate those ecological conditions that gave rise to the particular biodiversity of an area. The extent to which these conditions can be approximated depends on how well we understand the underlying mechanisms. For example, we cannot approximate the conditions that maintain guilds of seed dispersers without understanding the mechanisms by which the temporal, spatial, chemical, and size array of seeds, as resources, are utilized by their harvesters. In most cases, these mechanisms are too poorly known to permit more than the most tenuous generalizations.

Here, I emphasize management issues that concern natural areas—wildlands—in the context of the economic landscape in which they are embedded. I also identify research priorities that are needed to address these issues. A more complete treatment of research priorities for conservation biology is provided by Soulé and Kohm (1989).

Management should not be solely a reactive response to immediate crises—though it most frequently is—but should also be a proactive process to reduce the need for interventions in the future. I begin with a discussion of a "taxonomy" of those natural areas and processes that are likely to require active management. I conclude with a discussion of a daunting question: How can we make the proper management decisions now when the future seems so uncertain?

TAXONOMY OF SENSITIVE NATURAL AREAS AND PROCESSES

Management requires the allocation of limited resources: personnel, time, money, and expertise. Fortunately, not all natural areas or the ecological processes contained within them require intensive active management. The following represents a taxonomy of natural areas and ecological processes that are likely to require active management to prevent serious resource degradation and local extinction.

1. Rare Species, Habitats, and Resources

Conservation legislation emphasizes the protection of rare habitats and species. This is to be expected because environmental legislation is often the response to crises. In this instance, "rare" and "endangered" become synonyms with little regard given to causes of rarity (see chapter 2). Some species that are naturally widespread but everywhere uncommon presumably require less protection than species that have recently become uncommon due to human activity. Species that occur in rare but stable habitats presumably require less immediate attention than species whose ranges have recently shrunk to a few small sites. The concern with rarity per se, of course, is that the loss of any individual habitat or population represents a large total percentage loss with little opportunity for restoration. I would include here not just absolute losses but also many kinds of degradative effects such as fertilizer runoff into vernal pools or introgression of coyote genes into populations of the rare red wolf (*Vulpes fulva*).

Understanding the mechanisms of rarity and historical patterns is crucial to the development of a rational management policy. A few examples will serve to illustrate this need. Small isolated populations may result from the recent decline and fragmentation of a widespread population or may represent an ancient distributional pattern of small populations restricted to uncommon habitats. Many currently endangered species, such as grizzly bears (*Ursus horribilis*) and harpy eagle (*Harpia harpyja*), represent the former case whereas relictual species such as the Ozark collared lizard (*Cortaphytus collaris*) and Florida yew (*Taxus floridana*) would represent typical examples of the latter case. Stochastic processes, both demographic and environmental, place such species at increased risk, but the management considerations may be quite distinct. For example, populations that recently have become fragmented into small isolated demes may be losing alleles through fixation and thereby become more vulnerable to extinction if the local environment does not remain constant. In these cases, translocating individuals between demes may be a reasonable way to increase population heterozygosity. By contrast, species that have old disjunctive distributions may also have low within-population heterozygosity but in addition may have alleles functioning in very different genetic environments. In the terminology of Mayr (1966) and Dobzhansky (1970), the populations have different coadapted genomes. In these cases, translocating individuals carries the risk of introducing outbreeding depression in the recipient populations (Templeton 1986). For a discussion of the genetic considerations in the management of small populations, see Lande and Barrowclough (1987) and Lande (1988).

2. Small and Fragmented Habitats

Fragmentation of habitats per se is not necessarily bad. A population that is distributed among isolated habitats but with occasional migration may be less susceptible to extinction from demographic stochasticity or from localized catastrophes. Fragmentation becomes a problem, however, when migration is absent and the habitat quality is too poor or the area too small to sustain viable populations. Because fragmentation is now typically a consequence of human activities, remnant habitats are now generally in close proximity to areas of human disturbance, such as forest clear-cuts and agriculture, and may thereby be subjected to excessive stress from external invaders such as fire, pesticides, and weed species (Janzen 1986; Lovejoy et al. 1986).

In some cases it may be possible to create migration corridors between isolated areas with strategic acquisition of land. A case in point is the recent purchase of 6,804-ha Pinhook Swamp in north Florida as a corridor to connect the Okeefenokee Wildlife Refuge of southern Georgia with the Osceola National Forest in Florida (The Nature Conservancy 1989). There is also considerable opportunity to increase the effective size of a natural area through cooperative regional land-use planning. For example, interagency cooperation among land managers in the United States, such as the, National Park Service, Forest Service, and so on, could result in greatly increased networks and effectively increase the size of the reserves (Salwasser, Schonewald-Cox, and Baker 1987). Similarly, in Costa Rica, national parks are now administered as components of larger conservation units, and, in parts of east Africa, parts of larger game-management areas.

The minimal area required or the maximum habitat degradation tolerated, of course, varies greatly, even among closely related taxa. Collared peccaries (*Tayassu tajacu*) survive in smaller and more fragmented habitats than white-lipped peccaries (*Tayassu pecari*); howler monkeys (*Alouatta palliata*) are less sensitive to habitat disturbance than spider monkeys (*Ateles geoffroyi*); black bears (*Ursus americanus*) are less sensitive than grizzly bears; mountain lions (*Felis concolor*) are less sensitive than jaguars (*Felis onca*); and so on. For some taxa, the minimal amount of intact habitat that appears to be required is discouragingly large. Although most attention has been given to the great area requirements of large carnivores, many other species also appear to require very large areas. For example, Thiollay (1989) tentatively estimates that the minimum rain-forest reserve for raptors in French Guyana would have to be somewhere between 1 and 10 million ha to contain the full complement of forest species. Species that exist only in large intact habitats are, of course, extremely vulnerable to habitat loss and, given the rapid expansion of anthropogenic landscapes, face imminent extinction. An extremely important question, then, is this: How can management intervention compensate for loss of area? Interesting examples of the decision-making process that is required for conservationists to choose the most effective forms of intervention are provided by Maguire, Seal, and Brussard (1987) for management of the Sumatran rhino (*Rhinoceros sondaicus*) and Weaver, Escano, and Winn (1987) for the grizzly bear.

In general, to understand the extent to which it is possible to substitute intervention for area, we have to develop a better understanding of the relationship between habitat and demography—that is, we need field research that (1) emphasizes the linkage between the key population parameters of survivorship, mortality, and reproductive success and the characteristics of the habitat that determine its quality; and, (2) the significance of the spatial distribution of these habitats to the demography of the species. For the latter, Pulliam and Danielson (1991) develop a modeling approach based on mechanisms of habitat selection. They argue that for species with little selective ability, the spatial pattern of habitats may sometimes be more important than total area, whereas for highly vagile species, total area may be most critical. Thus, information about how species select habitats will be critical to decisions concerning the minimal size and spatial array of habitats that must be part of the management plan for the species.

3. Sites with Vulnerable Locations

Some areas are vulnerable because the likelihood that they will experience major disturbances is very high; that is, they are sites with locational or situational sensitivity. Some common examples of these areas include habitats near expanding urban populations, sites with high recreation development potential, sites with high agricultural potential, sites downwind or downstream from major sources of pollution (Figure 14.1), and so forth. However, proximity to intensive land use does not always threaten natural areas. A forested natural area close to agricultural development may be either endangered or conserved. In most climates, such as the east coast of North America during the last century and

Figure 14.1
San Joaquin Freshwater Marsh. One of the last freshwater marshes in southern California, the marsh is completely surrounded by the town of Irvine and acts as a biofilter, protecting upper Newport Bay. The marsh is also an important stopover for waterfowl using the Pacific flyway. (Photo source: I.K. Curtis, Inc., Burbank, California)

currently in most of the tropics, forests generally have been cut and replaced by agriculture. By contrast, in drier areas forests have often been conserved for their water catchment value. For example, it was the agricultural community in the semi-arid San Joaquin Valley of California that lobbied for the establishment of Sequoia National Park to protect its watershed and thereby conserve water for agriculture. In Costa Rica, the slopes above Lake Arenal long ago were converted to cattle pasture and crops, but they are now being reforested with native trees to protect Lake Arenal as a source of irrigation water to farms in the tropical dry forest lowlands.

Even though the need to protect agricultural development may sometimes lead to the conservation of natural areas, often agriculture and other development close to forest land may pose special danger from fires, pesticide runoff, and invasion by exotics. Terrestrial natural areas typically are embedded in landscapes that are highly anthropogenic; that is, they are characterized by agriculture, livestock rangelands, managed forests, and suburban and urban land uses. Productivity of the anthropogenic landscape surrounding any particular natural area is, to varying degrees, subsidized by external sources of energy and import of materials. Furthermore, anthropogenic landscapes are characteristically "leaky" systems, passively of their chemicals and actively of their weedy species. Because commercial agriculture and most kinds of industry use chemicals intensively and often inefficiently, chemicals leak into streams; groundwater; and, ultimately, natural wetlands, lakes, and coastal systems. Anthropogenic systems are also "leaky" in the sense that their characteristically weedy species are invasive and may become serious problems near the boundaries of natural areas. For these reasons it is important to capture entire watersheds in reserves and to maximize the average distance between the edge of a reserve and its center; therefore, for these purposes, round reserves are better than long, narrow ones.

4. Areas with Low Resilience to Perturbations

Systems may be particularly sensitive to perturbations for three reasons. First, areas with low net primary productivity (e.g., deserts, mountaintops, tundra, many tropical oxisol landscapes, etc.) generally will have slow rates of recovery. For many low-productivity habitats, management options are limited, and the best action is preventing disturbances. In some cases, most notably deserts, intervention may hasten the restoration of disturbed areas. Nitrogen may be more limiting than water for restoration of vegetation in some deserts. Where this occurs, application of slow-release nitrogen fertilizer may improve recovery rates, for example, of badly eroded lands. It may also be possible to manipulate desert rodent communities in ways that contribute to the restoration of desert shrub vegetation, for example, the strategic planting of shrubs for habitat cover and food may hasten dispersal of seeds and mycorrhizal inoculum.

Second, some systems have low thresholds of stability to particular kinds of perturbations, and persistent new species assemblages may become established. Tropical dry forests, for example, are converted to grassland as fires become frequent. In Southeast Asia the conversion to alang-alang (*Imperata cylindrica*) seems to be persistent. In Central America, tropical dry forest reinvades jaragua (*Hyparrhenia rufa*) grasslands within a few years after the elimination of fire (Janzen 1988), while tropical wet forest appears to reinvade these grasslands more slowly. Very little is known about the underlying mechanisms that explain the differing recovery rates that seemingly similar forest systems have in response to the same perturbation. It may well be that the explanation will be found in the nutrient linkages between higher plants and soil that is provided by the microbial community (see, for example, Perry et al. 1989).

Third, systems with low rates of exchange may degrade quickly (e.g., isolated islands and highly fragmented habitats may not easily recover lost species; isolated wetlands may accumulate pollutants resulting from separate runoff events). For systems that are distant from a source pool of native migrants, it may be necessary to both translocate native species where they have disappeared as well as actively remove highly competitive invasive species. For wetland sinks that accumulate pollutants, it may be necessary to flush out pollutants. This is probably the only long-term, albeit very expensive, solution to the contaminated wetlands of California's Central Valley.

In all three cases, the effects of perturbations may overlap in time, causing impacts from separate perturbations to become cumulative. Additionally, I should note that the relation between rate of recovery and rate of perturbation may be important. A very fast recovery system may overshoot and oscillate with a perturbation time course that is short, whereas a system with a long recovery period may never catch up with a series of rapid perturbations (Pimm 1986).

5. *Keystone Resources*

Most biological processes are characterized by considerable redundancy. For example, the elimination of one tree species from a forest may have no measurable effect on system processes such as average nitrogen mineralization or on the total abundance within functional guilds such as herbivores because the resources lost by the elimination of one species are compensated by other species. However, redundancy within biological systems is not universal, and there are important situations where the decline of one or more species can have an impact on many other species—that is, some resources and interactions play a keystone role in the sense that their elimination would have a major impact on community structure and processes. Some examples include seasonally limiting resources such as figs and palms for neotropical vertebrate frugivores (Terborgh 1986), dry-season pools for stream fishes (Zaret and Rand 1971), easily decoupled systems such as dependency between "trap-lining" pollinators and epiphytes (Janzen 1971), and top predators.

Management plans should explicitly take keystone resources into consideration, and research priority should be given to understanding (1) the mechanisms that determine the importance of keystone resources to the community and (2) the extent to which keystone resources might be managed to enhance the populations that depend on them. One of the major questions of conservation biology concerns the possibilities of using managed keystone resources to support populations in reserves that would otherwise be too small to contain viable populations. Success would depend on two factors: (1) the extent to which minimal area is simply a function of the spatial and temporal distribution of resources and hence might be manipulated, versus the more intractable behavioral use of space; and, (2) the secondary consequences of increasing the density of the managed populations. For example, in the Neotropics, tapirs (*Tapirus bairdii*) and white-lipped peccaries that feed primarily on large fruits and seeds appear to require very large areas of intact habitat. Because very few protected areas are extensive enough to provide adequate habitat for these species, they are especially vulnerable to extinction from habitat loss. One management option is to maintain zoo populations as a source for replenishing natural populations. Though probably essential, the use of captive populations to augment wild populations raises concerns about the effects of introducing captives into wild populations, especially the possibility of introducing disease. Another option for maintaining source populations is to attempt to support additional wild populations of tapirs and peccaries in smaller, more intensively managed reserves by manipulating the abundance of key resources (e.g., planting palms and other trees that produce food during times of seasonal scarcity). This is similar in principle to the management of game ranches in East Africa but differs by placing emphasis on the achievement of conservation needs of species and guilds by minimal manipulation of the natural habitat. Surplus individuals could then be used to replenish populations in larger and more natural reserves. It will be important to recognize that other significant trade-offs might occur. To continue with the example of the white-lipped peccaries, locally increasing their density by planting palms might impose significant seed mortality on other large-seeded trees, cause competition with other seed harvesters and dispersers, and thereby gradually change the forest composition of the small reserve. If it is possible and ecologically acceptable to use small reserves as managed source populations to augment other sites, then additional utility value accrues to small reserves.

MANAGING FOR AN UNCERTAIN FUTURE

Landscape functions change on several temporal scales that are of concern to conservation management. For areas such as U.S. national and state forests that are subject to "multiple-use" agreements, utilization of the landscape can change quickly in response to political pressures, budget allocation, and changing public priorities. On a somewhat longer time scale, land use may change in response to long-term economic trends, such as influencing agricultural and forestry practices, or in response to environmental conditions such as successional responses to disturbance, recovery from fire, and so on. On a longer time scale, land use will change in response to shifts in the global climate.

In this section, the focus is on three issues that pose especially difficult problems to the management of natural areas. They are (1) the management of "anthropogenic populations" or populations strongly influenced by human activity; (2) the changing economic landscape that surrounds natural areas, especially tropical regions; and (3) global climate change. They are difficult issues because the changes can occur rapidly, and the magnitude of the change at the local level is highly uncertain. The negative impact may be widespread, and the ecological effects of the changes, complex in themselves, may be further confounded by the social and economic responses that are made to the environmental change. For example, in terms of land-use change, the piedmont region of Georgia from the 1930s to the 1980s went from intensive farmland with very little forest to mostly forest with very little farmland. Thus, in approximately fifty years, the Georgia landscape has become strikingly less fragmented and the forest habitat more connected (Turner and Ruscher 1988). As discussed in the next section, deer (*Odocoileus virginiana*) populations rapidly expanded, due in part at least to the decline of farmland and the increase in secondary forest cover. In terms of climate change, some recent models suggest that decreasing moisture and increasing temperatures in eastern North America may result in the loss of yellow birch (*Betula alleghaniensis*), sugar maple (*Acer saccharum*), eastern hemlock (*Tsuga canadensis*) and beech (*Fagus grandifolia*) from the southern parts of their range (Davis 1988, reported in Roberts 1988). In Georgia, loblolly pine (*Pinus taeda*), economically the most important species, may disappear from the southern two-thirds of the state by 2080 (Miller et al. 1987).

1. Managing "Anthropogenic Populations": Invasive or Endangered Species

I restrict our use of the concept "anthropogenic populations" to those species whose population dynamics are strongly influenced by human activities. These may be highly invasive species that have a negative impact on natural communities and habitats (Figure 14.2) or species that are in rapid decline due to the impact of human activities.

Invasive species. Generally, terrestrial native species that recently have undergone great population expansions, in both geographic range and

Figure 14.2
*Hastings Natural History Reservation. This oak savanna reserve lies in California's coastal Carmel Valley. It is one of the best examples of Pacific Coast oak savanna, yet most of the grass cover consists of introduced Mediterranean annuals. In the absence of the native perennial grass cover, germination and seedling survival of valley (*Quercus lobata*) and blue (*Quercus douglasii*) oaks are poor. (Photograph by Galen Rowell/Mountain Light).*

local density, share at least one trait in common: They are ecotone specialists who therefore thrive in fragmented habitats.

For managers of natural areas, high priority should be given to populations that are rapidly increasing to the extent that they pose a significant threat to populations of other species or are degrading habitat. One example is the exotic Asian kudzu vine (*Pueraria lobata*) that aggressively invades hardwood forests of the southeastern United States. But not all problem species are exotics; there are long lists of native species that at times can pose significant management problems. Examples for North America include some of our most familiar species: eastern white-tailed deer, raccoons (*Procyon lotor*), opossum (*Didelphis marsupialis*), coyote (*Canis latrans*), crows, jays, cowbirds (Corvidae), mesquite (*Prosopis glandulosa*), honey locust (*Gleditsia triacanthos*) and many more. Food generalists such as corvids, opossums, and raccoons can impose serious mortality on nesting songbirds, especially along temperate forest edges (Wilcove, McLelland, and Dobson 1986). The expansion of mesquite in the southern plains of the United States has converted thousands of hectares of grassland into thorn scrub. Less dramatically in the Southeast, dense stands of honey locust often replace other successional tree species.

White-tailed deer (*Odocoileus virginianus*) have been the subject of hundreds of field studies, mostly dealing with ensuring that adequate supplies are available to hunters. In recent decades, largely due to successful stocking programs by state game-management programs and changes in land use from intensive farmland to successional forests and suburban woodlots, eastern white-tailed deer populations have greatly expanded. In Georgia, for example, deer have increased exponentially from approximately a few thousand individuals in 1940 to more than one million in 1985 (Odum 1987). High-density deer populations can impose severe herbivory pressure and influence the composition of plant communities. Table 14.1 lists a few of the many plants that are known to have declined in the presence of high deer populations.

Although deer populations have expanded as secondary forests have replaced farmland in the eastern United States, agricultural landscapes remain the principal source of invasive species. Agricultural practices increase the source pool, immigration rates, and colonization success of potentially invasive species. In general, there are four principal ways in which agriculture can increase invasive species (Carroll 1991):

a. By creating a greater frequency and extent of disturbed lands, agriculture creates a greater source pool of invasive species of natural habitats.

b. Through growth in international and regional commerce and tourism, propagules of potentially invasive species are more frequently transported to new areas.

Table 14.1
Plant Densities Influenced by White-Tailed Deer Browsing*

Direct Effects	Indirect Effects
Yew[1]	Leatherwood
Eastern hemlock	Wood sorrel
White oak[2]	
Live oak[3]	
Pin cherry[4]	
Sugar maple	
White ash	
Yellow birch	
Tree seedling species diversity	
Deciduous more than coniferous vegetation	
Orchids	
Lilies	
Blackberry	
Ferns (increased)	
Black cherry (increased)	

*All are declines except as noted. Indirect effects result from decline of eastern hemlock.
[1]Alverson, Waller, and Solheim 1988
[2]Bratton, Mathews, and White 1980
[3]Miller 1989
[4]Tilgham 1989

c.   By the spread of similar agricultural practices throughout the world, the chance of successful colonization by weeds is enhanced. For example, many of the most serious weeds have nearly global latitudinal distributions (Holm et al. 1977). From the perspective of a weed, cotton fields in southern Texas, Honduras, or Nigeria are largely similar resources.

d.   The expansion of agricultural lands results in the fragmentation of natural areas. Thus new additions to the pool of natural reserves tend to be small and the ratio of boundary to area is necessarily greater for smaller reserves. The increase in boundary effects increases the likelihood that agricultural pests will successfully invade small reserves and generally increases the impact of agriculture on natural areas.

In tropical regions, the invasion of weeds from disturbed lands into natural areas has received the most attention. Invasive weeds have two key effects on forests: competition with tree seedlings and the enhanced ability of dry weed biomass to carry destructive fires. These effects are illustrated by the pantropical weed *Chromolaena odorata* (syn. *Eupatorium odoratum*), Asteraceae, a highly branching, scandent perennial shrub that often reaches 7 m in height. A summary of attributes and ecological effects of this species is impressive: In Southeast Asia and West Africa it is often the greatest barrier to agroforestry; in Ghana it constitutes the largest component of second-growth seed banks and has the highest rate of germination; in moist forests of West Africa and India it constitutes the largest contributor to weed populations in forest clearings; in tree nurseries it has an allelopathic effect on tree seedlings; it strongly competes for soil nutrients; and it is protected from herbivory by very large quantities of essential oils and other secondary compounds (see Carroll 1991 and contained references).

Although *Chromolaena* and other invasive weeds may be important competitors of regenerating native vegetation in clearings, they often have far greater impact through fire. Weeds on adjacent agricultural lands and in forest clearings often develop large fuel loads, and dense populations of weeds can impose significant risk from damaging fires. In tropical regions, fire for pasture rejuvenation and weed control is a common and inexpensive management technique; consequently, wildfires pose a major and frequent threat to tropical forests.

Tropical wildfires are especially hazardous during long dry seasons and are most damaging to small forest patches and to forests that carry large amounts of downed wood and slash from logging operations. The dry biomass of *Chromolaena* with its high content of essential oils is highly flammable. In the savanna parks of east and southern Africa, *Chromolaena* carries fire from the regularly burned grasslands into forests and causes a continual attrition of forest patches (MacDonald and Frame 1988). Similar consequences of fire carried by weed biomass occur in Indonesia and Malaysia where *Imperata* carries fire into forest patches and becomes established in forests that have been degraded and opened by logging practices. In the dry forests of Costa Rica, *Hyparrhenia* carries fires into patches of deciduous forest during the dry season.

Just as weeds invade natural communities, so too may diseases move from agricultural animals and plants into natural habitats. One of the best-studied epidemics is the Great Rinderpest disease that entered tropical Africa in 1889 to 1890, probably from cattle introduced into Eritrea during the Italian occupation (Ford 1971). Less than a year after the disease reached Africa, F.D. Lugard in an 1891 communication to the Imperial British East Africa Company reported that in parts of Uganda, "almost all the game, including the small antelope, seem to have perished . . ." (cited in Ford 1971). In the savanna regions, the near-total loss of cattle resulted in famine, outbreaks of smallpox, and the collapse of these pastoral/agrarian societies.

Following the collapse of pastoralism and agriculture, large areas of cultivated lands returned to bush and acacia woodlands. The replacement of open savanna and cultivated fields by dense brush and woodlands created ideal habitat for the savanna tsetse fly (*Glossina morsitans*), an important vector of human and animal trypanosomiasis. In response to these habitat changes, the savanna tsetse fly greatly expanded its range during the early part of the twentieth century. Introduced diseases continue to have important ecological effects in national parks. Occasional outbreaks of distemper-like disease among wild canids in the Serengeti are thought to be spread from domestic dogs and cats kept by the Masai who reside in the conservation areas (Frame et al. 1979).

Endangered species. From the manager's perspective, two concerns are raised by declining small populations: (1) loss of genetic diversity and (2) risk of extinction from environmental and demographic stochasticity. There is general concern among conservation biologists that populations that have undergone precipitous declines may lose much heterozygosity through genetic drift and thereby lose their potential to respond to environmental changes (e.g., Frankel and Soulé 1981). Minimal levels of heterozygosity tolerated by populations without their experiencing measurably increased risk of extinction must be empirically determined and are largely unknown for virtually all natural populations. It is likely that these levels will vary considerably by taxa and environmental pattern.

Recently, some general guidelines for the management of small populations have been suggested. Franklin (1980) suggested that effective population sizes less than 50 are susceptible to inbreeding depression, and effective population sizes of 500 or greater are needed to prevent gene loss through drift. The guideline for avoiding inbreeding depression is based on the experience of animal breeders with domestic animals and so may be an underestimate for wild species in natural circumstances. The guideline for avoiding drift is based on application of a particular model to a laboratory study of bristle numbers in *Drosophila*. Thus, the 50 to 500 guidelines should be seen less as advice to managers than as a stimulus to field and theoretical research. In a comprehensive treatment of the role of genetics in population management, Lande and Barrowclough (1987) show that the 500 rule may be adequate for quantitative traits but single locus variation under neutral selection may require effective population sizes of 10^5 to 10^6.

Several models (e.g., Lacy 1987; Chesser 1983) suggest that genetic variation is better maintained in a metapopulation with migration among subunits than in a single population equivalent in size to the metapopulation. Simulation models also provide some evidence that a population subdivided into a metapopulation with migration between subunits will persist longer in a highly stochastic environment than a single population (Reddingius and den Boer 1970; Roff 1974). This has led to suggestions (e.g., Goodman 1987) that a series of small refuges with migration will lower extinction rates below that of a single large population. Thus, an important field research focus is the significance of small populations existing as subunits within a metapopulation. To meet conservation needs for the maintenance of genetic variation and for lowering risk of extinction due to environmental stochasticity, we need to emphasize effective population size and migration within any existing metapopulation. Furthermore, if such field studies are to contribute to the development of robust guidelines for managers, they need to be done in the context of simulation and analytical models. It is, of course, easier to "see" environmental effects on small populations than the consequences of genetic homogeneity. Mortality is generally more obvious than loss in reproductive fitness, and managers have therefore historically focused more attention on survivorship than on genetics (for a good treatment of the relationship of population biology to conservation, see Simberloff 1988).

Reintroduction of species into formerly occupied habitats or to augment precipitously small populations has been attempted, with mixed results, for a variety of animals and plants. Bobcats onto Cumberland Island National Park, wolves into the upper peninsula of Michigan, collared lizards onto Ozark Mountain "balds," and pupfish into desert refugia are indicative of translocation attempts in the United States. It is clear that some kinds of translocation will always be problematic (e.g., the reintroduction of wide-ranging top carnivores into limited habitat such as the reintroduction of wolves into northern Michigan or their proposed reintroduction into Yellowstone National Park). Approximately 700 translocations are attempted each year in Australia, Canada, New Zealand, and the United States combined. Of these, about 86 percent of the attempts with native game species are successful whereas only 44 percent are successful for threatened, endangered, or sensitive species (Griffith et al. 1989).

Before translocations are attempted, one should understand the mechanisms through which the species lost their old habitat. For example, it may be difficult and costly to keep a reintroduced species in a habitat that is now simply too small or that contains an inadequate mix of resources. In many cases, habitat restoration efforts may be needed before translocation efforts will be successful. It is also important to consider the probable consequences of reintroduction on the resident species, though this may be difficult to predict. Reintroduction of a plant with historically restricted ranges and patchy distributions may have little likelihood of creating adverse consequences. However, reintroduction of a generalist herbivore into a habitat, especially if resident predator populations are low or absent, could have major impacts on a plant community. Certain habitats, especially islands and other productive but species-poor sites, are especially vulnerable to invasive species, and great care should be taken before reintroductions of even native species are attempted. In all cases, it is important to have well-considered contingency plans to deal with any reintroduction programs that go awry before any attempt to establish populations is made.

2. The Changing Economic Landscape

Species, except for a handful of the more charismatic endangered ones, typically share their habitat with people. Certainly the vast majority of species, even in the tropics, are found in less-than-pristine environments. In the United States, multiple-use national forest lands contain approximately one-third of all federally listed threatened and endangered species. California's Inyo National Forest alone has more visitor days than Yellowstone, Grand Canyon, and Glacier national parks combined (Pister 1988a). Protection of sensitive species and habitats may include rigorous laws governing various kinds of reserve areas and prescriptive protection given to particular species and habitats. But in many cases, especially on private lands, protection is *laissez-faire* and fortuitous.

The majority of common property wildlands worldwide are by custom or by legislation governed by "multiple use," and though the concept is politically useful, it can be environmentally lethal. Many kinds of tenure rights are associated with renewable resources around the world (see, e.g., Fortmann and Bruce 1988 with regard to multiple forest use), and these patterns of use must be considered along with characteristics of the local ecosystem in any long-term attempt at management. Multiple use often results in user conflicts when the tenure of the use groups overlaps in the use of common property; for instance, the tenure of off-road vehicle use versus the tenure of hikers; the tenure of loggers in national forests versus the tenure of natural-history devotees.

Governmental policy concerning the users of one resource can create effects on another. For example, in tropical developing countries, liberal logging concessions open up forests that are then more accessible to shifting cultivators (Gillis and Repetto 1988). In the United States, liberal logging concessions open up forests that consequently support higher deer populations and encourage greater use by hunters. Indeed, the U.S. Forest Service includes recreational values to rationalize selling timber below market value (e.g., see U.S. Forest Service 1985). There is considerable debate whether the national accounting of costs and benefits is unbiased, leading critics to argue that the real costs of environmental degradation and wildlife habitat loss are generally understated and the benefits overvalued (Repetto 1988; Rice 1989). In the United States, energy laws passed in the 1970s (Geothermal Steam Act and the Public Utility Regulatory Act) provide incentives for the entrepreneurial development of energy resources, in particular requiring that public utilities purchase energy converted by these companies. The proliferation of these companies in the eastern Sierra Nevada of California, which in aggregate produce little contribution to regional energy needs, has created excessive environmental damage and degraded recreational and wildlife habitats (Pister 1988b).

In moist tropical countries, logging and forest conversion to plantations and cattle pastures have commonly been encouraged by governments through liberal concessions, tax benefits, and access to low-interest loans. At the national level it is understandable why such policies should be pursued, even though they may have short-term viability. Virtually all tropical countries are desperately short of large external debts. Thus cattle and timber exports provide a ready source of foreign trade with little investment required of national capital. In Indonesia, priority forestry projects receive long-term tax exemptions, rapid depreciation allowances, duty-free import of equipment, and liberal transfer abroad of profits in foreign currency (Allen, Straka, and Watson 1986). In Brazil, corporations may convert tax liabilities into venture capital for investment in government-approved Amazonian livestock projects. Various forms of subsidy provide economic support for Amazonian cattle production, with the net result that nearly four tons of foreign beef could be imported into Brazil for the real cost to the government of producing one ton of beef in the Amazon region (Browder 1988). In another study of cattle ranching in the Amazon region, overgrazing pasturelands is one way to short-term profitability that, by decreasing the useful life of pastures, creates incentives to convert cheaper forest land into new pasture (Hecht, Norgaard, and Possio 1988). Particularly disturbing are reports that demand for cocaine has resulted in the conversion of nearly 800,000 ha of Peru's rain forest into coca plantations (*The Atlanta Journal and Constitution* 1989). Gillis and Repetto (1988) provide an excellent assessment of the principal reasons why government policy for multiple use of common property resource has often led to forest destruction (Table 14.2).

At the local level, unsustainable land-use practices may yield short-term profits for large, well-capitalized absentee landowners but ultimately result in economic disaster for the small landholder and wage laborer. In a study of comparative yields from 1 ha of Peruvian rain forest, the sustainable-use value (fruits, latex, and highly selective logging) was estimated at $6,282 after subtracting collection and transportation costs (Peters, Gentry, and Mendelsohn 1989). By contrast, the more conventional uses yielded far less: clear-cut ($1,000), conversion to *Gmelina arborea* pulpwood and timber plantation ($3,184), and cattle ($2,960 without deducting the maintenance costs). Clearly in this case, a sus-

Table 14.2
Reasons Why Government Policy Often Leads to Forest Destruction*

1. The continuing flow of benefits from intact natural forests has been consistently undervalued by both policymakers and the general public.
2. Similarly, the net benefits from forest exploitation and conversion have been overestimated, both because the direct and indirect economic effects have been exaggerated and because many of the costs have been ignored.
3. Development planners have proceeded too boldly, exploiting tropical forests for commodity production without biological knowledge of their potential or limitations or awareness of the economic consequences of development policies.
4. Policymakers have attempted—without much success—to draw on tropical forest resources to solve fiscal, economic, social, and political conflicts elsewhere in society.
5. National governments have been reluctant to invest the resources that would have been required for adequate stewardship of the public resource over which they asserted authority.
6. National governments have undervalued the wisdom of traditional forest uses and the value of local traditions of forest management that they have overruled.

*From Gillis and Repetto 1988.

tainable extractive use of the forest is economically more rational for local economic development and could contribute to the long-term sustainability of a more or less natural forest ecosystem. Optimistic economic analyses of the extractive value of tropical forests have given support to the establishment of "extractive reserves" in Brazil and elsewhere. However, some caution concerning the relationship of extractive reserves to the conservation of biodiversity is warranted. First, it is not at all obvious that forest dwellers are willing to attempt a living by this modern market-oriented hunting and gathering. In Brazil, the extractive reserve concept was largely designed to assist rubber tappers, who engage in a difficult and full-time occupation that allows little extra time for significant harvesting of other natural products. Second, unmanaged harvesting could create serious habitat degradation. And third, many of these natural products are themselves important resources in the forest ecosystem. In particular, extractors would be competing for fruits with large herbivores, a group that is often endangered.

As the previous examples demonstrate, conservation is not just biology and must be an interdisciplinary effort. Conservation must become an integrative effort to treat the landscape as a life-support system for economic and biological diversity. Among the research priorities that address the interface of conservation and economic development, I would highlight the following:

a. Determine the real costs of degradation and regeneration of biodiversity and ecosystem functions without invoking excessive future discounting (Ledec and Goodland 1988).

b. Determine the complete "flow of benefits" from intact wildlands.

c. Find appropriate uses of buffer zones for economic yields that neither degrade the nearby wildlands nor create incentives for local human population growth (e.g., see Carroll 1991).

3. Global Warming and the Changing Environment

"The problem of global environmental change is crucial and urgent" (National Academy of Sciences 1988, 1). Among the principal concerns for conservation I particularly note the following: (i) Change may occur more rapidly, within 50 to 100 years, than organisms can be expected to adapt; (ii) fragmentation and isolation of habitats have restricted migration as an option for natural communities; (iii) the economic need to adjust to the effects of climate change may force dramatic conservation in land use in ways that could further degrade natural communities; (iv) expected increases in weather extremes suggest that environmental stochasticity may become a greater contributor to local extinction rates; and, (v) highly invasive weedy species may be particularly favored by the disruptive influences of climate change. Many of these concerns have been developed more fully in recent symposia and workshops, such as the recent symposium sponsored by The World Wildlife Fund ("Consequences of the Greenhouse Effect for Biological Diversity," 4–6 October 1988) and the workshops of the US-IGBP Committee on Global Change (National Academy of Sciences 1988).

Biological interactions are strongly influenced by temperature and moisture, and, in many cases, members of the interaction will not share the same sensitivity. Some examples illustrate how interactions with significant ecological consequences might be influenced by climate change.

In the deserts of the southwestern United States, densities of the scorpion *Paruroctonus luteolus* appear to be limited by predation from the sand scorpion *P. mesaensis*. The sand scorpion's activity is largely restricted to the warmer months. Thus, when *P. luteolus* is mating during the cool nights of fall and is otherwise vulnerable, the sand scorpion remains in its burrow. On unusually warm fall nights the sand scorpion remains active and preys upon the vulnerable *P. luteolus* (Polis 1989).

For some species (e.g., some pierid butterflies [Kingsolver 1983; Cappuccino and Kareiva 1985]), body temperatures necessary to permit growth and reproduction seldom may be reached. A generally warmer climate presumably would be difficult for the scorpion *P. luteolus* and beneficial to some pierid butterflies. There are many cases where predators and parasitoids respond differently than their prey to climate variation (Messenger, Billotti, and van den Bosch 1976).

Climate change may induce unexpected and complex community responses when these interactions are variously intensified or relaxed. For example, mortality of parasitoid and predatory insects that prey upon caterpillars in temperate forests is strongly affected by the severity of winter. During mild winters when mortality from environmental stress is low, the higher springtime densities of predators and parasitoids may sometimes depress the populations of their caterpillar prey (e.g., Evans 1982; Frazer and Gilbert 1976; Power and Kareiva 1989). The influence of weather on these predator-prey interactions is by no means certain, and mild winters may in some cases result in large increases in caterpillar abundances in spite of considerable mortality from predators and parasitoids. Successful reproduction by insectivorous birds is strongly dependent on food density, and mild winters may either force greater competition between birds and predatory insects or result in an abundance of food. Of particular significance is that most climate-change scenarios portray milder temperate zone winters as an average condition but also predict greater year-to-year variance in local weather with greater extremes. Thus, climate change frequently may result in greater competition for food, but the greater year-to-year variance also suggests that population booms and crashes will occur more frequently.

Although rapid changes in climate may influence important species interactions, there is considerable historical evidence to suggest that communities, at least temperate zone communities, are not tightly coupled species assemblages and that the geographic ranges of species have largely changed independently (Davis et al. 1986; Huntley and Webb 1989). Thus, what we designate as a community may be a useful heuristic concept for some purposes, but it probably has limited usefulness as a management unit.

Current models of climate change are far too broad-scaled to provide more than a rather murky window into the future. But they do suggest that we need to focus research in several key areas:

a. Identification of key interactions and resources that are likely to be significantly changed and development of modeling approaches to investigate the probable results from those changes.

b. Identification of the spatial array of habitats and their relation to the demographies of representative taxa and, again through modeling, investigation of the effects of changing those spatial and habitat quality relationships. To accomplish the latter we must place more emphasis on field studies to understand the various mechanisms species use to locate and exploit habitats. To accomplish the former, we will need to depend opportunistically on "natural experiments" (e.g., unusual weather) or undertake large-scale experimental manipulations.

In the absence of latitudinal migration corridors it will be particularly important to focus research on protected areas that are likely to contain refugia for those taxa unable to cope with the changing climate. For example, many large watersheds contain multiple habitats, wet refugia, and altitudinal zonation. Because watersheds also are critical for the environmental services they supply to the economy, such landscapes may become the most significant conservation units in a future of changing climate.

CONCLUSIONS

The long-term security of natural areas that are surrounded by various kinds of economic landscapes will strongly depend on the way land surrounding natural areas is used. In particular, patterns of use will have to be economically sustainable, socially equitable, and explicitly linked with management objectives in the embedded natural area.

The decreasing size of many natural areas and the increasing distances between them greatly contribute to the vulnerability of natural biodiversity. The small size of most natural areas results in increasing rates of invasion of weedy species and greater perturbations from nearby agricultural and other development practices. The increasing isolation of natural areas results in a decrease in rates of recolonization and thereby greater vulnerability to species and gene extinction processes.

These problems will only intensify as economic pressures increase and as the changing climate creates greater uncertainty and at the same time forces us to take action. Thus, ecologists must learn to broaden their academic habitats and move from the relative intellectual security of pristine equilibrium systems to the realities of disequilibrium (see chapter 4), and to incorporate the economic landscape as part of the life-support system for natural biodiversity.

LITERATURE CITED

Allen, R.C., T.J. Straka, and W.F. Watson. 1986. Indonesia's developing forest industry. *Environ. Manage.* 10:735–59.

Alverson, W.S., D.M. Waller, and S.L. Solheim. 1988. Forests too deer: Edge effects in Northern Wisconsin. *Cons. Biol.* 2:348–58.

Atlanta Journal and Constitution. 1989. Cocaine makers fell Peru's forest, flood rivers with toxic chemicals. 13 August 1989.

Bratton, S.P., R.C. Mathews, Jr., and P.S. White. 1980. Agricultural area impacts within a natural area: Cades Cove, a case history. *Environ. Manage.* 4:433–48.

Browder, J.O. 1988. The social cost of rain forest destruction: A critique and economic analysis of the "hamburger debate." *Interciencia* 13:115–20.

Cappuccino, N., and P. Kareiva. 1985. Coping with a capricious environment: A population study of a rare pierid butterfly. *Ecology* 66:152–61.

Carroll, C.R. 1991. The interface between natural areas and agroecosystems. In *Agroecology*, ed. C.R. Carroll, J.H. Vandermeer, and P. Rosett, 365–84. New York: McGraw-Hill.

Chesser, R.K. 1983. Isolation by distance: Relationship to the management of genetic resources. In *Genetics and conservation*, ed. C.M. Schonewald-Cox, S.M. Chambers, B. MacBryde, and W.L. Thomas, 66–77. Menlo Park, Calif.: Benjamin/Cummings.

Davis, M.B. 1988. Consequences of the greenhouse effect for biological diversity. Presentation at a symposium sponsored by the World Wildlife Fund, 4–6 October 1988. As reported by Roberts, L. 1988. Is there life after climate change? *Science* 242:1010–12.

Davis, M.B., K.D. Woods, S.L. Webb, and R.P. Futyman. 1986. Dispersal versus climate: Expansion of *Fagus* and *Tsuga* into the Upper Great Lakes region. *Vegetatio* 67:93–104.

Dobzhansky, Th. 1970. *Genetics of the evolutionary process.* New York: Columbia University Press.

Evans, E. 1982. Influence of weather on predator/prey relations: Stinkbugs and caterpillars. *J. N.Y. Entomol. Soc.* 90:241–46.

Ford, J. 1971. *The role of the trypanosomiasis in African ecology.* Oxford: Clarendon.

Fortmann, L., and J.W. Bruce, eds. 1988. *Whose trees? Propriety dimensions of forestry.* Boulder, Colo., and London: Westview.

Frame, L.H., J.R. Malcolm, G.W. Frame, and H. van Lawick. 1979. Social organization of African wild dogs *Lycaon pictus* on the Serengeti plains, Tanzania 1967–1978. *Z. Tierpsychol.* 50:225–49.

Frankel, O.H., and M.E. Soulé. 1981. *Conservation and evolution.* Cambridge: Cambridge University Press.

Franklin, I.R. 1980. Evolutionary changes in small populations. In *Conservation biology: An evolutionary-ecological perspective*, ed. M.E. Soulé and B.A. Wilcox, 135–49. Sunderland, Mass.: Sinauer.

Frazer, B.D., and N. Gilbert. 1976. Coccinellids and aphids: A quantitative study of the impact of adult ladybirds preying on field populations of pea aphids. *J. Entomol. Soc. British Columbia* 73:33–56.

Gillis, M., and R. Repetto. 1988. Conclusion: Findings and policy implications. In *Public policies and the misuse of forest resources*, ed. R. Repetto and J. Gillis, 385–410. Cambridge: Cambridge University Press.

Goodman, D. 1987. Consideration of stochastic demography in the design and management of biological reserves. *Nat. Resource Model.* 1:205–34.

Griffith, B., J.M. Scott, J.W. Carpenter, and C. Reed. 1989. Translocation as a species conservation tool: Status and strategy. *Science* 245:477–80.

Hecht, S.B., R.B. Norgaard, and G. Possio. 1988. The economics of cattle ranching in eastern Amazonia. *Interciencia* 13:233–40.

Holm, L.G., D.L. Plucknett, J.V. Pancho, and J.P. Herberger. 1977. *The world's worst weeds: Distribution and biology.* Honolulu: University Press of Hawaii.

Huntley, B., and T. Webb III. 1989. Migration: Species response to climatic variations caused by changes in the earth's orbit. *J. Biogeog.* 16:5–19.

Hutchinson, G.E. 1965. *The ecological theater and the evolutionary play.* New Haven: Yale University Press.

Janzen, D.H. 1971. Euglossine bees as long-distance pollinators of tropical plants. *Science* 171:203–6.

———. 1986. The eternal external threat. In *Conservation biology: The science of scarcity and diversity,* ed. M.E. Soulé, 286–303. Sunderland, Mass.: Sinauer.

———. 1988. Guanacaste National Park: Tropical ecological and biocultural restoration. In *Rehabilitating damaged ecosystems. Volume II,* ed. J. Cairns, 143–92. Boca Raton, Fla.: CRC Press.

Kingsolver, J.G. 1983. Ecological significance of flight activity in *Colias* butterflies: Implications for reproductive strategy and population structure. *Ecology* 64:546–61.

Lacy, R.C. 1987. Loss of genetic diversity from managed populations: Interacting effects of drift, mutation, immigration, selection and population subdivision. *Cons. Biol.* 1:143–58.

Lande, R. 1988. Genetics and demography in biological conservation. *Science* 241:1455–60.

Lande, R., and G.F. Barrowclough. 1987. Effective population size, genetic variation, and their use in population management. In *Viable populations for conservation,* ed. M.E. Soulé, 87–124. Cambridge: Cambridge University Press.

Ledec, G., and R. Goodland. 1988. *Wildlands: Their protection and management in economic development.* Washington, D.C.: The World Bank.

Lovejoy, T.E., R.O. Bierregaard, Jr., A.B. Rylands, J.R. Malcolm, C.E. Quintela, L.H. Harper, K.S. Brown, Jr., G.V.N. Powell, H.O.R. Schubart, and M.B. Hays. 1986. Edge and other effects of isolation on Amazon forest fragments. In *Conservation biology: The science of scarcity and diversity,* ed. M.E. Soulé, 257–85. Sunderland, Mass.: Sinauer.

MacDonald, I.A.W., and G.W. Frame. 1988. The invasion of introduced species into nature reserves in tropical savannas and dry woodlands. *Biol. Cons.* 44:67–93.

Maguire, L.A., U.S. Seal, and P.F. Brussard. 1987. Managing critically endangered species: The Sumatran rhino as a case study. In *Viable populations for conservation,* ed. M.E. Soulé, 543–64. Cambridge: Cambridge University Press.

Mayr, E. 1966. *Animal species and evolution.* Cambridge: The Belknap Press of Harvard University.

Messenger, P.S., E. Billotti, and R. van den Bosch. 1976. The importance of natural enemies in integrated control. In *Theory and practice of biological control,* ed. C.B. Huffaker and P.S. Messenger, 543–64. New York: Academic Press.

Miller, S.K. 1989. Reproductive biology of white-tailed deer on Cumberland Island, Georgia. National Park Service Cooperative Study Unit Technical Report no. 52. Athens, Ga.: University of Georgia.

Miller, W.F., Dougherty, and G.C. Switzer. 1987. Effects of rising carbon dioxide and potential climate change on loblolly pine distribution, growth, survival, and productivity. In *The greenhouse effect, climate change, and U.S. forests*, ed. W.E. Shands and J.S. Hoffman, 157–87. Washington, D.C.: The Conservation Foundation.

National Academy of Sciences. 1988. Toward an understanding of global change: Initial priorities for U.S. contributors to the International Geosphere-Biosphere Program. Washington, D.C.: National Academy Press.

The Nature Conservancy Magazine. 1989. New projects, Florida, Pinhook Swamp. 22–24. July-August.

Odum, E.P. 1987. The Georgia landscape: A changing resource. Final report of the Kellogg Physical Resources Task Force, E.P. Odum, Chair. Institute of Ecology, University of Georgia, Athens.

Perry, D.A., M.P. Amaranthus, J.G. Borchers, S.L. Borchers, and R.E. Brainerd. 1989. Bootstrapping in ecosystems. *BioScience* 39:230–37.

Peters, C.M., A.H. Gentry, and R.O. Mendelsohn. 1989. Valuation of an Amazonian rainforest. *Nature* 339:655–56.

Pimm, S.L. 1986. Community stability and structure. In *Conservation biology: The science of scarcity and diversity*, ed. M.E. Soulé, 309–29. Sunderland, Mass.: Sinauer.

Pister, P. 1988a. The Congressional Record, 10/4/88.

———. 1988b. Saving the Eastern Sierra: Recreation or generation? *Los Angeles Times*, 6 November 1988.

Power, A.G., and P. Kareiva. 1989. Herbivorous insects in agroecosystems. In *Agroecology*, ed. C.R. Carroll, J.H. Vandermeer, and P. Rosett. New York: McGraw-Hill.

Polis, G.A. 1989. The unkindest sting of all. *Natural History* 34–39. July 1989.

Pulliam, H.R., and B.J. Danielson. 1991. Sources, sinks and habitat selection: A landscape perspective on population dynamics. *Am. Nat.* 137 Supplement: 550–66.

Reddingius, J., and P.J. den Boer. 1970. Simulation experiments illustrating stabilization of animal numbers by spreading of risk. *Oecologia* 5:240–84.

Repetto, R. 1988. Subsidized timber sales for national forest lands in the United States. In *Public policy and the misuse of forest resources*, ed. R. Repetto and M. Gillis, 353–84. Cambridge: Cambridge University Press.

Rice, R.E. 1989. *National forests. Policies for the future. Vol. 5: The uncounted costs of logging*. Washington, D.C.: The Wilderness Society.

Roberts, L. 1988. Is there life after climate change? *Science* 242:1010–12.

Roff, D.A. 1974. Spatial heterogeneity and the persistence of populations. *Oecologia* 15:245–58.

Salwasser, J., C. Schonewald-Cox, and R. Baker. 1987. The role of interagency cooperation in managing for viable populations. In *Viable populations for conservation*, ed. M.E. Soulé, 159–74. Cambridge: Cambridge University Press.

Simberloff, D. 1988. The contribution of population and community biology to conservation science. *Ann. Rev. Ecol. Syst.* 19:473–511.

Soulé, M.E., and K.A. Kohm. 1989. *Research priorities for conservation biology*. Critical Issues series. Washington, D.C.: Island Press.

Templeton, A.R. 1986. Coadaptation and outbreeding depression. In *Conservation biology: The science of scarcity and diversity*, ed. M.E. Soulé, 105–6. Sunderland, Mass.: Sinauer.

Terbourgh, J. 1986. Keystone plant resources in the tropical forest. In *Conservation biology: The science of scarcity and diversity*, ed. M.E. Soulé, 330–44. Sunderland, Mass.: Sinauer.

Thiollay, J.M. 1989. Area requirements for the conservation of rain forest raptors and game birds in French Guiana. *Cons. Biol.* 3:128–37.

Tilghman, N.G. 1989. Impacts of white-tailed deer on forest regeneration in northwestern Pennsylvania. *J. Wildl. Manage.* 53:524–32.

Turner, M.G., and C.L. Ruscher. 1988. Changes in landscape patterns in Georgia, USA. *Landsc. Ecol.* 1:241–51.

U.S. Forest Service. 1985. Annual report, for fiscal year 1984. Washington, D.C.: U.S. Government Printing Office.

Weaver, J.L., R.E.F. Escano, and D.S. Winn. 1987. A framework for assessing cumulative effects on grizzly bears. *Trans. 52nd N.A. Wildl. Nat. Res. Conf,* 364–76.

Wilcove, D.S., C.H. McLelland, and A.P. Dobson. 1986. Habitat fragmentation in the temperate zone. In *Conservation biology: The science of scarcity and diversity*, ed. M.E. Soulé, 237–56. Sunderland, Mass.: Sinauer.

Zaret, T.M., and A.S. Rand. 1971. Competition in tropical stream fishes: Support for the competitive exclusion principle. *Ecology* 52:336–42.

CHAPTER
15

Park Protection and Public Roads

CHRISTINE SCHONEWALD-COX

and MARYBETH BUECHNER

ABSTRACT

Landscape fragmentation is increasingly subdividing natural areas into semi-isolated remnants. Although such subdivision has benefits for some species, it often reduces the ability of an area to protect species that are sensitive to disturbance or prone to local extinction. Within protected areas, such as parks, landscape fragmentation results from road development. Roads both subdivide continuous habitat and act as corridors for the entry of materials, edge-adapted species, and disturbances into natural areas. We review the general effects of landscape fragmentation on sensitive species and the ecological effects of roads. We then combine this information with a pilot analysis of the extent to which our largest national parks are fragmented by roads. Our results indicate that substantial areas of our national parks are close enough to paved roads that they may be impacted by road effects. We consider the implications of this for park planning processes.

INTRODUCTION

In the United States the alteration of habitat for human uses is progressively subdividing native habitats (Harris 1984). Landscape fragmentation often comes about when new areas of human-altered habitat are produced between sections of a formerly continuous community or ecosystem (Wilcox 1980; Forman and Godron 1986; Wilcove 1988; Yahner 1988). As human modification of an area progresses, the landscape becomes increasingly different from its ancestral state. The overall aspects of this process are illustrated in Figure 15.1. As humans extract resources

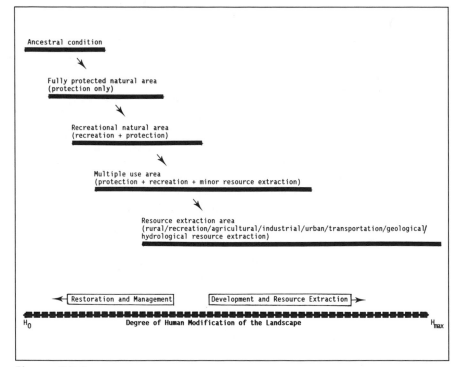

Figure 15.1
In this figure the degree of human modification of the landscape is illustrated for several types of land use. Development and resource extraction result in increasing modification of the landscape, whereas restoration and management result in reduced modification.

from a natural area, the degree of human domination of the landscape increases, and the ability of the area to support species sensitive to human impacts decreases. Restoration and management of human-modified landscapes can to some extent reverse this process. The preservation of natural areas in a relatively undisturbed state is vital to the conservation of numerous species of plants and animals. Although many species thrive or survive in human-altered habitats, remnant natural areas, the largest of which are our national parks, are the sole refuges for many other species.

The subdivision of a landscape by human-modified areas changes ecosystem patterns, and fragmentation of ecosystems constrains spatial patterns and affects the rate and frequency of geomorphic and nongeomorphic processes (Swanson et al. 1988). Such processes alter landforms, causing changes in air, water, temperature, and nutrient flow patterns and modifying the movement of organisms, propagules, energy, and materials (Swanson et al. 1988). These changes can result in positive feedback loops that continue or accelerate the process of fragmentation (Schonewald-Cox and Stohlgren, 1989). The challenge that landscape fragmentation presents for conservation of natural resources is most serious when generated edges that are normally found outside of protected areas, e.g., those associated with roads, also occur within park or reserve boundaries (Schonewald-Cox and Bayless 1986; Schonewald-Cox 1988).

Much of the anthropogenic fragmentation occurring within parks results from the development of paved roads within park boundaries. Within parks, the edges generated along roads and park boundaries by humans responding to access opportunities and protection regulations are superimposed on any preexisting natural edges in the habitat, thereby increasing the exposure of park areas to various disturbances (geological, microclimatic, etc.) (Schonewald-Cox and Bayless 1986; Schonewald-Cox 1988). Edge effects, natural or anthropogenic, including changes in microclimate and species composition, can extend for long distances into the interior of patches (Temple 1986; Forman and Godron 1986). An analysis of the effects of roads is especially critical to our understanding of landscape fragmentation because roads serve as both barriers and as corridors. Roads at once subdivide landscapes, removing habitat and inhibiting species dispersal and migration, and facilitate the movement of disturbances (e.g., pollutants or exotic species) into natural areas (e.g., van der Zande, ter Keurs, and van der Weiden 1980). Fragmentation of parks by roads is most likely to negatively affect species that (1) do not do well in edge habitat; (2) are sensitive to human contact; (3) exist at low densities; (4) are unlikely or unable to cross roads; (5) seek roads for heat or food; and, (6) require considerable space (more than is available in any single road-created fragment).

National parks face a dilemma with respect to the development of paved roads within the parks. The dual priority interpretation of the National Park Service Organic Act (1916) has resulted on the one hand in a directed effort toward the preservation of biodiversity within parks for future generations. On the other hand, implementation of the alternate agency mandate, public enjoyment, has resulted in the physical subdivision of parks by visitor access routes and conveniences. A technical conflict results concerning how best to protect parks while providing for public enjoyment (Brown 1989; Sudia 1989). Dr. T. Sudia, former Chief Scientist of the National Park Service, has stressed that landscape fragmentation within parks should be examined as a potential threat to fragile resources, perhaps more threatening than what goes on in adjacent areas. While we benefit from the public access provided by roads, we also increase exposure of park interiors to road-related impacts. If we are to effectively protect the natural and cultural resources of our wildlands, we need to document and mitigate the effects of this division of a once-extensive natural landscape into a series of smaller, semi-isolated remnant areas.

In this chapter we review the general effects of landscape fragmentation on sensitive species or systems and the ecological effects of roads. We then report the results of a pilot analysis of the extent to which the largest U.S. national parks in the contiguous forty-eight states are fragmented by roads. A consideration of our results in light of information on the ecological effects of landscape fragmentation by roads has important implications for conservation and park planning.

EFFECTS OF LANDSCAPE FRAGMENTATION ON SENSITIVE SPECIES

Subdivision of ecosystems also produces changes in the size and shape of continuous areas of habitat, the degree of isolation of remnant areas (Wilcox and Murphy 1985), the influx of materials and species found in the disturbed habitat, and the rate of further induced fragmentation. Table 15.1 outlines some major effects of landscape fragmentation on sensitive species or systems. In general, these changes manifest themselves in an increasing amount of edge habitat relative to patch interior habitat in the landscape (Wilcove, McClellan, and Dobson 1986; Yahner and Scott 1988). Many species will obviously benefit from an increased amount of edge habitat. In this analysis, however, we are concerned with those species and habitats that are both negatively impacted by the effects of landscape fragmentation and in need of protection.

Table 15.1

Summary of some major effects of landscape fragmentation on sensitive species or systems, particularly as determined by roads within parks. Other species or systems (not directly considered here) may be insensitive to or may benefit from fragmentation processes. This list is based on the current understanding of landscape processes and thus is subject to change.

I. MODIFICATION OF HABITAT
    A. Changes in the size and shape of landscape elements
        1. Decreased size of continuous habitat in remnant patches
        2. Altered shape of continuous areas of patch interior habitat
        3. Altered geometry of edges
        4. Increased perimeter:area ratios of remnant patches
    B. Changes in the connectivity and isolation of landscape elements
        1. Increased degree of isolation of remnant patches for species, materials, or effects restricted to patch interior habitats
        2. Increased connectivity of remnant patches for species, materials, or effects following edge or modified habitat
        3. Increased access for logging, mining, hunting, and other resource-extraction activities
        4. Increased access for poachers and other illegal activities
    C. Changes in habitat types
        1. Increased amount of edge and modified habitats
        2. Decreased amount of patch interior habitats
        3. Changes in the composition and geometry of edge habitats
        4. Loss of sensitive species from small remnant patches
        5. Altered balance of exotic and native species
        6. Altered balance of weedy or edge and patch interior species
        7. Increased spatial and temporal variation in habitat quality for patch interior species
        8. Increased habitat homogeneity within small remnant patches
        9. Changes in the capacity of the reserve for populations of sensitive species
II. MODIFICATION OF THE QUALITY OF PROTECTION PROVIDED
    A. Changes in balance of patch interior versus edge species and native versus exotic species
    B. Increased exposure of internal areas and further subdivision of landscape
        1. Direct removal of habitat
        2. Increased amount of edge in landscape
        3. Increased exposure to edge effects
        4. Increasing fluctuation of microclimate and related processes
        5. Influx of foreign materials (pollen, insects, toxins, garbage, etc.)
        6. Disturbance of habitat (soil compaction, direct destruction of vegetation or substrate, etc.)
    C. Declines of populations of species that
        1. occur naturally at low densities
        2. have large area requirements
        3. do not do well in edge habitats
        4. are sensitive to human contact
        5. are unlikely or unable to cross roads

(continued)

Table 15.1 (*continued*)

6. are frequently killed on roads (e.g., seek out roads for heat or food)
7. are otherwise sensitive to extinction resulting from habitat fragmentation or disturbance

III. MAJOR OBSERVED CHANGES
   A. Peninsular effects and some island effects
   B. Altered population dynamics of many species
   C. Possible increased probability of further fragmentation
   D. Increase in absolute amount of edge in the landscape
   E. Decrease in the amount of edge that can support sensitive species
   F. Subdivision of protected habitats and forced metapopulation structure of patch interior species
   G. Altered patch dynamics; for example, loss of species for which patch colonization rates are lower than local patch extinction rates
   H. Increased instability of ecological processes and increased frequency of fluctuation in habitat quality
   I. Predisposition of local extinction of some species

Fragmentation of natural areas may have direct influences on the number of species of interest that are protected in a given area. This is most obvious when landscape subdivision substantially increases the insularity of populations within a given area. There have been a number of studies of species loss from reserves following their separation from other areas of similar habitat (Leck 1979; Soulé, Wilcox, and Holtby 1979; Western and Ssemakula 1981; Karr 1982; Willis 1979, 1980; Humphreys and Kitchener 1982; Glenn and Nudds 1989). Although there have been some problems with such studies (Boecklen and Gotelli 1984; Glenn and Nudds 1989), they have generally indicated that the extinction of species occurs relatively rapidly in small reserves surrounded by sharp barriers or by heavily human-dominated habitats (Leck 1979; Willis 1979, 1980; Karr 1982; Diamond, Bishop, and van Bolen 1987). On the other hand, loss of species may occur only very slowly in large reserves or in those that remain continuous with surrounding areas (Glenn and Nudds 1989).

Subdivision of a landscape does not necessarily completely isolate populations within remnant patches. Subdivision within a landscape can instead result in a series of local subpopulations linked by migration, that is, a metapopulation (Levins 1970; Ehrlich et al. 1980; Gilpin 1987; Opdam 1988). In theory the survival of a metapopulation depends on three factors: (1) dynamics of subpopulations (e.g., extinction rates), (2) dispersal flow of colonists between subpopulations, and (3) spatial and temporal variation in habitat quality and dispersal among patches. If the mean colonization rate of new patches is smaller than the mean extinction rate of patches for a metapopulation, it will eventually become extinct (Gilpin 1987; Opdam 1988). For metapopulations existing in a subdivided habitat, both the size and degree of isolation of remnant patches are important influences on subpopulation dynamics, such as colonization or extinction rates (review in McCollin, Tinklin, and Storey 1988; Opdam 1988), and, hence, on the survival of the overall metapopulation.

ECOLOGICAL EFFECTS OF ROAD-GENERATED EDGES

Roads can have major influences on adjacent ecosystems. They remove habitat, create ecotones, act as sources of pollution, are barriers to animal and plant dispersal, are sources of mortality, and function as corridors for the movement of plants, animals, and disturbances (e.g., Ferris 1979; van der Zande et al. 1980; Baudry and Merriam 1988). Such effects move varying distances into the habitat adjacent to the road (Table 15.2), depending (in part) on the nature of the adjacent habitat. Certain types of habitats have been identified as particularly sensitive to road effects. Hynson et al. (1982) suggest that some raptor nest sites; big-game wintering areas; habitats for rare, threatened, or endangered species; wading-bird rookeries; streams with high wildlife value; areas with many snags or overmature trees; and habitats of species important for local ecosystem function should be avoided whenever new roads are developed in an area.

Roads remove habitat directly by occupying space within the ecosystem; a 30-m wide road covers 30,000 m^2 for each kilometer of its length. Roads alter habitat by creating ecotones and other types of edges on their margins. The modified habitat adjacent to roads is often substantially different from the habitat that it replaced (Michael 1975; Michael, Ferris, and Haverlack 1976; Ferris 1979; van der Zande et al. 1980). Edges created by roads are often abrupt, contrasting with the wider and more diffuse edges common to many natural landscapes (Gates and Gysel 1978; Ratti and Reese 1988). Some authors have speculated that numerous bird species, even those that do well in less-abrupt edge habitats, may be poorly adapted to the abrupt edge created by roads (e.g., Ratti and Reese 1988).

Materials may be carried by air or water from roads into habitats adjacent to a road. Airborne pollution, especially contamination by heavy metals, is a significant problem near many roadways (Hamilton and Harrison 1987a, 1987b). Airborne pollution may produce secondary effects via its effect on plant/insect interactions or predation. Bolsinger and Flueckiger (1987) found a significant increase in aphid infestation in plants raised in the ambient air near roadways compared to plants raised in filtered air. Scanlon (1987) found that small mammals and earthworms

Table 15.2

Estimates of the proportion of park area that could be affected by road-generated edges of varying widths. For each park we show (1) length of paved roads within the park (R_T), (2) paved road length plus the perimeter length of the park ($P + R_T$), (3) area of the park, and (4) estimated number of square kilometers and percent of park area that lies within the specific distance from paved roads.

	Park Characteristics				Estimated Area and Percentage of Park that Occurs Less than the Given Distances from a Paved Road									
Park	R_T	$P+R_T$	Acres	km²	100 m	%	250 m	%	500 m	%	1 km	%	10 km	%
Badlands	40	444	243,244	984	8	1	20	2	40	4	80	8	796	81
Big Bend	159	461	802,541	3,248	32	1	79	2	159	5	317	10	3,172	98
Big Cypress	115	325	716,000	2,898	23	1	57	2	115	4	229	8	2,294	79
Canyonlands	5	216	337,570	1,366	1	0	2	0	5	0	10	1	50	4
Cape Cod	19	202	43,557	176	4	2	9	5	19	11	38	22	380	100
Cape Hatteras	117	390	30,319	123	23	19	58	48	117	95	233	100	2,334	100
Crater Lake	119	233	183,224	741	24	3	59	8	119	16	237	32	2,374	100
Death Valley	319	825	2,067,628	8,368	64	1	160	2	319	4	638	8	6,384	76
Everglades	85	476	1,398,938	5,661	17	0	43	1	85	1	171	3	1,706	30
Glacier	136	528	1,013,572	4,102	27	1	68	2	136	3	272	7	2,722	66
Grand Canyon	80	1,048	1,218,375	4,931	16	0	40	1	80	2	160	3	1,596	32
Grand Teton	94	298	309,993	1,254	19	2	47	4	94	7	189	15	1,886	100
Great Smoky	156	401	520,269	2,105	31	1	78	4	156	7	311	15	3,112	100
Isle Royale	0	413	571,790	2,314	–	–	–	–	–	–	–	–	–	–
Kings Canyon	20	284	461,901	1,869	4	0	10	0	20	1	39	2	392	21
Mt. Rainier	115	210	235,404	953	31	3	58	6	115	12	311	33	3,106	100
North Cascade	?	?	504,781	2,043	?	?	?	?	?	?	?	?	?	?
Olympic	61	627	921,942	3,731	12	0	30	1	61	2	121	3	1,214	32
Organ Pipe	35	180	330,689	1,338	7	0	18	1	35	3	71	5	710	53
Redwood	31	261	110,132	446	6	1	16	3	31	7	62	14	624	100
Rocky Mountain	88	287	265,200	1073	17	2	44	4	88	8	175	16	1,752	100
Sequoia	55	272	402,482	1,629	11	1	28	2	55	3	111	7	1,110	68
Shenandoah	114	494	195,382	791	23	3	57	7	114	14	228	29	2,278	100
Sleeping Bear	15	171	71,132	288	3	1	7	3	15	5	30	10	298	100
Yellowstone	37	479	2,219,791	8,983	75	1	188	2	376	4	751	8	7,512	84
Yosemite	155	420	761,170	3,080	31	1	77	2	155	5	309	10	3,092	100
Zion	8	171	146,598	593	2	0	4	1	8	1	17	3	168	28

adjacent to highways showed increased levels of heavy metal contamination compared to those in control areas. Shrews, the only predatory small mammals sampled, showed the highest levels of contamination. Other predators, such as raptors that hunt near roads, could also be at risk from heavy metal contamination of their prey (Scanlon 1987).

Movement of water-borne sediment and chemicals from roads can be a major impact on adjacent habitat (Helsel et al. 1979; Hamilton and Harrison 1987a; Hynson et al. 1982). Bauer (1985) reports that seven of twenty-five forest operations surveyed in Idaho were considered to have major impacts on salmonid habitat due to the direct delivery of sediment from roads or skid trails into nearby streams. Amaranthus et al. (1985) reported that erosion due to debris slides near roads and landings can be as much as 100 times that of undisturbed areas. In addition to the movement of sediment, increased levels of lead, zinc, and copper have been found in streams near roads (Hamilton and Harrison 1987a).

Road edges represent new habitats, dispersal corridors, or other resources for some species (Bider 1968; Michael et al. 1976; Ferris 1979; Getz, Cole, and Gates 1978; Verkaar 1988; Baudry and Merriam 1988; Melman, Verkaar, and Heemsbergen 1988). A typical "roadside flora" has been identified for numerous regions in North America (e.g., Frankel 1970; Wills 1981; Belzer 1984; Edsall 1985; Mason 1987). Butterflies have repeatedly been used as examples of species benefiting from forest gaps (e.g., Lovejoy et al. 1986 for Brazil) (although roads can also be barriers to butterflies; see van der Zande et al. 1980). Breeding-bird communities within 100 m of a highway in northern Maine have been shown to contain a larger component of edge species than did communities at greater distances from the roadway (Ferris 1979). Although overall bird abundance and species number were similar from 0 to 400 m from the roadside, the abundances of three bird species were positively correlated with distance from the road over a range of 400 m from the highway right-of-way (Ferris 1979). Dispersal corridors for rodents often occur along road edges (e.g., references cited in van der Zande et al. 1980). Bradley and Fagre (1988) working in south Texas found that bobcats use dirt roads and fence lines within their home ranges more frequently than other search paths for foraging. The movement of mammalian or avian predators into an area fragmented by roads can affect various prey species (Wilcove 1985; Wilcove et al. 1986; Small and Hunter 1988; Yahner and Scott 1988).

A number of animals avoid roads or are found at lower-than-expected densities near roadways. Van der Zande et al. (1980) showed that the densities of several grassland bird species were reduced at distances of up to 1,200 m meters from roads in the Netherlands. Ferris's (1979) study showed that the abundances of five bird species were negatively correlated with distance to the edge of a Maine highway. Witmar and deCalseta

(1985) found Roosevelt elk in the Oregon Coast Range occurred at only 50 percent of the expected frequency in a 500-m band around paved roads; elk sightings also were reduced within 125 m of spur roads open to vehicles (but see Edge, Marcum, and Olson-Edge 1987). In some parts of their range, mountain lions avoid roads and are found in relatively small numbers in areas with high road densities (Van Dyke, Brocke, and Shaw 1986; VanDyke, Brocke, Shaw, Ackerman, et al. 1986). Wolves (e.g., Thiel 1985) and grizzly bears (McLellen and Shackleton 1988) have been shown to avoid roads in some areas. The presence of roads may alter animal activity patterns as well as change densities. Murphy and Curatolo (1987) showed that in some seasons the activity budgets of Alaskan caribou were altered within 600 m of a pipeline road with relatively heavy traffic, or within 300 m of a pipeline road without heavy traffic. Animals near the road spent less time feeding and used more energy in nonforaging activities than did animals further from the road.

Roads act as barriers for the dispersal of many species (Mader 1984, 1988; Baudry and Merriam 1988), and even large mammals may find roads to be barriers to movement. Curatolo and Murphy (1986) showed that caribou cross oil pipeline–road combinations with traffic only 50 percent as often as they move across control areas. Mountain lions in Utah and Arizona cross improved dirt roads and paved roads less often than they cross unimproved dirt roads (Van Dyke, Brocke, Shaw, Ackerman, et al. 1986). Animals that do cross roads freely often die on them. In 1974 state officials counted 146,229 road-killed white-tailed deer on U.S. highways in spite of various fences and warning devices designed to prevent deer from approaching the highways (Feldhamer et al. 1986). Van der Zande et al. (1980) cite reports indicating that an average of 653,000 birds and 159,000 mammals die on roads in the Netherlands every year.

The increased amount of edge habitat relative to patch interior habitat, the movement of disturbances associated with generated edges, and the increased isolation of continuous areas resulting from subdivision by park roads reduce the population sizes of the inhabitants of remnant patches and increase the rarity or unpredictability of resources within the interior habitats in subdivided parcels. Unpredictable events can have substantial impacts on the viability of populations in park fragments. Models have been developed suggesting that, in a varying environment, population persistence is more sensitive to environmentally produced fluctuations in population size and growth rate than to the original size of the population (Leigh 1981; Wright and Hubbell 1983; Goodman 1987). Variation in key parameters can be caused by fluctuations in the resource base of the population, by changes in the level of environmental toxins, by local climatic perturbations, or by disturbance from human activities. Many of these factors are increased by the presence of roads.

A PILOT STUDY OF PARK FRAGMENTATION BY ROADS AND ADMINISTRATIVE BOUNDARIES

We conducted a pilot study to obtain a rough estimate of the extent to which some of the largest national parks in the lower forty-eight states are subdivided by roads. We are interested in the potential of this subdivision to reduce the level of protection provided to park species and communities. Although areas adjacent to roads obtain protection as part of the park, the degree of protection is not the same as for the less-accessible areas as a result of the edges often produced along roads (Schonewald-Cox and Bayless 1986). Generated edges existing along roadways represent abrupt transitions in ecological elements as well as transitions in human behavior (Schonewald-Cox 1988).

We utilized AutoCAD to digitize maps from a Rand-McNally Travel Guide for twenty-seven of the largest national parks in the lower 48 states. Parks consisting of physically separate areas (based on the administrative boundaries) were analyzed as whole parks and as sets of separate parcels. Within each park we digitized the perimeter of the park along the administrative boundary and the entire length of each road. We counted as a "fragment" each area completely surrounded by roads. For each park we estimated total area, length of park perimeter, total length of roads within the park, number of fragments within the park, area of each fragment, and fragment perimeter length. The limiting factor throughout our analysis was the very coarse level of resolution available from our maps. As a result of this limitation, the values reported here are to be taken as illustrative only and are not meant as exact measurements.

Many of our largest parks are substantially fragmented by paved roads. Figure 15.2 shows some examples of the subdivision of several parks. Roads tend to cut through the center of a park and to subdivide it into several fragments. The vast majority of park fragments are small, with most falling below 100 km^2; very few fragments are greater than 1,000 km^2 (Figure 15.3a). Most of the parks that we analyzed in this pilot study have between two and ten fragments (Figure 15.3b). Interestingly, the smallest parks tend to be divided into more fragments than the larger parks. The upper limit to the number of fragments within a park fell steadily with park size, with the exception of the two largest parks, Death Valley and Yellowstone (Figure 15.3b). This trend also was seen when

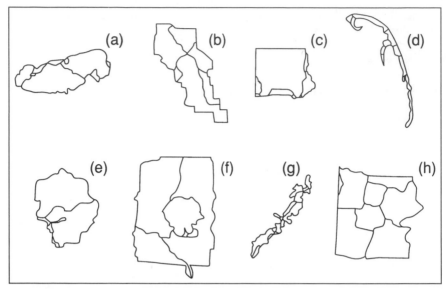

Figure 15.2
Examples of the pattern of subdivision of U.S. national parks by paved roads. For each example, we have shown the park administrative boundary and the paved roads within the park. A = Great Smoky Mountains National Park, B = Death Valley National Monument, C = Mount Rainier National Park, D = Cape Cod National Seashore, E = Yosemite National Park, F = Crater Lake National Park, G = Shenandoah National Park, H = Yellowstone National Park. Parks are not shown to the same scale.

park area was plotted against the total "edge" length of parks as measured by park perimeter length (P) plus total length of roads within the park (R_T) (Figure 15.3c). Though the length of park perimeter plus roads (P + R_T) was greater for large parks than for small ones, the slope of the relationship falls far short of that needed to produce equal "edge" (P + R_T) -to-area ratios for parks of all sizes. The observation that small parks are more highly fragmented than large ones holds true whether separate parcels of parks are plotted separately (as they are in Figure 15.3b) or summed so that the park may be plotted as a single area (as done in Figure 15.3c).

Assuming, conservatively, that generated edges lie on the boundary of each fragment (treating each fragment as an individual unit) and that the effects of roads are variable in width, what amount of habitat is potentially affected in each park? In other words, how much of the park potentially could be impacted by road effects? How much less than the total park area is estimated to function in protection? To address these

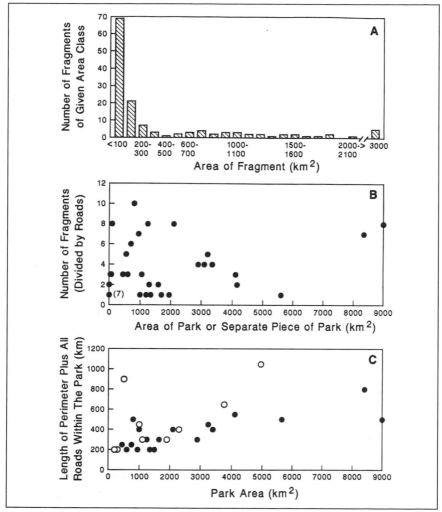

Figure 15.3
Results of a coarse-scale map analysis of the subdivision of twenty-seven U.S. national parks by paved roads. (a) Distribution of fragment sizes. The number of road-surrounded fragments of various areas occurring within the parks is shown as a frequency histogram. (b) The number of road-surrounded fragments into which each park is divided versus the area of the park. In this graph, if a park is separated into two or more physically separate parcels, these are treated independently and shown as separate points. (c) The total "edge" of a park (perimeter length plus road length) versus the total area of the park. Open circles show parks that are divided into more than one physical parcel; solid circles show parks that consist of single units.

questions, we calculated the percent of park area that could be impacted by effects moving various distances into each park from the center of a road (Table 15.3). Table 15.2 gives examples of the distances road-generated effects have been measured to move into adjacent habitats. Road effects may extend even greater distances into a park if they affect animals with large home ranges, such as large carnivores. Many carnivores and raptors have home ranges with diameters exceeding 1 km, some exceeding 10 km (Harris 1984; Gittleman 1989). If these species react either positively to roads (e.g., bobcats using roads as foraging routes) or negatively (e.g., cougars or wolves being killed on roads), then this aspect of the generated edge carries into the park for kilometers.

We asked what percentage of park area occurs further than specified distances from park roads. The results of our pilot analysis suggest that substantial areas of our national parks are close enough to roads that they may be impacted by effects like those shown in Table 15.3. Areas of over 100 km² in several of the parks we surveyed are within 500 m of paved roads. Highly subdivided parks may have 30 percent or more of their land within 1 km of paved roads. Even the largest parks had relatively little area further than 10 km from park roads (Table 15.2). Again, the trend for smaller parks to be more highly subdivided by roads resulted in a relatively high percentage of park area near roads in most

Table 15.3

Examples of the Distances that Road-Generated Edge Effects Extend into Adjacent Habitat

Road Effect	Distance	Reference and Location
High levels of heavy metals:	Up to 48 m	Scanlon 1987; Virginia
In soil near highways	30 m	Warren and Birch 1987; England
In plants	40 m	Foner 1987; Israel
	120 m	Deroanne-Bauvin et al. 1987; Netherlands
In animals	Up to 48 m	Scanlon 1987; Virginia
Sediment reaching areas needing protection	9 m–89 m	Hynson et al. 1982; U.S.
Increase in edge-species component of bird community	100 m	Ferris 1979; Maine
Plant damage from de-icing salts	120 m	Simini and Leone 1986; U.S.
Decrease in abundance of Roosevelt elk	125 m–500 m	Witmar and deCalseta 1985; Oregon
Altered activity budget of caribou	300 m–600 m	Murphy and Curatolo 1987; Alaska
Altered abundance of some bird species	400	Ferris 1979; Maine
	200 m–1200 m	Van der Zande et al. 1980; Netherlands

relatively small parks surveyed (Table 15.3). Increase in park subdivision by roads may be influenced by park age, proximity to urban areas, ease of access via major travel routes, or the popularity of major park features (e.g., beaches).

DISCUSSION AND CONCLUSIONS

Our preliminary analysis indicates that many U.S. national parks may be sacrificing some of their ability to protect sensitive species in order to be able to provide access for visitors. This appears to be part of a more general process by which natural areas face a gradual erosion of their ability to protect sensitive species, occurring first from the fragmentation of the landscape surrounding the natural area and second from the internal subdivision of the natural area itself (Figure 15.4).

Most ecological research conducted in national parks for use in the detection and management of damaging impacts on focal species and natural processes treats parks as single continuous units facing threats originating outside the park. Research on focal species, for example, may concern declining species or nonnative species that cross park perimeters (e.g., Buechner 1987). Declining species that cross over park boundaries into adjacent unprotected areas have been a concern in Rocky Mountain, Yellowstone, Shenandoah, and Great Smoky Mountains national parks (Elfring 1986; Lee 1986; Buechner 1987; Schonewald-Cox 1988). Species unwanted by parks are themselves considered impacts if they move into the park landscape, as has been seen for Great Smoky Mountains, Hawaii Volcanoes, Grand Canyon, Death Valley, and Bandelier national parks (Elfring 1986; Lee 1986; White 1987). Studies of both declining and unwanted species tend to focus on the park as a single unit while obtaining knowledge on how external habitat is conducive to their survival and dispersal across park boundaries.

Study of the impacts of habitat alteration on species, communities, or ecosystem processes and translation of such studies for use in parks are not new (see review in Agee and Johnson 1988). However, integration of an analysis of anthropogenic effects with the use of basic ecological models is relatively recent. This has been accomplished in park studies of the role of prescribed fire in natural succession, the effect of acid rain on water quality, and the management of small populations. During the last decade, attention has been given to the role of anthropogenic and naturally occurring impacts in determining degree of insularity, functional size, and protection effectiveness of parks (DiSilvestro 1986; Gregg, Krugman, and Wood 1989; Schonewald-Cox 1988; Schonewald-Cox and Stohlgren 1989; Agee and Johnson 1988; Waller 1988). Most of this work

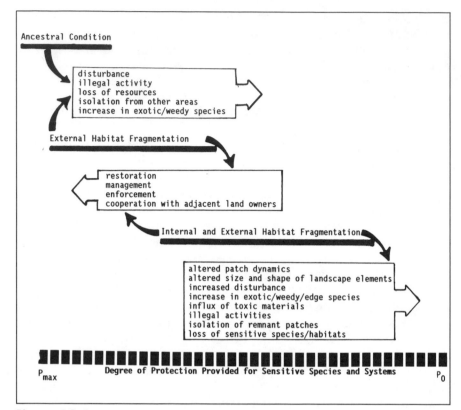

Figure 15.4
This figure illustrates the range of protection provided by the ancestral condition of a hypothetical natural area, a natural area with external habitat fragmentation, and a natural area facing internal plus external landscape fragmentation. The stresses imposed on the area by external and internal landscape fragmentation and the actions that can be taken to relieve these stresses are shown in the boxes.

concerns documentation of effects of habitat alteration or subdivision that occurs outside park boundaries. Visitor impacts are addressed in management of resources, landscape, enforcement, maintenance, and interpretation; however, they are not routinely incorporated in studies of ecology and park function. Yet visitation is the basis of most internal road development in parks. Parks are highly subdivided by roads and other human access routes. It is worth asking whether parks should be assumed to function as single cohesive units for species and communities.

Although fragmenting features in the landscape traditionally have been studied as localized impacts, it was not until a few years ago that the perspective began to focus on the subdividing and erosive properties of these features. New studies are under way at Organ Pipe Cactus National Monument, addressing the fragmentation properties of Arizona State Highway 85 bisecting the park. Though the highway provides no obstacles to ubiquitous or vagile species, the two fragments created by the road are disconnected for species that do not cross roads or die on them. Lowe (pers. com.) suspects that the Organ Pipe Cactus National Monument is losing significant numbers of reptiles to road traffic (roads operating as fatal sinks) and currently is measuring effects of the highway on fragmentation and dynamics of Monument reptile populations.

Perhaps for some species we should view parks such as those shown in Figure 15.1 as collections of smaller, tightly clustered units rather than as single cohesive large areas. Populations of species for which roads are substantial barriers to movement or important sources of mortality would then consist of several subpopulations, each significantly smaller than estimated for the whole park. Fragmentation of parks by roads may force such populations into a highly subdivided metapopulation structure. The demographic populations of species for which roads are not serious impediments to movement or major sources of mortality could still be (positively or negatively) affected by the generated edges associated with roads.

We have noted numerous negative effects of road-generated edges on protected habitats and species. However, we must also note that road-generated edges also produce positive effects. The first and most obvious is the visitor contact and education that comes from the movement of people into parks. This can translate into public support for park policies and has the potential to change values held by the public. Edge species benefit directly from the increased availability of habitat along roads.

Another positive effect of road-generated edges is the protection from spread of local catastrophic events. Species subdivided by roads into relatively isolated subpopulations may benefit from the fact that roads will also be barriers to detrimental local events such as small fires or some diseases. However, although roads may be solid enough barriers to decouple the demographics of populations in adjacent fragments and stop the spread of slow-moving catastrophes on a local scale, they still may be too permeable to protect populations from the spread of regional catastrophes. Roads are not wide enough to insulate fragments from the spread of nonnative species, rapidly spreading parasites, or crown fires and exceptionally high winds. This has been illustrated by the spread of 1988 fires in the Yellowstone region, the intrusion of pigs into Pinnacles National Monument, and the movement of southern pine beetles through white fir in Great Smoky Mountains National Park. Roads may protect some park populations from local threats by acting as barriers at one scale while at the same time failing to protect these populations from widespread threats by *not* acting as barriers at another.

Internal fragmentation of parks resulting from patterns in the natural landscape (or in special instances—cultural landscape) is not typically considered a threat. It is assumed to be a part of the normal ecosystem processes occurring in the park landscape (e.g., Southern pine beetle, White 1987). However, fragmentation resulting from internal park development or disjunct park property jurisdictions increases the surface exchange of visitors and sites, fragile species and sinks, and impacts and park interiors. Road-created fragments and their edges comprising the park are likely to vary independently in hospitability to endemic species. Roads can serve as corridors connecting sources and park interiors for nonnative species and as sinks for protected species (e.g., van der Zande 1980 and references therein).

Our analysis underscores the need for the documentation of internal fragmentation of parks. It suggests that ecological studies of the minimum requirements for population sustainability, the intrusion distances of road-generated edges, and the minimal critical areas for ecosystem processes should be tied to knowledge of how landscape fragmentation affects parks and other protected habitats (Pickett and Thompson 1978; Harris 1984; Salwasser, Schonewald-Cox, and Baker 1987; White 1987; Schonewald-Cox 1988). As has been analyzed for Australia by Kuiken (1988), we also need to integrate our road-planning process with our attempts to conserve valued natural landscapes more systematically than we do at present. Planning for the expansion of parks, development of new parks, and management and planning of existing parks for visitation should be tied to how landscape subdivision affects both people and parks.

ACKNOWLEDGMENTS

We wish to thank Stephanie Blume, Peter Hunter, and Marilu Carter for their help. This work was funded by the U.S. National Park Service. The statements made herein do not necessarily reflect the opinions or policies of the U.S. National Park Service but are the sole responsibilities of the authors.

LITERATURE CITED

Agee, J.K., and D.R. Johnson, eds. 1988. *Ecosystem management for parks and wilderness*. Institute for Forest Resources Contribution no. 65. Seattle: University of Washington Press.

Amaranthus, M.P., M.R. Rice, N.R. Barr, and R.R. Ziemer. 1985. Logging and forest roads related to increased debris slides in southwestern Oregon. *J. For.* 1985: 229–33.

Baudry, J., and H.G. Merriam. 1988. Connectivity and connectedness: Functional versus structural patterns in landscapes. In *Connectivity in landscape ecology,* ed. K.F. Schreiber, 23–28. Proceedings of the 2nd International Seminar of the International Association for Landscape Ecology (IALE), Munster, Germany, 1987. Munstersche Geographische Arbeiten 29.

Bauer, S.B. 1985. Evaluation of nonpoint source impacts on water quality from forest practices in Idaho: Relation to water quality standards. In *Perspectives in nonpoint source pollution,* 455–58. Proceedings of a National Conference. Kansas City, Missouri, May 1985.

Belzer, T.J. 1984. *Roadside plants of southern California.* Missoula, Mont.: Mountain Press.

Bider, J.R. 1968. Animal activity in uncontrolled terrestrial communities as determined by a sand transect technique. *Ecol. Monogr.* 38:269–308.

Boecklen, W.J., and N.J. Gotelli. 1984. Island biogeographical theory and conservation practice: Species-area or specious-area relationships? *Biol. Cons.* 29:63–80.

Bolsinger, M., and W. Flueckiger. 1987. Enhanced aphid infestation at motorways: The role of ambient air pollution. *Entomol. Exp. Appl.* 45:237–243.

Bradley, L.C., and D.B. Fagre. 1988. Coyote and bobcat responses to integrated ranch management practices in south Texas. *J. Range Manage.* 41:322–27.

Brown, W.E. 1989. Values and purposes of the National Park system. *Forum* 6:2–9.

Buechner, M. 1987. Conservation in insular parks: Simulation models of factors affecting the movement of animals across park boundaries. *Biol. Cons.* 41:57–76.

Curatolo, J.A., and Murphy, S.M. 1986. The effects of pipelines, roads, and traffic on movements of caribou, *Rangifer tarandus. Can. Field Nat.* 100:218–24.

Deroanne-Bauvin, J., E. Delcarte, and R. Impens. 1987. Monitoring lead deposition near highways: A ten-year study. *Sci. Total Environ.* 59:257–66.

Diamond, J.M., K.D. Bishop, and S. van Balen. 1987. Bird survival in an isolated Javan woodlot: Island or mirror. *Cons. Biol.* 1:132–42.

DiSilvestro, R.L., ed. 1986. *Audubon wildlife report 1986.* New York: National Audubon Society.

Edge, W.D., C.L. Marcum, and S.L. Olson-Edge. 1987. Summer habitat selection by elk in western Montana: A multivariate approach. *J. Wildl. Manage.* 51:844–51.

Edsall, M.S. 1985. *Roadside plants and flowers: A traveler's guide to the Midwest and Great Lakes area.* Madison: University of Wisconsin Press.

Ehrlich, P.R., D.D. Murphy, M.C. Singer, C.B. Sherwood, R.R. White, and I.L. Brown. 1980. Extinction, reduction, stability and increase: The responses of checkerspot butterfly populations to the California drought. *Oecologia* 46:101–5.

Elfring, C. 1986. Wildlife and the National Park Service. In *Audubon wildlife report 1986,* ed. R.L. DiSilvestro, 462–94. New York: National Audubon Society.

Feldhamer, G.A., J.E. Gates, D.M. Harman, A.J. Loranger, and K.R. Dixon. 1986. Effects of interstate highway fencing on white-tailed deer activity. *J. Wildl. Manage.* 50:497–503.

Ferris, C.R. 1979. Effects of Interstate 95 on breeding birds in northern Maine. *J. Wildl. Manage.* 43:421–27.

Foner, H.A. 1987. Traffic lead pollution of some edible crops in Israel. *Sci. Total Environ.* 59:309–15.

Forman, R.T.T., and M. Godron. 1986. *Landscape ecology.* New York: John Wiley & Sons.

Frankel, R.E. 1970. *Ruderal vegetation along some California roadsides.* University of California Publications in Geography Volume 20. Berkeley: University of California Press.

Gates, J.E., and L.W. Gysel. 1978. Avian nest dispersion and fledgling success in field-forest ecotones. *Ecology* 59:871–83.

Getz, L., F.R. Cole, and D.L. Gates. 1978. Interstate roadsides as dispersal routes for *Microtus pennsylvanicus. J. Mammal.* 59:208–12.

Gilpin, M.E. 1987. Spatial structure and population vulnerability. In *Viable populations for conservation,* ed. M.E. Soulé, 125–39. Cambridge: Cambridge University Press.

Gittleman, J.L., ed. 1989. *Carnivore behavior ecology and evolution.* Ithaca: Cornell University Press.

Glenn, S.M., and T.D. Nudds. 1989. Insular biogeography of mammals in Canadian parks. *J. Biogeogr.* 16:261–68.

Goodman, D. 1987. The demography of chance extinction. In *Viable populations for conservation,* ed. M.E. Soulé, 11–34. Cambridge: Cambridge University Press.

Gregg, W.P., Jr., S.L. Krugman, and J.D. Wood, Jr., 1989. *Fourth world wilderness congress worldwide conservation: Proceedings of the symposium on biosphere reserves.* Washington, D.C.: U.S. Department of the Interior.

Hamilton, R.S., and R.M. Harrison, eds. 1987a. *Highway pollution.* Proceedings of the Second International Symposium, London, July 1986. *Sci. Total Environ.* Volume 59.

———. 1987b. Heavy metal pollution in roadside urban parks and gardens in Hong Kong. *Sci. Total Environ.* 59:325–28.

Harris, L.D. 1984. *The fragmented forest: Island biogeography theory and the preservation of biotic diversity.* Chicago: University of Chicago Press.

Helsel, D.R., J.I. Kim, T.J. Grizzard, C.W. Randall, and R.C. Hoehn. 1979. Land use influences on metals in storm drainage. *J. Wat. Pollut. Control Fed.* 51:709–17.

Humphreys, W.F., and D.J. Kitchener. 1982. The effect of habitat utilization on species-area curves: Implications for optimal reserve area. *J. Biogeogr.* 9:391–96.

Hynson, J., P. Adamus, S. Tibbetts, and R. Darnell. 1982. *Handbook for protection of fish and wildlife from construction of farm and forest roads.* Report 82/18. U.S. Fish and Wildlife Service, Office of Biological Services.

Karr, J.R. 1982. Population variability and extinction in the avifauna of a tropical landbridge island. *Ecology* 63:1975–78.

Kuiken, M. 1988. Consideration of environmental and landscape factors in highway planning in valued landscapes: An Australian survey. *J. Environ. Manage.* 26:191–201.

Leck, C.F. 1979. Avian extinctions in an isolated tropical wet forest preserve, Ecuador. *Auk* 96:343–52.

Lee, W.S., ed. 1986. The National Wildlife Refuge system. In *Audubon wildlife report 1986*, R.L. DiSilvestro, 410–60. New York: National Audubon Society.

Leigh, E.G. 1981. The average lifetime of a population in a varying environment. *J. Theor. Biol.* 90:213–39.

Levins, R. 1970. Extinction. In *Some mathematical questions in biology*, 75–108. Providence: American Mathematical Society.

Lovejoy, T.E., R.O. Bierregaard, Jr., A.B. Rylands, J.R. Malcolm, C.E. Quintela, L.H. Harper, K.S. Brown, Jr., A.H. Powell, G.V.N. Powell, H.O.R. Schubart, and M.B. Hays. 1986. Edge and other effects of isolation on Amazon forest fragments. In *Conservation biology: The science of scarcity and diversity*, ed. M.E. Soulé, 257–85. Sunderland, Mass.: Sinauer.

Mader, H.J. 1984. Animal habitat isolation by roads and agricultural fields. *Biol. Cons.* 29:81–96.

———. 1988. The significance of paved agricultural roads as barriers to ground-dwelling arthropods. In *Connectivity in landscape ecology*, ed. K.F. Schreiber, 97–100. Proceedings of the 2nd International Seminar of the International Association for Landscape Ecology (IALE), Munster, Germany, 1987. Munstersche Geographische Arbeiten 29.

Mason, C.T. 1987. *A handbook of Mexican roadside flora*. Tucson: University of Arizona Press.

McCollin, D., R. Tinklin, and R.A.S. Storey. 1988. The status of island biogeographic theory and the habitat diversity hyphothesis. In *Connectivity in landscape ecology*, ed. K.F. Schreiber, 29–34. Proceedings of the 2nd International Seminar of the International Association for Landscape Ecology (IALE), Munster, Germany, 1987. Munstersche Geographische Arbeiten 29.

McLellen, B.N., and D.M. Shackleton. 1988. Grizzly bears and resource-extraction industries: Effects of roads on behavior, habitat use and demography. *J. Appl. Ecol.* 25:451–60.

Melman, P.J.M., H.J. Verkaar, and H. Heemsbergen. 1988. The maintenance of road verges as possible ecological corridors of grassland plants. In *Connectivity in landscape ecology*, ed. K.F. Schreiber, 131–33. Proceedings of the 2nd International Seminar of the International Association for Landscape Ecology (IALE), Munster, Germany, 1987. Munstersche Geographische Arbeiten 29.

Michael, E.D. 1975. *Effects of highways on wildlife*. West Virginia Department of Highways Report WVDOH42.

Michael, E.D., C.R. Ferris, and E.G. Haverlack. 1976. Effects of highway rights-of-way on bird populations. *Symp. Environ. Concerns Rights-of-way Manage.* 1:253–61.

Murphy, S.M., and J.A. Curatolo. 1987. Activity budgets and movement rates of caribou encountering pipelines, roads, and traffic in northern Alaska. *Can. J. Zool.* 65:2483–90.

Opdam, P. 1988. Populations in fragmented habitat. In *Connectivity in landscape ecology*, ed. K.F. Schreiber, 75–78. Proceedings of the 2nd International Seminar of the International Association for Landscape Ecology (IALE), Munster, Germany, 1987. Munstersche Geographische Arbeiten 29.

Pickett, S.T.A., and J.N. Thompson. 1978. Patch dynamics and the design of nature reserves. *Biol. Cons.* 13:27–37.

Ratti, J.T., and K.P. Reese. 1988. Preliminary test of the ecological trap hypothesis. *J. Wildl. Manage.* 52:484–91.

Salwasser, H., C. Schonewald-Cox, and R. Baker. 1987. The role of interagency cooperation in managing for viable populations. In *Viable populations for conservation,* ed. M.E. Soulé, 159–73. Cambridge: Cambridge University Press.

Scanlon, P.F. 1987. Heavy metals in small mammals in roadside environments: Implications for food chains. *Sci. Total Environ.* 59:317–23.

Schonewald-Cox, C.M. 1988. Boundaries in the protection of nature reserves. *BioScience* 38:480–86.

Schonewald-Cox, C.M., and J.W. Bayless. 1986. The boundary model: A geographic analysis of design and conservation of nature reserves. *Biol. Cons.* 38:305–22.

Schonewald-Cox, C.M., and T.J. Stohlgren. 1989. Biological diversity and global change: Habitat fragmentation and extinction. In *Proceedings U.S./U.S.S.R. Symposium: Air Pollution and Vegetation Change,* ed. R. Noble and Shriner, U.S. Department of Energy.

Simini, M., and I.A. Leone. 1986. Studies on the effects of de-icing salts on roadside trees. *Arboric. J.* 10:221–31.

Small, M.F., and M.L. Hunter. 1988. Forest fragmentation and avian nest predation in forested landscapes. *Oecologia* 76:62–64.

Soulé, M.E., B.A. Wilcox, and C. Holtby. 1979. Benign neglect: A model of faunal collapse in the game reserves of East Africa. *Biol. Cons.* 15:259–72.

Sudia, T.W. 1989. National parks in the 21st century. *Forum* 6:38–40.

Swanson, F.J., T.K. Kratz, N. Caine, and R.G. Woodmansee. 1988. Landform effects on ecosystem patterns and processes. *BioScience* 38:92–98.

Temple, S.A. 1986. Predicting impacts of habitat fragmentation on forest birds: A comparison of two models. In *Wildlife 2000: Modeling habitat relationships of terrestrial vertebrates.* ed. A. Keast, M.L. Morrison, and C.J. Ralph, 301–4. Madison: University of Wisconsin Press.

Thiel, R.P. 1985. Relationship between road densities and wolf habitat suitability in Wisconsin. *Am. Midl. Nat.* 113:404–7.

van der Zande, A.N., W.J. ter Keurs, and W.J. van der Weiden. 1980. The impact of roads on the densities of four bird species in an open field habitat: Evidence of a long-distance effect. *Biol. Cons.* 18:299–321.

Van Dyke, F.G., R.H. Brocke, and H.G. Shaw. 1986. Use of road track counts as indices of mountain lion presence. *J. Wildl. Manage.* 50:102–9.

Van Dyke, F.G., R.H. Brocke, H.G. Shaw, B.A. Ackerman, T.H. Hemker, and F.G. Lindzey. 1986. Reactions of mountain lions to logging and human activity. *J. Wildl. Manage.* 50:95–102.

Verkarr, H.J. 1988. The possible role of road verges and river dikes as corridors for the exchange of plant species between natural habitats. In *Connectivity in landscape ecology,* ed. K.F. Schreiber, 79–84. Proceedings of the 2nd International Seminar of the International Association for Landscape Ecology (IALE), Munster, Germany, 1987. Munstersche Geographische Arbeiten 29.

Waller, D.M. 1988. Sharing responsibility for conserving diversity: The complementary roles of conservation biology and public land agencies. *Cons. Biol.* 2:398–99.

Warren, R.S., and P. Birch. 1987. Heavy metal levels in atmospheric particulates, roadside dust and soil along a major urban highway. *Sci. Total Environ.* 59:253–56.

Western, D., and J. Ssemakula. 1981. The future of savannah ecosystems: Ecological islands or faunal enclaves? *Afr. J. Ecol.* 19:7–19.

White, P.S. 1987. Natural disturbance, patch dynamics and landscape pattern in natural areas. *Nat. Areas J.* 7:14–22.

Wilcove, D.S. 1985. Nest predation in forest tracts and the decline of migratory songbirds. *Ecology* 66:1211–14.

———. 1988. Changes in the avifauna of the Great Smoky Mountains: 1947–1983. *Wilson Bull.* 100:256–71.

Wilcove, D.S., C.H. McClellan, and A.P. Dobson. 1986. Habitat fragmentation in the temperate zone. In *Conservation biology: The science of scarcity and diversity,* ed. M.E. Soulé, 237–56. Sunderland, Mass.: Sinauer.

Wilcox, B.A. 1980. Insular ecology and conservation. In *Conservation biology: An evolutionary-ecological perspective,* ed. M.E. Soulé and B.A. Wilcox, 95–117. Sunderland, Mass.: Sinauer.

Wilcox, B.A., and D.D. Murphy. 1985. Conservation strategy: The effects of fragmentation on extinction. *Am. Nat.* 125:879–87.

Willis, E.O. 1979. The composition of avian communities in reminiscent woodlots in southern Brazil. *Papeles Avulsos Zool.* 33:1–25.

———. 1980. Species reduction in reminiscent woodlots in southern Brazil. In *Proceedings of the Seventeenth International Ornithological Congress,* Berlin 1978.

Wills, M.M. 1981. *Roadside flowers of Texas.* Austin: University of Texas Press.

Witmar, G.W., and D.S. deCalseta. 1985. Effect of forest roads on habitat use by Roosevelt elk. *Northwest Sci.* 59:122–25.

Wright, S.J., and S.P. Hubbell. 1983. Stochastic extinction and reserve size: A focal species approach. *Oikos* 41:466–76.

Yahner, R.H. 1988. Changes in wildlife communities near edges. *Cons. Biol.* 2:333–39.

Yahner, R.H., and D.P. Scott. 1988. Effects of forest fragmentation on depredation of artificial nests. *J. Wildl. Manage.* 52:158–61.

CHAPTER
16

From Conservation Biology to Conservation Practice: Strategies for Protecting Plant Diversity

DONALD A. FALK

ABSTRACT

This chapter explores linkages between studies of the biology of rare plants and strategies for their conservation. Because of their low numbers and consequent vulnerability to destruction, rare plant species provide an important test of the current state of the art in conservation, particularly important in an era of biological management. The primary threats and patterns of endangerment to the flora of the United States are summarized, with special reference to causes of decline beyond outright destruction of habitat. The chapter addresses the major biological considerations in rare plant conservation and management, including endemism and narrow distribution, and demographic or genetic effects in small populations. Finally, integrated strategies for rare plant conservation are discussed, emphasizing the interactions among land conservation, biological management, offsite research and propagation, and introduction and habitat restoration. A bibliography on rare plant biology and conservation is also provided.

INTRODUCTION

Real-world conservation does not take place in tidy packages, within the neat confines of intellectual models. Nor is it primarily governed by rational discourse, systematic investigation of the structure and workings of natural systems, or carefully reasoned plans for sustainable use of natural resources. The daily practice of conservation is as different from the world of theory and scholarly research as is the blackboard at a military academy from the battlefield. As every conservationist knows, decisions in the field are as likely to be influenced by real-estate transactions, land use, the economics of resource extraction, state and federal taxation, political expediency, and the vagaries of public opinion as they are by careful planning grounded in sound conservation biology.

Nonetheless, what distinguishes the current era of biological conservation from earlier periods is the attempt to build a foundation for action in sound scientific principles. The underlying axiom of conservation biology, of which this volume is an expression, is that conservation should be based on detailed understanding of the nature and dynamics of biological systems and, correspondingly, that research should be organized to shed light on aspects of biological diversity useful for conservation. This ongoing dialectic between conservation and research is one of the defining characteristics of the current era.

A second element in conservation practice is the prevailing orientation to biological management. Conservationists today engage in, and approve of, a degree of intervention into the functioning of ecosystems that would have been considered unthinkable or even unethical less than half a century ago. The modern land steward's toolbox includes plows, bulldozers, chain saws, herbicides, fire equipment, and even the occasional shotgun—ironically, the very tools of destruction in other hands. To refer to a "managed natural area" is not conservation Newspeak; it is the prevailing mode of conservation in North America, and increasingly in other parts of the world as well. The ascendancy of journals such as *Conservation Biology*, *Natural Areas Journal*, and *Restoration and Management Notes* and professional societies such as the Society for Conservation Biology, Natural Areas Association, and the Society for Ecological Restoration attest to the wide acceptance of the management orientation in the conservation profession.

Land management also has become a substantial fraction of the budgets and programs of private conservation groups. For instance, in 1988 nearly one-quarter of all program expenditures by The Nature Conservancy were for land stewardship (The Nature Conservancy 1988). The increasing fragmentation and isolation of remaining habitat, along with the intensifying need to control invasive exotic vegetation and herbivores, compel greater attention to management of preserves. As a result, many organizations now embrace a philosophy of conservation management that entails calculated intervention into the processes of ecological succession, competition, and species distribution. Fire management, for example, has become a standard and necessary practice in the maintenance of prairies, grasslands, savannas, chapparal, and many other community types. Active removal of feral and exotic animals is an essential function in maintaining ecological quality in many preserves, as is control of invasive exotic vegetation. Thus, although conservationists rightly argue that the preservation of large areas of habitat is essential for the "natural" functioning of ecosystems, much of the daily practice of preserve management consists of intentional human intervention into those very processes, in order to create what we conceive would be the natural bioscape (Falk 1991a).

Conservation of rare plants provides a useful case study of these principles. Because of their extreme rarity in the wild, they are a test of our ability to maintain the quality of ecosystems. Many species hover near minimum viable population size and are vulnerable to stochastic demographic, genetic, and environmental variation, as well as destruction or alteration of remaining habitat. Their very rarity and fragility reduces the margin for error in efforts to protect them. Many also are edaphic endemics or otherwise naturally restricted in distribution, and therefore are of considerable scientific interest. Ironically, until recently endangered plants as a group have been underrepresented in the scientific literature, further complicating their recovery (Falk and Holsinger 1991; Given in press). Significant interest from the perspective of evolutionary biology attaches to the increasing number of rare taxa that are believed to be recently evolved, and hence rare because they have not yet radiated geographically; far from representing phylogenetic "senescence," rare plants may be among the best opportunities to observe the speciation process itself (Falk in press).

This chapter will explore current strategies to protect endangered plants in an integrated fashion, reflecting their biology and the particular threats they face. In general, the approach advocated will be that of *integrated conservation*, which has been described in detail elsewhere (Falk 1987; Falk 1990a; Falk and McMahan 1988). Integrated conservation strategies are based on the calculated interweaving of multiple conservation methods, with the shortcomings of one complemented by the strengths of another. Typically involving collaboration among several

agencies, integrated conservation strategies are inherently multidisciplinary, drawing as needed on legal protection, land acquisition, site management, species reintroduction, and ecological restoration. The approach
is characteristically information-intensive, especially with regard to understanding of reproductive ecology, seedling growth, competition, and
seed dispersal.

Ultimately, the test of conservation is the successful management
and maintenance of diversity at all levels of biological organization. Thus,
in making the transition from conservation biology to conservation practice we ultimately pass from the realm of asking "What is . . . ?" to determining "What should be done?"

THREATS TO PLANT DIVERSITY IN THE UNITED STATES

The imperative for an integrated, multidisciplinary approach to conservation derives in large part from the variety and multiplicity of threatening processes to which plant populations are subjected. Moreover,
many of these destructive processes are only partially alleviated or prevented by land acquisition or other means of avoiding outright destruction of the habitat. Although land protection per se is an essential prerequisite, effective long-term conservation will require intervention and
biological management practices that draw on a wide range of tools, skills,
and knowledge.

The dimensions of plant endangerment in the United States have
been well documented (The Nature Conservancy 1989a; U.S. Fish and
Wildlife Service 1990; Center for Plant Conservation 1988a; Roberts
1988, 1989a; Lewin 1986). Of the approximately 20,000 vascular plant
taxa native to the United States, over 4,400 (22%) are currently of conservation concern and monitored for status by the U.S. Fish and Wildlife
Service, The Nature Conservancy, or the Center for Plant Conservation.
Of these, approximately 1,300 taxa are known from five or fewer populations or have 1,000 or fewer extant individuals. Moreover, an estimated 745 taxa are judged to be severely threatened and at risk of imminent extinction (Center for Plant Conservation 1988, 1991a). The
greatest concentration of rare plants within the U.S. is in tropical, subtropical, or Mediterranean areas: more than three-quarters of all critically
threatened U.S. species are endemic to Hawaii, Puerto Rico, Florida,
Texas, and California (Myers 1985; Center for Plant Conservation 1991a;
Huenneke 1989). The overall pattern of endangerment in the U.S. is thus
comparable to worldwide trends, mirroring increasing threats to plant
diversity, particularly in tropical areas (Davis et al, 1986; Roberts 1988;
Lewin 1986; Myers 1985).

Endangered species in the United States also reflect considerable taxonomic and ecogeographic diversity. Over 1,125 genera are represented in the Center for Plant Conservation endangered species database (1991a). Interestingly, certain genera comprise a substantial percentage of the total; endangered species in just eleven genera account for 14 percent of the entire list (Table 16.1).

Unfortunately, the statutory basis for species protection still leaves much to be desired. Although plant conservation provisions of the U.S. Endangered Species Act of 1973 have recently been strengthened somewhat (as amended, 16 U.S.C. 1531 *et seq.*), plant species are still more difficult to protect than animals under the law owing to the different sources of legal precedent for plants and animals (McMahan 1980). Moreover, the historic rate of plant listings under the Act has been low; as of August 1990, only 237 plant taxa (approximately 5.4% of U.S. species of conservation concern) had received protection under the Act (U.S. Fish and Wildlife Service 1990). Of the 745 plant taxa identified as at critical risk of endangerment, only one-fourth (184) have legal protection under the Act; the remaining three-quarters have *no* federal protection except by voluntary policies adopted by land-managing agencies. The number with approved recovery plans is even lower: only 108 species have such plans, fewer than half of the listed species, and only 2 percent of total species of conservation concern.

Table 16.1

Plant genera with the highest number of taxa of conservation concern in the United States. These eleven genera account collectively for 14.6% of all U.S. endangered species. Source: Center for Plant Conservation, 1991a.

Genus	Family	Number of Taxa of Conservation Concern	Percent of Total
Astragalus	Fabaceae	156	3.5
Eriogonum	Polygonaceae	101	2.3
Penstemon	Scrophulariaceae	64	1.5
Erigeron	Asteraceae	55	1.3
Lupinus	Fabaceae	46	1.1
Arctostaphylos	Ericaceae	44	1.0
Phacelia	Hydrophyllaceae	39	0.9
Draba	Brassicaceae	37	0.8
Carex	Cyperaceae	32	0.7
Arabis	Brassicaceae	32	0.7
Castilleja	Scrophulariaceae	32	0.7
Total in top 11 genera		638	14.5
All other genera (842)		3,774	85.5
Total U.S.		4,412	100.0

Less than half of the endangered species in the United States occur on private conservation or public lands; the remainder are under private ownership and are vulnerable to development. Certain regions of the U.S. with high rates of plant endangerment, such as Texas and Florida, also have a significant number of rare species on private lands threatened by immediate development (Wallace and McMahan 1988). Many species, however, are threatened by factors that are not necessarily controlled simply by acquisition of land. Though public or private conservation ownership generally is a prerequisite to effective long-term conservation, the diversity of threats to plant populations and species suggests that a multifaceted approach emphasizing management will be required.

As Table 16.2 illustrates, the diversity of plant species in the U.S. is matched by a diversity of threats to their survival. The dominant—and most visible—cause of plant endangerment is unquestionably destruction or conversion of habitat. Surprisingly, however, outright habitat destruction is only part of the picture; a wide range of factors can contribute to the decline of a population or species. In Hawaii, for example, the primary

Table 16.2
Major threats to plant species in the United States. Endangered taxa are often subject to more than one cause of endangerment or decline. Note that many of these threatening processess are only partially reduced or avoided by land acquisition per se.

Habitat Destruction
  Commercial development
  Highway construction or widening
  Hydroelectric projects
Habitat quality degradation
  Alteration of hydroecology
  Acid precipitation
  Soil erosion
Successional changes or arrestation
Competitive exclusion by exotic vegetation
Seed predation by wild and feral animals
Overgrazing
Logging
Hybridization
Fire suppression
Overcollecting
Disease
Inbreeding depression and subminimal population size
Loss of pollinator, dispersal agent, or other symbiont
Recreational land use, off-road vehicles
Vandalism
Climate change

impacts on upland forests at present are invasive exotic vegetation, feral introduced herbivores and seed predators, and disease, even on public and private conservation land (Stone and Scott 1985; Kimura and Nagata 1980). Control of such threats would be impossible without adequate ownership and protection of the land itself, but agencies such as The Nature Conservancy devote increasing effort to the management of these biological impacts. Likewise, throughout the Intermountain Region and the southwestern United States, a significant number of endangered species occur on public lands used for commercial grazing or logging. For instance, two species of *Agave* (*A. arizonica* and *A. murpheyi*) are found entirely on U.S. National Forest land but suffer ongoing pressure from cattle grazing (Delamater and Hodgson 1987). The same is true for the Clay phacelia (*Phacelia argillacea*) and many other Great Basin endemics.

The impact of feral or exotic animals is frequently a decisive factor in plant endangerment. *Pritchardia munroi*, a Hawaiian endemic known from only a single remaining individual in the wild, has been prevented from reproducing by nearly complete predation of each year's seed crop (Stone and Scott 1985; W. Garnett, pers. comm.). Rieseberg et al. (1989) have related the decline of Catalina mahogany (*Cercocarpus traskiae*) on the Channel Islands off the coast of California in large measure to introduced goats, sheep, pigs, and deer (along with the genetic threat of hybridization with a more common member of the same genus). Loss of riparian habitat along the San Marcos River in Texas for Texas wild rice (*Zizania texana*) has been compounded by grazing damage by nutria (*Myocaster coypu*), an introduced South American herbivore (Nabhan 1989).

Other causes of endangerment can be even more insidious and resistant to control. The extinction of a highly coevolved seed-dispersal agent, the dodo, (*Raphus cucullatus*) is thought to contribute to the present lack of reproduction in an endemic species, *Calvaria major* (Temple 1977). A corresponding North American example is the Florida torreya (*Torreya taxifolia*), which occurs largely on land protected by The Nature Conservancy, U.S. Army Corps of Engineers, and the state of Florida. The species, thought to originally have had a more cool-temperate Pleistocene distribution, failed to reradiate following glacial recession and now survives only in relatively cool microclimate refugia near the Apalachicola River in southern Georgia and the Florida panhandle (U.S. Fish and Wildlife Service 1988). These refugia, however, are still too warm to represent optimal climate for the species, which evidently persists in a condition of chronic physiological stress. *In situ* the species is attacked by a suite of six pathogenic fungi that kill trees before they reach reproductive age; most of the remaining trees are stump sprouts. Thus, without sexual reproduction and genetic recombination that might offer an ev-

Figure 16.1
Pritchardia munroi *is a Hawaiian endemic known from only a single individual remaining in its native habitat. (Photo by W. Garnett)*

Figure 16.2
*Loss of riparian habitat along the San Marcos River and grazing by an exotic herbivore (*Myocaster coypu*) threatens the endangered Texas wild rice (*Zizania texana*). (Photo by J. Poole)*

olutionary prospect of adaptation, the species appears ultimately doomed (McMahan 1989b). Possible conservation measures include control of the fungal pathogens and experimental placement of specimens in cooler locations to the north to break the cycle of physiological stress, pathology, reproductive failure, and genetic depauperization (Figure 16.3).

A discussion of threats to species survival would be incomplete without consideration of the long-term impact of climatic change. Narrowly-distributed species are thought to be among the most vulnerable to disruption and local extirpation in periods of rapid climatic change (Tangley 1988; World Wildlife Fund 1989). This is due in part to narrow environmental tolerances of many endemic species; as temperature or precipitation zones shift geographically, the absence of suitable habitat in new areas can place an ecological specialist at a competitive disadvantage. The increasing fragmentation of areas of natural habitat exacerbates this phenomenon; without continuous migratory corridors to accommodate shifting species distributions, many species simply will not find the suitable combination of soil, precipitation, insolation, temperature, and other ecological factors that permit them to survive (Peters

Figure 16.3
*Tissue culture of the Florida torreya (*Torreya taxifolia*) provides specimens for reintroduction in the northern portions of its range. (Photo by L.S. DeBruyn)*

and Darling 1985). Individuals, of course, do not "migrate"; shifts in species ranges take place over generations of sexual reproduction and seed dispersal, processes that are themselves easily disrupted by sudden changes in climate or in the behavior and distribution of ecological symbionts such as pollinators, dispersal agents, or mycorrhizal fungi. Species of montane and alpine communities are thought to be particularly vulnerable in a warming period, since upslope or northward migration of lowland communities will gradually displace them. These considerations are of great significance for conservation strategies because they influence the long-term viability of populations and communities, even on currently protected land (Tangley 1988; Peters 1988; Colinvaux 1989).

To a great extent, the impact of climate change on species survival must be viewed in light of short-term successional changes and long-term ecosystem evolution. As the local values of abiotic (soil, temperature, precipitation) and biotic (competitors, symbionts) factors change on a given site, the suite of species that can survive inevitably changes as well (Hammond 1972; Grime 1977). Further, as Harper (1977), MacMahon (1980), and others have observed, colonization of a new (or newly habitable) site is as much the result of chance distribution and dispersal factors as it is of predictable parameters influencing the suitability of local habitat. Moreover, succession appears to be particularly sensitive to precipitation and temperature regimes (MacMahon 1980), factors that are predicted to change dramatically in most climate-change scenarios. Dispersal and migration processes typically are matched to rates of successional and ecosystem change on the order of hundreds to thousands of years, and evolutionary adaptation on an even longer time scale. Projected rates of climate change, however, will occur tens to hundreds of times faster than these processes can accommodate. Thus there is every reason to expect that narrowly distributed species, including habitat endemics, will be threatened severely by rapid climate change.

BIOLOGICAL CONSIDERATIONS IN RARE PLANT CONSERVATION

Many threats to rare plants, other than outright destruction of habitat, are closely linked to their biology. Though a comprehensive survey of the biology of rare plants is beyond the scope of this chapter, a few salient factors deserve mention (see also chapter 2; Falk and Holsinger 1991; review in Falk 1990a).

Ecological Characteristics of Rare Plants

From the perspective of conservation biology, insights into species ecology provide clues that can inform strategies to preserve them. Many rare United States plants are endemic to edaphically limited communities such as limestone, serpentine, or shale outcrops (Huenneke 1991; Kruckeberg and Rabinowitz 1985). The actual mechanisms of endemism and the relationship between endemic distribution and particular edaphic factors, are still matters of controversy. Traditionally, plant species restricted to particular rock outcrop types were considered to be either obligate for a certain soil chemistry component or genetically depauperate, and thus with narrow niche breadth and reduced competitive ability. In their review of edaphic endemics of the eastern U.S., Baskin and Baskin (1988) concluded that the weight of evidence now does not universally support this model (see also Fiedler 1986 and chapter 2). Many endemics are found on more than one rock or soil type, often with wide divergence of pH or physical characteristics. Moreover, many plant species grow as well or better under cultivation in nonspecialized commercial potting soil as they do in their native soil, further indicating the lack of a requirement for an unusual soil chemical component.

Baskin and Baskin's review (1988) of studies of genetic or genetically controlled variation concluded that many (but not all) endemic taxa show substantial morphological, anatomical, autoecological, reproductive, and cytological variability, as well as allozyme variation. This suggests that limited genetic variation per se is not an adequate explanation of endemic restriction. In many cases, it is the habitat itself that affects the distribution of populations.

The mechanism for rarity may well be other than edaphic restriction. The rare *Pedicularis furbishiae*, for instance, occurs in chronically disturbed riparian habitat subject to periodic scouring by ice and floods. These disturbances extirpate whole populations, each of which may represent a founder event initiated by only one or a few dispersed seeds. The intermittent and ephemeral nature of *P. furbishiae* populations may well inhibit gene flow among populations, but the species' limited range is a matter of habitat disturbance, not edaphic restriction due to low genetic variation per se. Studies comparing patterns of variation among congeners (Karron, 1987, 1989, 1991) may help to elucidate other related issues, such as differing competitive abilities and ecological amplitudes among closely related plant lineages.

Genetic Characteristics of Rare Plant Species

The distribution of genetic variation in rare plant species is a key consideration in conservation strategies and is treated in detail elsewhere in this volume (chapters 2 and 13; see also Falk and Holsinger 1991). Studies by Hamrick and others (Hamrick et al. 1979; Hamrick 1983; Hamrick and Godt 1989; Hamrick et al. 1991) have assessed a wide range of correlate factors, including geographic range, mode of reproduction, breeding system, pollination and dispersal mechanisms, successional stage, population characteristics, and others. Strong as many of these correlations are, ecological characteristics account for only 28 percent of within-population variation and 47 percent of the variability observed among populations (Hamrick and Godt 1989).

Studies of genetic variation in endemic species reveal a divergent pattern. Loveless and Hamrick (1984) characterized "endemic" plants (sensu narrowly distributed geographically) as generally depauperate of allozyme variation. Interestingly, however, endemic species were second only to widespread plants in mean diversity at the species level (Ht = .272 for endemics, compared to .316 for widespread plants), and higher than narrowly (.255) or regionally distributed (.218) plants. Likewise, the partitioning of mean diversity *within* populations (Hs) was actually highest of all groups among endemic plants, perhaps reflecting the tendency of some rare species to be locally abundant (Rabinowitz, Cairns, and

Dillon 1986). Many narrow endemics are, of course, low or lacking entirely in genetic variation (Hamrick et al. 1979; Hamrick et al. 1991). Studies of extremely rare species such as *Pedicularis furbishiae* (Waller, O'Malley, and Gawler 1987), *Pinus torreyana* (Ledig and Conkle 1983), *Chrysoplenium iowense* (Schwartz 1985), and *Sullivantia* ssp. (Soltis 1982) revealed virtually no electrophoretically detectable variation.

Although some rare species can be surprisingly heterozygous, studies aggregating data from many genera have on the whole shown endemic and/or narrowly restricted species to have lower levels of variability then widespread species. Whether this lower observed variation is a cause or an effect of rarity, however, is less clear. Many U.S. plant taxa are geographically as well as ecologically restricted; of the 745 native taxa thought to be most critically threatened, 575 (77%) are found in only a single state (Center for Plant Conservation 1988a, 1991b). Endangered species are frequently found in no more than five populations, and might thus be expected to manifest lower ecological amplitude—and less genetic variation overall—than widespread species. Brown and Briggs (1991) report that 81 percent (38 of 47 taxa) of endangered species in Australia are known from five or fewer populations, including 47 percent (22 taxa) known from only a single population. Indeed, some careful studies of congeners eliminating the uncertainty of cross-genus comparison) do confirm lower levels of genetic variation in some narrowly distributed species (Karron 1987, 1991).

The ecological and evolutionary significance of this low observed variation is far from clear, however. For instance, classic population genetics predicts severe inbreeding depression in extremely small populations of normally out crossing plants, especially where there is little gene flow from outside (Wright 1977). Inbreeding depression is of particular concern because it is thought to affect central components of individual fitness and evolutionary adaptation early in the plant's life history: seed production, germination success, seed viability, seedling growth rates (Charlesworth and Charlesworth 1987). Interestingly, Barrett and Kohn (1991) observe that many of the strongest effects of inbreeding actually occur late in a plant's life cycle—such as flower and fruit production—rather than in early stages such as seed germination. Naturally rare species may have been so thoroughly exposed to inbreeding that depression is unlikely to occur (Stebbins 1950). According to this model, species with populations that are normally small and disjunct have already expressed their lethal or deleterious alleles and have eliminated them from the genome (Lande and Schemske 1985). This is corroborated by some studies such as Karron's work with *Astragalus*, which found no evidence in rates of seed set or embryo abortion for inbreeding depression in two highly restricted species of locoweed (1989).

A number of species, such as *Physostegia correllii, Hedyotis parvula, Pritchardia munroi, Arctostaphylos hookeri* ssp. *ravenii*, and *Castilleja uliginosa*, have only one surviving individual known in the wild (McMahan 1989a). In such cases the notion of intraspecific genetic diversity is obviously irrelevant; survival of the individual is congruent with survival of the species, and special measures are required. *Castilleja uliginosa*, for example, is a California endemic species that diverged from the more widespread *C. miniata*; at present there is only one remaining plant. Furthermore, because the plant is an obligate outcrosser, increase in the wild seems highly unlikely. One possible strategy for *C. uliginosa* may be to cross-pollinate with the two closely related species and then backcross the F_1 progeny to produce genetically pure individuals of the species (P. Raven, R. Ornduff, pers. com.).

Demographic Effects in Small Populations

As noted earlier, over 1,300 rare and endangered U.S. plant species are known from five or fewer populations, or 1,000 or fewer individuals. This degree of rarity exposes these species to chance events that may precipitate a population crash. All populations in nature normally fluctuate in size from year to year due to variation in weather, pollination, seed production, dispersal, seedling success, and a host of other variables. Extreme perturbations in these factors can precipitate dramatic fluctuations in population size from year to year. If a species' population is already small, a combination of several of these impacts in a negative direction can cause a population to fall below minimum viable population size, resulting in local extirpation of the species. Moreover, if a species exists only in a few such populations, each population lost brings the species as a whole significantly closer to extinction. Thus, rare species, which tend to have few populations and frequently few individuals per population to begin with, are particularly susceptible to this pattern of decline.

The study of these factors generally revolves around the determination of minimum viable population levels and the species susceptibility to stochastic (from Greek *stochastikos*, to aim an arrow at a target: the superimposition of a random event on an orderly process). This phenomenon has been extensively studied in animals (Pimm, Jones, and Diamond 1988; Soulé and Simberloff 1986) but has received little empirical work in plants (Lewin 1989; Menges 1991). Theoretically, for a given species there exists a minimum threshold below which the population will be susceptible to stochastic events, any one of which can cause a drastic decline or even local extinction. Rare species are particularly vulnerable to Shaffer's (1981) four types of stochasticities (genetic, demographic, environmental, catastrophic) because in each case, the smaller the population, the greater the probability that stochastic fluctuations will result in decline of the population to zero.

INTEGRATED CONSERVATION STRATEGIES FOR ENDANGERED SPECIES

The biological characteristics of the taxon of concern, and its principal threats, are two of the primary determinants of sound conservation strategy. The third central element consists of the conservation resources available and the particular approach that is to be taken to address the problem. The juxtaposition of these three elements in a manner that is responsive to the particular characteristics of a given situation constitutes the essence of integrated conservation strategies (Falk 1987, 1990a, 1991b). The integrated model is based on the precept that the full range of conservation measures—from land acquisition and management to offsite research, propagation, and genebanking—will be necessary to conserve biological diversity in the coming decades. The need for such an approach arises from the diversity, scale, and pace of threats to species in the wild and of the severe limitations of resources available for conservation programs. In such an environment, conservationists no longer have the luxury of relying on a single approach regardless of the cost or time involved. They must use every means at their disposal to protect species and communities from extinction, in an optimally cost-effective manner.

This section will assess several conservation strategies that may be components of the integrated approach, including three of the most important for endangered species: land protection, biological management and restoration, and the role of offsite conservation and research.

Land Protection

Land acquisition is such a visible mainstay of conservation programs in the United States that we often take its presence and scale for granted. In fact, less than one-tenth of the land area of the United States is specifically protected and firmly dedicated for conservation purposes in parks, wilderness areas, and private conservation land (U.S. Department of Interior 1988). The Nature Conservancy, by far the largest and most successful private land conservation effort in the history of the United States, has cumulatively protected over two million ha of land since its inception (The Nature Conservancy 1988). In sheer hectarage, however, even this impressive accomplishment is dwarfed by other types of land ownership, such as the more than ten million ha reserved as Native American tribal lands in the Southwestern United States (Nabhan 1988). In the states from the Rocky Mountains westward, approximately 47 percent of the land is owned by the U.S. government, with another 21 percent owned by state and local governments (U.S. Geological Service 1970). Most of this land area is managed by the U. S. Bureau of Land Management (BLM), Forest Service (USFS), and other agencies with multiple-use mandates that include strong commitments to resource extraction and recreational use. For example, on a national scale the National Forest System contains approximately 77 million ha in 42 states and two territories, accounting for 8.4 percent of the land area of the United States (Wilcove 1988).

Only 17 percent of the system's total hectarage is set aside as "Wilderness Areas" or "Research Natural Areas," but because of the huge extent of National Forest lands this amounts to over 13 million ha. By contrast, the federal agencies most typically associated with biological conservation, the National Parks and Monuments, control only 32 million ha, or less than 3.5 percent, of the land area of the country. This pattern is even more significant for certain community types. For instance, in the state of Oregon 57.9 percent of all forested lands are owned and managed by the Forest Service, Bureau of Land Management, and Indian tribal lands; 38.2 percent are in private or corporate hands, and less than 4 percent are controlled by state or local government (Black 1989).

Thus, from a strategic and policy point of view, management of public lands in the United States is as significant as, and in many areas more significant than, private conservation acquisition of land. The latter mode of ownership, however, retains the important advantage of having conservation as its principal purpose, with other uses clearly secondary. The principal constraint on private efforts to conserve land has traditionally been the limited availability of funds, especially in contrast to sums expended for federal government programs. The difficulties of acquisition may be compounded by the refusal of landowners or developers to sell to a conservation organization for fear that doing so will inhibit further growth in the value of adjacent real estate. Moreover, the very lands facing the greatest threat (and therefore naturally the highest priorities for conservation) frequently are those in the path of commercial development, where the price of land may be extremely high and the willingness of a developer to relinquish control of a parcel of land extremely low. Land costs in such instances may be tens of thousands of dollars per hectare, so that the acquisition of even a small preserve may require millions of dollars of capital.

Under such trying circumstances, land-conservation organizations have increasingly diversified their approach. The Nature Conservancy, for instance, follows a complex strategy for protecting land even when outright acquisition is not feasible or cost-effective (The Nature Conservancy 1987, 1988). Alternatives to purchase include management agreements and covenants, long-term leases, and conservation easements. Even where land is to be acquired, transactions may be made more palatable with deed restrictions, conditional transfer agreements, and favorable tax considerations. Despite funding constraints, the Conservancy has compiled a strong record of land protection; from 1981 to 1988 the Conservancy added some 614,000 ha of conservation land, as compared to 229,000 ha protected by the U.S. Fish and Wildlife Service (Farney 1989).

The Conservancy also is increasingly adopting a preserve complex or "megasite" approach to land conservation, utilizing regional clusters of smaller protected areas in a region rather than relying on the acquisition of one or two huge, but unattainable, preserves (Jenkins 1989). A well-chosen constellation of preserves may contain a greater variety of habitat types and communities and help to buffer the system from impacts of climatic change by providing sites with different microclimates to which species may migrate or be moved. The megasite approach also permits the establishment of migratory corridors and buffer zones, and allows for greater flexibility in management regimes.

Biological Management

As the field of conservation has increased in biological sophistication, it has become evident that acquisition and protection of habitat are only the first steps in a long-term strategy to protect biological diversity. In part this reflects the application of relevant scientific models to conservation practice. For instance, island biogeographic theory (MacArthur and Wilson 1967) has been used widely to provide a theoretical basis for understanding the effects of habitat fragmentation, by using an analogy between oceanic islands and "islands" of protected land and habitat. Likewise, models of the distribution of allelic variation among and within populations have been used to guide sampling strategies for genetic conservation (Brown and Briggs 1991).

A deeper influence on conservation thinking, however, has been the recognition that conserving biological diversity is a subtler and more difficult goal than simply a matter of accumulating acreage. As noted above, in many parts of the United States there are severe threats to native species even on protected or publicly owned land. Many of these threatening influences (e.g., invasive exotic vegetation, feral herbivores, overgrazing, successional changes, illegal collecting, hybridization, fungal pathogens) are not eliminated or mitigated by acquisition of the land, although acquisition remains in many cases a necessary precondition to further management regimes. Such threats compel intervention into the ecology of a site and the dynamics of the constituent species—hence the term *biological management*.

Several practices in the realm of biological management are particularly germane to the conservation of rare plants. Perhaps the simplest is the propagation and rerelease of plant material for *enhancement* of a damaged but extant population. Enhancement projects have the benefit of working with an existing population in its natural ecological setting. Ideally, plant material is taken from the site itself, so as to minimize the likelihood of introducing foreign genetic material. For example, an enhancement project carried out for *Erysimum menziesii* involved the collection, propagation, and replanting of 3,500 seedlings of a rare Pacific coastal dune species (Ferreira and Smith 1987). A similar project was carried out for a Vermont population of *Hudsonia tomentosa* damaged by off-road vehicles on a Nature Conservancy preserve. Propagules taken from the site were rooted at a nearby nursery and then replanted (Des Meules, pers. com.).

Introduction and *reintroduction* increasingly are used as tools in the management of rare species (Falk and McMahan 1988; Falk 1987). Both methods involve the release of plant material onto a site not currently inhabited by the target species in order to establish a breeding population. The distinction between the two approaches is significant, however, and deserves to be retained. *Reintroduction* denotes the release of material onto a documented historical location for the taxon, including reestablishment after a catastrophic event. *Introduction*, by contrast, involves placement of plants at a location with no direct evidence of former presence on the site. Reintroduction is thus the more conservative of the two options from the perspective of population genetics and biogeography in that it approaches more closely the ideal of management of natural systems as they would exist without human interference. However, reintroductions may be hampered if the conditions that caused the initial decline of the population, such as a fungal pathogen or invasive exotic competitor, are still present onsite. Introduction of material to a new site, on the other hand, must always be regarded as an empirical test of our understanding of the biology of a species, because soil, light, water, and temperature parameters may differ from the original site, and pollinators, mycorrhizal symbionts, or seed dispersal agents may be absent or altered (Lewin 1989). Likewise, introduction by definition involves manipulation of the distribution of genetic variation among populations and thus is opposed by some biologists.

Unfortunately, where all other suitable sites for a species have been destroyed, reintroduction and introduction may represent the only hope for maintaining a population in the wild. As a consequence, such measures have been widely used for rare species, including the recovery of *Agave arizonica* (Delamater and Hodgson 1987; Figure 16.4), *Arctostaphylos uva-ursi* var. *leobreweri* (S. Edwards, pers. com.), *Chrysopsis floridana* in Florida (Wallace and McMahan 1988), *Pediocactus knowltonii* (Olwell et al. 1987), *Penstemon barrettiae* (Thompson 1988b; Native Plants Society of Oregon 1989), *Stephanomeria malheurensis* in Oregon (Parenti and Guerrant 1990; Brauner 1988; U.S. Fish and Wildlife Service 1982), *Styrax texana* in Texas (Cox 1987) and others. Moreover, as Webb (1985) has observed, the "native" species complement we associate with a particular site may in fact represent an earlier or undocumented introduction, so that the distinction between native and exotic species is not always supported by biological or historical reality. In the face of continued destruction of existing populations of rare species and the increasing fragmentation of preserves and other protected areas, it is likely that these methods will be even more widely used.

Many management practices for rare species fall under the rubric of *successional management*. Huenneke (1988) describes several considerations in management of rare plant sites, including habitat quality, competition and other interactions with co-occurring species, relationships with pollinators and dispersal agents, and community ecology. The latter in particular may involve direct manipulation of seral stages and disturbance factors such as fire (The Nature Conservancy 1989b). Fire-adapted communities like chaparral, prairie, and savanna may be impossible to maintain without periodic burns. Some species have seeds that will germinate only after scarification by fire; seeds of some species of *Phacelia*, *Emmenanthe*, and many other herbs and grasses germinate only after exposure to high temperatures and charred wood left by a chaparral fire. Moreover, in the absence of periodic fires, shrubs and overstory plants invade, and early successional communities are shaded out. Populations of species endemic to these communities may be lost as a consequence, and if these populations are few in number then a species may become extinct. Species endemic to other disturbance-adapted communities can be threatened by land-management practices that stabilize the ecosystem. For example, dune stabilization is thought to have contributed to the decline of *Erysimum capitatum* var. *angustatum* (U.S. Fish and Wildlife Service 1984) and *E. teretifolium* (Huenneke 1988). Successional changes also may be a factor in the survival of rare species even where the community per se does not appear to be disturbance-adapted overall. An example is *Iliamna corei*, a rare member of the Malvaceae known only from a single site in western Virginia

Figure 16.4
Agave arizonica, *the Arizona agave, is cultivated at the Desert Botanical Garden as part of its recovery plan. (Photo by F.R. Thibodeau)*

(Thompson 1988a; U.S. Fish and Wildlife Service 1985). Only four ra-
mets remain extant on the site, along with an unknown quantity of un-
germinated seed in the soil. Seed germination appears to be inhibited by
accumulation of leaf litter associated in part with successional changes
in the forest canopy. Other endangered species affected by successional
changes or their suppression include *Astragalus agnicidus* (Berg and Bittman 1988) and *Penstemon haydenii* (Stubbendieck, pers. com.).

A conceptual distinction traditionally has been made between management of existing communities and the establishment of new ones.
However, as the preceding discussion illustrates, the boundary between
management and restoration is sometimes largely academic, especially
for regimes that involve active manipulation of successional phases and
introduction of new species to a site. These forms of "management" may
entail such extensive modification of the habitat that they are indistin-
guishable from what is customarily referred to as "restoration." The essential continuity between management and restoration has been recognized (Jordan, Gilpin, and Aber 1987), although the relationship
between the two practices is still evolving. Linkages between community
restoration and endangered species conservation continue to develop, but
rare plants already have been incorporated into restoration projects in a
number of instances. These include *Pogogyne abramsii* (Zedler and Black
1990), *Cordylanthus palmatus* (Coats, Showers, and Pavlik 1990), *Cordylanthus maritimus* ssp. *maritimus* (Fink and Zedler 1990), *Erysimum
menziesii* (Sawyer and Andre 1990), *Monardella linoides* ssp. *viminea*
(Scheid 1987), and a number of others.

Restoration is becoming a crucial strategy for biodiversity conservation as remaining natural habitat becomes more fragmented. In a natural state, disturbances (and even local extinctions) create a successional
mosaic on the landscape, with the result that at any given time some
areas were recently disturbed and others relatively stable. Successional
species generally had somewhere to migrate to; local extirpations were
probably relatively common, but there were always new populations
emerging (Harris 1984). Now that the few parcels of undisturbed land
are increasingly fragmented, it will become necessary to restore migratory
pathways and to use site management to ensure that adequate successional diversity is maintained.

The Role of Off-site Conservation

Botanical gardens traditionally have been viewed with skepticism or in-
difference by the conservation community, largely because they were seen
as irrelevant to the core work of protecting natural habitat. The most
tangible expression of this attitude was the dichotomy drawn between *ex*

situ and *in situ* conservation. With the emergence of integrated conservation methods in a more management-oriented era, and recognition of the need for the full range of conservation resources, this false dichotomy has begun to fade. Gardens, arboreta, university greenhouses, seed banks, and a wide range of other off-site resources now are used regularly by habitat-managing agencies and may indeed be considered part of the conservation community overall (Given in press; Falk 1987, 1990a, b; Falk and Thibodeau 1986, 1987).

Off-site programs can make especially significant contributions at certain levels of biological hierarchy, most notably in efforts to conserve diversity at the levels of the allele, gene, individual, and population (after Millar and Ford 1988; Falk 1991b). These levels of organization represent the best match between the technical capabilities of off-site conservation programs and the needs of biodiversity conservation. Some of the criticisms traditionally leveled against such programs in fact have been grounded in misapprehension of their potential role; gardens and seed banks are not appropriate vehicles by which to address the conservation of communities or ecosystems, nor are they designed to do so. Although many botanic gardens manage significant natural areas, this is by and large an adjunct of their primary institutional activity. Offsite programs do, however, provide an effective and cost-efficient means to capture and preserve genetic variation found in the wild and, in some cases, to preserve whole species.

The work of off-site programs may be considered in four general categories: genepool maintenance, propagation, research, and education. The most direct service that gardens may perform in conserving plant diversity is in maintenance of living genetic reserves, typically in the form of living collections or seed banks. Botanical gardens and arboreta are, in fact, specifically established for the long-term maintenance of living plant material, and often have extensive staff and infrastructure for this purpose. Though the viability of conservation collections over extremely long periods of time has not yet been tested (Elias 1987), there is evidence that gardens can play a significant role in protecting species from extinction. A number of species now exist only cultivation, including *Franklinia alatamaha*, *Kokia cookei*, and, before its reestablishment in the wild as part of a habitat-restoration project, *Arctostaphylos uva-ursi* var. *leobreweri*. *Stephanomeria malheurensis* was virtually eliminated from its only known site in central Oregon; its reintroduction was made possible by use of seeds stored at the University of California, Davis. Several species known from only a single surviving individual or with populations below sustainable levels in the wild have been maintained in botanic gardens, including *Betula murrayana* (Figure 16.5), *Castilleja uliginosa*, *Pritchardia munroi*, *Arctostaphylos hookeri* ssp. *ravenii*, and *Prunus mar-*

Figure 16.5
Dr. Burton Barnes of the University of Michigan examines the single individual of Betula murrayana. *(Photo by B. Parsons, The Holden Arboretum)*

itima var. *gravesii* (McMahan 1989a). The Center for Plant Conservation now maintains over 400 endangered native U.S. plant species in protective cultivation, utilizing a national network of 22 botanical gardens and arboreta. The highest priority for accession is accorded to the most critically threatened species in a given year, with emphasis on those species that have been identified as experiencing the most catastrophic decline.

Increasing attention is being given to the genetic aspects of off-site collections (Center for Plant Conservation 1991b; see also McMahan and Guerrant 1991; Holsinger and Gottlieb 1991). Although no offsite collection can hope to capture 100 percent of the genetic diversity present in a widespread species, endangered plants frequently are so rare in the wild that a relatively complete genetic sample may be made. Current work in population genetics indicates that for many species, a majority of the polymorphic portion of the genome may be captured in an off-site collection taken from up to 50 individuals from each of no more than five populations (Brown and Briggs 1991; Holsinger and Gottlieb 1991). Sampling of multiple populations enables interpopulational diversity to be captured, and models of variation within populations suggests that a sample taken from the recommended number of individuals will be adequate to capture all but the rarest alleles with a high degree of confidence. The role of such collections also has been considered for the preservation of genetic variation for agricultural purposes (see, for instance, Keystone Center 1988). These recommendations have been published by the Center for Plant Conservation (1991b) as "Genetic Sampling Guidelines for Conservation Collections of Endangered Plants."

The role of off-site facilities in propagation and research have been mentioned earlier in this chapter (see also Raven 1981). Reintroduction, introduction, and restoration programs need reliable sources of healthy, well-documented, genetically diverse material in order to be successful. Thus, the living collections of the Center for Plant Conservation, in addition to serving as a backup against extinction in the wild, may also serve as a source of living plants to assist in the management and recovery of natural populations. Propagated plant material has been used in the recovery of *Chrysopsis floridana, Helianthus schweinitzii, Pediocactus knowltonii, Penstemon barrettiae, Stephanomeria malheurensis, Styrax texana* (Figure 16.6), and others, with promising results (Falk 1990a). These projects are carried out in partnership with land-managing organizations such as The Nature Conservancy, the U.S. Fish and Wildlife Service, and land-managing agencies of federal and state governments.

Increasingly, major federal landholders such as the Bureau of Land Management and the Forest Service are incorporating integrated conservation measures into Recovery Plans and site-management plans. For instance, biological research programs for *Iliamna corei, Agave arizonica,*

Figure 16.6
Reintroduction of propagated material of Styrax texana. *(Photo by P. Leslie, SABG)*

Torreya taxifolia, and others have yielded information useful in the management of wild populations. Research and recovery projects at the Center for Plant Conservation member gardens have covered more than 20 endangered species, including *Schwalbea americana, Muhlenbergia torreyana, Thalictrum cooleyi, Zizania texana, Polygala smallii, Torreya taxifolia, Argyroxiphium virescens, Hedeoma todsenii, Opuntia spinosissima, Cucurbita okeechobeensis, Oxypolis canbyi,* and *Lobelia boykinii.*

Finally, gardens and arboreta are an effective means of communicating the importance of conservation to the public. This is especially true now that more gardens have established living collections of rare species and have incorporated conservation into their institutional missions. Gardens, after all, are largely designed to bring people and plants together; at their best, they can mount dramatic and influential displays and programs seen by millions of visitors each year. Even in affluent nations like the United States, few people ever have more than a fleeting glimpse of rare species habitats across the continent. Gardens and arboreta, which are frequently located in major urban areas, are often the only opportunity to reach people directly with a conservation message backed by living examples of rare species and plant community types.

STRATEGIC CONSIDERATIONS IN RARE-PLANT CONSERVATION

The diverse threats to U.S. plant species present technical, philosophical, and strategic problems for the conservation field. From a technical point of view, the emerging need for management raises many presently unanswered questions about the biology of rare species. Biological management is an inherently information-intensive activity, especially with regard to species' reproductive biology, physiology, population biology, genetics, demography, and other dimensions. Unfortunately, such information is largely absent in the current scientific literature on rare organisms. A recent study sponsored by the National Science Foundation (Wildt and Seal 1988) identified several critical areas of information that are presently underrepresented in the endangered species biological literature. These include population biology (including the biology of release and reintroduction especially in small populations), reproductive biology and breeding systems, and the impact of stress and disease. An understanding of these and other aspects of species biology will be critical if biological management is to succeed, in part because the populations and ranges of so many organisms have already been so severely reduced that there remains little margin for error (Roberts 1988).

The field of biological management and restoration is still in a period of rapid development. For example, a survey of intentional releases of birds and mammals for conservation purposes found wide variability in technique, success rate, and documentation (Griffith et al. 1989). Genetic aspects of restoration and reintroduction work have only recently been considered (Lewin 1989; Millar and Libby 1989; Guerrant, this volume) and many elements of successful management, such as fire ecology, are still being developed by trial and error (The Nature Conservancy 1989b). In many cases basic survey and inventory work still is needed to determine the extent of natural populations and the habitats they occupy (Bowles 1988; Bowles et al. 1986). Many rare species' populations are still unknown, and species once thought extinct are regularly rediscovered in the wild (Homoya, Aldrich, and Jacquart 1989). Field work also may generate basic ecological information, such as fire adaptation, essential for community or species management.

Biological management also ultimately raises a series of ethical questions regarding our relationship to nature (Nabhan 1988). If we seek to preserve nature "as it is," without the influence of the human hand, then management is in some respects an inherently contradictory activity, for in order to "manage" nature we must tamper with it. Indeed, biological management requires a degree of intervention into the functioning and distribution of natural systems and species that would have been unacceptable less then a century ago. The array of management practices used by the contemporary land manager necessitates deep involvement in the processes of those very entities—ecosystems, communities, species— whose "naturalness" we wish to preserve. The manager may at one time or another intervene in the forces of ecological succession and disturbance, population dynamics and genetic structure, species distribution, competition, herbivory and predation, pollination, soil and microclimate dynamics, disease, and many other factors. By the time all this has occurred, we may well wonder what "natural" aspect it was that we really valued in the first place (see Falk 1991a).

Strategically, where rare species are concerned there is frequently no practical alternative to an active management program. Precisely because there are so few individuals and sites for these organisms, each one must be treated as an irreplaceable evolutionary expression and be protected if at all possible. Loss of a population of a species known from only two other sites unquestionably exposes the species to a serious risk of extinction from stochastic forces alone (see chapter 11). Moreover, if there is any biological element to the decline (such as invasion by aggressive exotic species), or a systematic threat to the entire community type (for instance, inundation of desert riparian communities by hydropower projects), then the conservative assumption must be that the remaining populations will be exposed to similar impacts.

Because of their rarity and vulnerability in the wild, in the long run endangered species also force consideration of the optimal allocation of scarce conservation resources. Some strategic options, such as triage (Millar and Libby 1991), are highly charged emotionally as well as uncertain conceptually. For instance, should it be axiomatic that conservation resources always be directed to the most critically threatened species and communities, those in the most need? Or are some species so far gone that the enormous investment of resources required for their recovery would be better spent on other taxa less severely reduced in the wild, where there may be some hope of protecting a broader range of ecotypic diversity? Even within a species, should we endeavor to protect all populations if the species is sufficiently rare, to preserve what little genetic variability and demographic cushion there may be? Or, having protected several such populations of a given species, are there diminishing returns to the protection of additional populations, compelling additional populations to be abandoned in favor of providing protection for the first population of the next species (Brown and Briggs 1991; Falk 1991b; Holsinger and Gottlieb 1991)? Should scarce dollars be used to investigate in depth the pollination ecology of a rare and declining species of *Agave*, or used as a small fraction of the down payment on the acquisition of its habitat? Should one of five remaining populations (all on private land) of an endangered Florida species of *Dicerandra* be dug and "rescued" by removal to another site, or does the possibility of such activity soften the legal imperative for conservation of natural populations? Should the restoration and re-creation of damaged communities be considered an integral part of strategies to conserve diversity, or does restoration open the way for economically driven trade-offs and mitigation projects?

Whatever the answers are in any particular case, it is increasingly clear that long-term conservation strategies must utilize the full range of resources available. Even as land acquisition and protection efforts continue, there will have to be greater resources devoted to the management of populations and communities and a greater willingess to engage in reintroduction and restoration programs (Jenkins 1989).

The strategic issues surrounding conservation of endangered plants mirror similar policy questions that have arisen in the protection of animal species. In some cases—such as African elephants (Cherfas 1989), chimpanzees (Booth 1989), and giant pandas (Roberts 1989b)—there is no clear consensus on the proper course of action, and the conservation and research communities may even be at odds regarding strategy. Many of the relevant considerations are ultimately ethical and economic as well as biological in content and thus require thoughtful dialogue both within and beyond the conservation community. The answers are by no means obvious, and where endangered plants are concerned the fate of many species may hang in the balance.

ACKNOWLEDGMENTS

The author thanks the staff of the Center for Plant Conservation, particularly Margaret Olwell and Michael J. O'Neal, for substantive contributions to this chapter. Thanks also to former CPC staff Linda R. McMahan, Russell D. Stafford, and Kerry S. Walter. Linda DeBruyn, Jennifer Klein, and Grace Padberg provided manuscript preparation. Grant support provided to the Center for Plant Conservation by the Andrew W. Mellon Foundation, the John D. and Catherine T. MacArthur Foundation, the W. Alton Jones Foundation, the David and Lucile Packard Foundation, the Pew Charitable Trusts, the William and Flora Hewlett Foundation, and the National Science Foundation was instrumental in the development of many of the programs described herein.

LITERATURE CITED

Barrett, S.C.H., and J.R. Kohn. 1991. Genetic and evolutionary consequences of small population size in plants: Implications for conservation. In *Genetics and conservation of rare plants*, ed. D.A. Falk and K.E. Holsinger. New York: Oxford University Press.

Baskin, J.M., and C.C. Baskin. 1988. Endemism in rock outcrop plant communities of unglaciated eastern United States: An evaluation of the roles of the edaphic, genetic, and light factors. *J. Biogeogr.* 15:829–40.

Berg, K., and R. Bittman. 1988. Rediscovery of the Humboldt milk-vetch. *Fremontia* 16:13–14.

Black, C. 1989. Last stand for 1,000 year-old trees. *Boston Globe*, 31 July, B1.

Booth, W. 1989. Chimps and research: Endangered? *Science* 241:777–78.

Bowles, M.L. 1988. *Research and recovery objectives of the Morton Arboretum rare plant program.* Lisle, Ill.: Morton Arboretum.

Bowles, M.L., W.J. Hess, M.M. Demauro, and R.D. Hiebert. 1986. Endangered plant inventory and monitoring strategies at Indiana Dunes National Lakeshore. *Nat. Areas J.* 6:18–26.

Brauner, S. 1988. Malheur wire-lettuce (*Stephanomeria malheurensis*) biology and interaction with cheatgrass: 1987 study results and recommendations for a recovery plan. Report to Bureau of Land Management, Burns, Oregon District, U.S. Dept. of Interior.

Brown, A.H.D., and J.D. Briggs. 1991. Sampling strategies for genetic variation in *ex situ* collections of endangered plant species. In *Genetics and conservation of rare plants.*, ed. D.A. Falk and K.E. Holsinger. New York: Oxford University Press.

Center for Plant Conservation. 1988. A survey of the most critically threatened plant species of the United States. St. Louis, Missouri.

————. 1991a. Database of the Center for Plant Conservation. St. Louis, Missouri.

————. 1991b. Genetic sampling guidelines for conservation collections of endangered plants. In *Genetics and conservation of rare plants*, ed. D.A. Falk and K. E. Holsinger. New York: Oxford University Press.

Charlesworth, D., and C. Charlesworth. 1987. Inbreeding depression and its evolutionary consequences. *Ann. Rev. Ecol. Syst.* 18:237–68.

Cherfas, J. 1989. Decision time on African ivory trade. *Science* 246:26–27.

Coats, R., M.A. Showers, and B. Pavlik. 1990. A management plan for the alkali sink and its endangered plant, *Cordylanthus palmatus*. In *Restoration '89: The new management challenge*, ed. H.G. Hughes and T. Bonnicksen. Madison, Wisc.: Society for Ecological Restoration.

Colinvaux, P.A. 1989. The past and future Amazon. *Sci. Am.* 261:102–8.

Cox, P. 1987. Chasing the wild Texas snowbells. *Plant Conserv.* 2:1, 8.

Davis, S.D., S.J.M. Droop, P. Gregerson, L. Henson, C.S. Leon, J. Lambein Villa-Lobos, H. Synge, and J. Zantovska. 1986. *Plants in danger: What do we know?* International Union for Conservation of Nature, Gland, Switzerland.

Delamater, R., and W. Hodgson. 1987. *Agave arizonica*: An endangered species, a hybrid, or does it matter? In *Conservation and management of rare and endangered plants*, ed. T.S. Elias, 305–10. Sacramento: California Native Plant Society.

Elias, T.S. 1987. Can threatened and endangered species be maintained in botanic gardens? In *Conservation and management of rare and endangered plants*, ed. T.S. Elias and J. Nelson. Sacramento: California Native Plant Society.

Falk, D.A. 1987. Integrated conservation strategies for endangered plants. *Nat. Areas J.* 7:118–23.

————. 1990a. Integrated strategies for conserving plant genetic diversity. *Annals Missouri Bot. Gard.* 77:38–47.

————. 1990b. The theory of integrated conservation strategies for biological diversity. In *Ecosystem management: Rare species and significant habitats*, ed. R.S. Mitchell, C.J. Sheviak, and D.J. Leopold. Proc., 5–10. 15th Ann. Nat. Conf. Natural Areas Association. Albany: New York State Museum.

————. 1991a. Restoring the future, discovering the past. *Restoration and management notes* 8:71–72.

————. 1991b. Joining biological and economic models for conserving plant genetic diversity. In *Genetics and conservation of rare plants*, ed. D.A. Falk and K.E. Holsinger, 209–23. New York: Oxford University Press.

————. In press. An evolutionary perspective on plant conservation. *Bot. J. Linnean Soc.* London.

Falk, D.A., and K.E. Holsinger, eds. 1991. *Genetics and conservation of rare plants*. New York: Oxford University Press.

Falk, D.A., and L.R. McMahan. 1988. Endangered plant conservation: Managing for diversity. *Nat. Areas J.* 8:91–9.

Falk, D.A., and F.R. Thibodeau. 1986. Saving the rarest. *Arnoldia* 46:3–17.

————. 1987. Building a national *ex situ* network: The U.S. Center for Plant Conservation. In *Botanic gardens and the world conservation strategy*, ed. D. Bramwell, O. Hamann, V. Heywood, and H. Synge, 285–94. International Union for the Conservation of Nature. London: Academic Press.

Farney, D. 1989. Nature Conservancy–led land preservation plan outperforms and out–innovates Federal program. *Wall Street Journal*, 24 May, A16.

Ferreira, J., and S. Smith. 1987. Methods of increasing native populations of *Erysimum menziesii*. In *Conservation and management of rare and endangered plants*, ed. T.S. Elias, 507–11. Scaramento: California Native Plant Society.

Fiedler, P.L. 1986. Concepts of rarity in vascular plant species, with special reference to the genus *Calochortus* Pursh (Liliaceae). *Taxon* 35:502–18.

Fink, B., and J.B. Zedler, 1990. Endangered plant recovery: Experimental approaches with *Cordylanthus maritimus* ssp. *maritimus* In *Restoration '89: The new management challenge*. Madison, Wisc.: Society for Ecological Restoration.

Given, D.R. In press. *Principles and practices of plant conservation*. Christchurch: N.Z.: Department of Scientific and Industrial Research.

Griffith, B., J.M. Scott, J.W. Carpenter, and C. Reed. 1989. Translocation as a species conservation tool: Status and strategy. *Science* 245:477–80.

Grime, J.P. 1977. *Plant strategies and vegetation processes*. Chichester: Wiley-Interscience.

Hammond, A.L. 1972. Ecosystem analysis: A bionic approach to environmental science. *Science* 175:46–48.

Hamrick, J.L. 1983. The distribution of genetic variation within and among natural plant populations. In *Genetics and conservation*, ed. C.M. Schonewald-Cox, S.M. Chambers, B. MacBryde, and L. Thomas, 335–48. Menlo Park: Benjamin/Cummings.

Hamrick, J.L., and M.J. GODT. 1989. Allozyme diversity in plant species. In *Population genetics and germplasm resources in crop improvement*, ed. A.H.D. Brown, M.T. Clegg, A.L. Kahler, and B.S. Weir, 43–63. Sunderland, Mass.: Sinauer.

Hamrick, J.L., M.J. Godt, D.A. Murawski, and M.D. Loveless. 1991. Relationships between species characteristics and the distribution of allozyme variation. In *Genetics and conservation of rare plants*, ed. D.A. Falk and K.E. Holsinger. New York: Oxford University Press.

Hamrick, J.L., Y.B. Linhart, and J.B. Mitton. 1979. Relationships between life history characteristics and electrophoretically-detectable variation in plants. *Ann. Rev. Ecol. Syst.* 10:173–200.

Harper, J.L. 1977. *Population biology of plants*. New York: Academic Press.

Harris, L. 1984. *The Fragmented Forest*. Chicago: University of Chicago Press.

Holsinger, K.E., and L.D. Gottlieb, 1991. Conservation of rare and endangered plants: A synthesis. In *Genetics and conservation of rare plants*, ed. D.A. Falk and K.E. Holsinger. New York: Oxford University Press.

Homoya, M.A., J.R. Aldrich, and E.M. Jacquart. 1989. The rediscovery of the globally endangered clover, *Trifolium stoloniferum*, in Indiana. *Rhodora* 91:207–12.

Huenneke, L.F. 1988. Managing land to protect rare plant populations. *Fremontia* 16:3–8.

———. 1989. Conservation needs close to home. *Science* 244:854–55.

——. 1991. Ecological implications of genetic variation in plant populations. In *Genetics and conservation of rare plants.*, ed. D.A. Falk and K.E. Holsinger. New York: Oxford University Press.

Jenkins, R.E. 1989. Long-term conservation and preserve complexes. *The Nature Conservancy Magazine* 39:4–7.

Jordan, W., III, M. Gilpin, and J.D. Aber, eds. 1987. *Restoration ecology: A synthetic approach to ecological research.* New York: Cambridge University Press.

Karron, J.R. 1987. A comparison of levels of genetic polymorphism and self-compatibility in geographically restricted and widespread plant congeners. *Evol. Ecol.* 1:47–58.

——. 1989. Breeding systems and levels of inbreeding depresssion in geographically restricted and widespread species of *Astragalus* (Fabaceae). *Am. J. Bot.* 76:331–40.

——.1991. Patterns of genetic variation and breeding systems in rare plant species. In *Genetics and conservation of rare plants*, ed. D.A. Falk and K.E. Holsinger. New York: Oxford University Press.

Keystone Center. 1988. Final report of the Keystone International Dialogue on Plant Genetic Resources. Session I: *Ex situ* conservation of plant genetic resources. Keystone, Colo.: Keystone Center.

Kimura, B.Y., and K.M. Nagata. 1980. Hawaii's vanishing flora. Honolulu: Oriental Publishing Co.

Kruckeberg, A.R., and D. Rabinowitz. 1985. Biological aspects of endomysia in higher plants. *Ann. Rev. Ecol. Syst.* 16:447–79.

Lande, R., and D.W. Schemske. 1985. The evolution of self-fertilization and inbreeding depression in plants. I. Genetic Models. *Evolution* 39:24–40.

Ledig, F.T., and M.T. Conkle. 1983. Gene diversity and genetic structure in a narrow endemic, Torrey pine (*Pinus torreyana* Parry ex Carr.). *Evolution* 37:79–85.

Lewin, R. 1986. A mass extinction without asteroids. *Science* 234:14–15.

——. 1989. How to get plants into the conservationists' ark. *Science* 244:32–33.

Loveless, M.D., and J.L. Hamrick. 1984. Ecological determinants of genetic structure in plant populations. *Ann. Rev. Ecol. Syst.* 15:65–95.

MacArthur, R.H., and E.O. Wilson. 1967. *The theory of island biogeography.* Princeton, N.J.: Princeton University Press.

MacMahon, J.A. 1980. Ecosystems over time: Succession and other types of change. In *Forests: Fresh perspective from ecosystem analysis*, ed. R.H. Waring, 27–58. Proceedings from the 40th Annual Biological Colloquium, 1979. Corvallis: Oregon State University Press.

McMahan, L.R. 1980. Legal protection for rare plants. *Am. Univ. Law Rev.* 29:515–69.

——. 1989a. Rarest U.S. plants are literally one of a kind. *Plant Conserv.* 4:6.

——. 1989b. Conservationists join forces to save Florida torreya. *Plant Conserv.* 4:1, 8.

McMahan, L.R., and E.O. Guerrant. 1991. Practical pointers for conserving genetic diversity in botanical gardens. *The Public Garden* 6:20–25.

Menges, E.S. 1991. The application of minimum viable population theory to plants. In *Genetics and conservation of rare plants*, ed. D.A. Falk and K.E. Holsinger. New York: Oxford University Press.

Millar, C.I., and L.D. Ford. 1988. Managing for nature conservation: From genes to ecosystems. *BioScience* 38:456–57.

Millar, C.I., and W.J. Libby. 1989. Restoration: Disneyland or a native ecosystem? *Fremontia* 17:3–10.

———. 1991. Strategies for conservation of clinal, ecotypic and disjunct population diversity in widespread species. In *Genetics and conservation of rare plants*, ed. D.A. Falk and K.E. Holsinger. New York: Oxford University Press.

Myers, N. 1985. The end of the lines. *Nat. Hist.* 94:2–12.

Nabhan, G.P. 1988. Southwest project cuts across cultural, national boundaries. *Plant Conserv.* 3:1, 8.

———. 1989. *Enduring seeds: Native American agriculture and wild plant conservation*. San Francisco: North Point Press.

Native Plant Society of Oregon. 1989. Penstemon transplant: So far, so good. *Bull. Native Plant Soc. Ore.* 22(6):58.

Nature Conservancy, The. 1987. *Memorandum: Options for protecting your land*. Arlington, Va.: The Nature Conservancy.

———. 1988. Annual report. *The Nature Conservancy.*

———. 1989a. Natural heritage database. Arlington, Va.: The Nature Conservancy.

———. 1989b. Fire ecology report, summer 1989. Tallahassee, Fla.: The Nature Conservancy and Tall Timbers Research Station.

Olwell, M., A. Cully, P. Knight, and S. Brack. 1987. *Pediocactus knowltonii* recovery efforts. In *Conservation and management of rare and endangered plants*, ed. T.S. Elias, 519–22. Sacramento: California Native Plant Society.

Parenti, R.L., and E.O. Guerrant. 1990. Down but not out: Reintroduction of the extirpated Malheur wirelettuce, *Stephanomeria malheurensis. Endang. Spec. Update* 8(1):62–3.

Peters, R.L. 1988. Effects of global warming on species and habitats: An overview. *Endang. Spec. Update* 5(7):1–8.

Peters, R.L., and D.S. Darling. 1985. The greenhouse effect and nature reserves. *BioScience* 35(11):707–17.

Pimm, S.L., H.L. Jones, and J. Diamond. 1988. On the risk of extinction. *Am. Nat.* 132:757–85.

Rabinowitz, D., S. Cairns, and T. Dillon. 1986. Seven forms of rarity and their frequency in the flora of the British isles. In *Conservation biology: The science of scarcity and diversity*, ed. M. Soulé, 182–204. Sunderland, Mass.: Sinauer.

Raven, P.H. 1981. Research in botanical gardens. *Bot. Jahrb. Syst.* 102:53–72.

Rieseberg, L.H., S. Zona, L. Aberom, and T.D. Martin. 1989. Hybridization in the island endemic, Catalina Mahogany. *Cons. Biol.* 3:52–58.

Roberts, L. 1988. Beyond Noah's ark: What do we need to know? *Science* 241:1247.

———. 1989a. Extinction imminent for native plants. *Science* 242:1508.

———. 1989b. Conservationists in Panda-modium. *Science* 241:529–31.

Sawyer, J., and J. Andre. 1990. The Menzies' wallflower: An integrated approach to enhancing rare plant populations using restoration. In *Restoration '89: The new management challenge*. Madison, Wisc.: Society for Ecological Restoration.

Scheid, G.A. 1987. Habitat characteristics of Willowy monardella in San Diego County: Site selection for transplants. In *Conservation and management of rare and endangered plants*, ed. T.S. Elias, 501–6. Sacramento: California Native Plant Society.

Schwartz, O.A. 1985. Lack of protein polymorphism in the endemic relict *Chrysosplenium iowense* (Saxifragaceae). *Can. J. Bot.* 63:2031–34.

Shaffer, M.L. 1981. Minimum population sizes for species conservation. *BioScience* 31:131–34.

Soltis, D.E. 1982. Allozymic variability in *Sullivantia* (Saxifragaceae). *Syst. Bot.* 7:26–34.

Soulé, M.E., and D. Simberloff. 1986. What do genetics and ecology tell us about the design of nature preserves? *Biol. Cons.* 35:19–40.

Stebbins, G.L. 1950. *Variation and evolution in plants.* New York: Columbia University Press.

Stone, C.P., and J.M. Scott, eds. 1985. *Hawaii's terrestrial ecosystems: Preservation and management.* Cooperative National Park Resources Study Unit, Honolulu: University of Hawaii.

Tangley, L. 1988. Preparing for climate change. *BioScience* 38:14–18.

Temple, S. 1977. Plant-animal mutualism: Coevolution with dodo leads to near extinction of plant. *Science* 197:885–86.

Thompson, P.J. 1988a. Peter's mountain mallow. *Plant Conserv.* 3(4):7.

———. 1988b. *Penstemon barrettiae:* Renewed plant on the block. *Plant Conserv.* 3(2):7.

U.S. Department of the Interior. 1988. *Public land statistics.* Washington, D.C.: U.S. Governmental Printing Office.

U.S. Fish and Wildlife Service. 1982. Malheur wire-lettuce listed with critical habitat. *Endang. Spec. Tech. Bull.* 7:1.

———. 1984. Revised recovery plan for three endangered species endemic to Antioch Dunes, California. Region 1, Portland, Ore.

———. 1985. Protection sought for four vulnerable plants. *Endang. Spec. Tech. Bull.* 10:1, 8.

———. 1988. Florida torreya (*Torreya taxifolia*) recovery plan. Atlanta: Southeast Regional Office.

———. 1989. Box score of listings and recovery plans. *Endang. Spec. Tech. Bull.* 14:12.

———. 1990. Box score of listings and recovery plans. *Endang. Spec. Tech. Bull.* 15:8.

U.S. Geological Service. 1970. *The National Atlas of the United States of America.* Washington, D.C.: U.S. Dept. of the Interior.

Wallace, S.R., and L.R. McMahan. 1988. A place in the sun for plants. *Garden* 12:20–23.

Waller, D.M., D.M. O'Malley, and S.C. Gawler. 1987. Genetic variation in the extreme endemic *Pedicularis furbishiae* (Scrophulariaceae). *Cons. Biol.* 1:335–40.

Webb, D.A. 1985. What are the criteria for presuming native status? *Watsonia* 15:231–36.

Wilcove, D.S. 1988. *National forests: Policies for the future. Vol. 2: Protecting biological diversity.* Washington, D.C.: The Wilderness Society.

Wildt, D.E., and U.S. Seal. 1988. *Research priorities for single species biology.* Washington, D.C.: National Science Foundation.

World Wildlife Fund. 1989. *The consequences of global warming for biological diversity.* World Wildlife Fund Letter no. 5. Washington, D.C.: World Wildlife Fund.

Wright, S. 1977. *Evolution and the genetics of populations. Vol. 3: Experimental results and evolutionary deductions.* Chicago: University of Chicago Press.

Zedler, P.H., and C. Black. 1990. The creation of artificial habitat for the rare plant *Pogogyne abramsii* in San Diego County, California. In *Restoration '89: The new management challenge.* Madison, Wisc.: Society for Ecological Restoration.

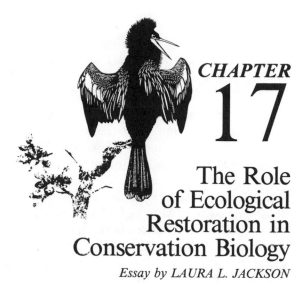

CHAPTER
17

The Role
of Ecological
Restoration in
Conservation Biology

Essay by LAURA L. JACKSON

INTRODUCTION

It is hard to imagine the sense of vacancy that exists in a place where nature has been completely removed, not just for city blocks, but for thousands of square miles. While attending college in Iowa, my friends and I would borrow a car and make the ten-mile pilgrimage to Turner Station, one of the handful of prairies left in southeastern Iowa. By accident and luck, the railroad right-of-way had protected a strip of unturned sod 6 m wide and 30 m long. Kneeling, we would search for single specimens of once-common forbs pushing up through the tall grasses and the thick mat of leaf litter. The prairie was infested with sumac, buckthorn, and tree seedlings because it had not burned in over ten years, and we worried that it would soon disappear entirely.

One night my friends went out and illegally set the prairie on fire. Two weeks later, on our next pilgrimage, a new growth of prairie species had emerged from the ashes, and the encroaching shrubs were dead. A long-forgotten rite of spring had been resurrected, saving (temporarily) the 0.027 hectares of prairie that were still left in the county.

Conservation biology has until recently focused on preserving un-besmirched habitat. But, as in the North American prairies, primary habitat has disappeared in so many places that it seems the only options left are to tend, in museum fashion, "postage-stamp" remnants such as the Turner Station prairie or to invest large sums to restore land to a hypothetical historic state. Restoration of entire historical communities may become an important component of the conservation of biodiversity worldwide (Jordan, Gilpin, and Aber 1987; Jordan, Peters, and Allen 1988). But because restoration is very expensive and would convert land that is in use by people, it has the potential to be even more controversial than setting aside primary habitat. Prime Iowa farmland costs over $2,000 an acre and feeds many people. In poorer countries, the ethical dilemmas are worse.

Can restoration contribute to the conservation of biodiversity in tropical developing countries, where the stakes are highest? Or is it merely an outgrowth of an affluent society that is willing and able to pay for it? The answer is inseparable from the issue of how we value nature.

CONSERVATION OF RESOURCES OR NATURE FOR ITS OWN SAKE?

Ecological restoration, which could be defined succinctly as "making nature," straddles the widely acknowledged philosophical rift in our conception of nature as a pool of natural resources for human use versus something with intrinsic value (Oelshlaeger 1991; other characterizations can be found in, for example, Leopold 1949; Worster 1977; Ehrenfeld 1988; Callicot 1990). The International Union for the Conservation of Nature and Natural Resources (IUCN) defines conservation as "the management of human use of the biosphere so that it may yield the greatest sustainable benefit to present generations, while maintaining its potential to meet the needs and aspirations of future generations. Thus conservation is positive, embracing preservation, maintenance, sustainable utilization, restoration and enhancement of the natural environment" (IUCN 1980, 8).

This definition exemplifies the culturally dominant view of nature, which might be called "utilitarian" or "resourcist." According to Oelschlaeger (1991, 286–87), resourcists believe that "the wilderness . . . [is] a stockpile of matter-energy to be transformed through technology, itself guided by the market and theoretical economics, into the wants and needs of the consumer culture . . . natural systems are no more than collections of parts, and the ecomachine [nature] can be engineered to produce desired outcomes and prevent undesired consequences." Harper (1987, 35), a preeminent plant ecologist and champion of reductionism in ecology, says that "the *raison d'etre* for a science of ecology is presumably the development of an understanding of the workings of nature that would enable us to predict its behavior and to manage and control (conserve or change) it to our liking." Thus, not only is nature a resource, but the sole purpose of ecological science is to enable us to manipulate natural resources.

Conservation and restoration borne out of this viewpoint focus on the *products* of nature such as timber, soil, pharmaceutical chemicals, or genes for disease resistance. Even "amenity value" is a product that can be quantified in terms of tourist dollars. More recently, *ecological function* has been added to the list of useful and therefore valuable natural attributes. Wetland law in the United States concentrates on avoiding the loss of or replacing lost wetland functions such as filtration of sediments and watershed protection (Kusler and Kentula 1990). Protection of tropical rain forests is justified partly because of the role these forests may play in global biogeochemical cycles. Recognition of the importance of ecosystem functions and processes, as well as products, is but a more sophisticated form of utilitarianism.

Utilitarianism has been criticized on the anthropomorphic grounds that it is impossible to foresee what will be useful or even necessary to humans in the long term and therefore we cannot make adequate judgments about what is useful. Dr. David Ehrenfeld's (1988, 214) criticism of the utilitarian stance is less anthropocentric: Other species have an intrinsic right to exist on legal principles because of their long-established existence and on religious principles because diversity is God's property. Furthermore, valuation of nature is bad political strategy.

> In the long run, basing our conservation strategy on the economic value of diversity will only make things worse, because it keeps us from coping with the root cause of the loss of diversity. It makes us accept as givens the technological socioeconomic premises that make biological impoverishment of the world inevitable. . . . [E]conomic criteria of value are shifting, fluid and utterly opportunistic in their practical application. This is the opposite of the value system needed to conserve biological diversity over the course of decades and centuries.

Ehrenfeld (1988) points out that most organisms do not have economic value, especially the rarest species. Should the California condor become extinct, he maintains, no ecosystem processes will collapse, and the tourist industry will not suffer. Yet something will be lost.

In contrast to a utilitarian view of nature is the belief that nature has an intrinsic right to exist, regardless of its current or potential use to humans. Biologists Paul R. Ehrlich and Edward O. Wilson (1991) give three reasons to care about the loss of biodiversity, the last two of which are utilitarian. The first reason given is that "[b]ecause *Homo sapiens* is the dominant species on Earth . . . people have an absolute moral responsibility to protect what are our only known living companions in the universe. Human responsibility is deep, beyond measure. . . ." Despite his experience as a resource manager, Aldo Leopold came to a view of nature that both acknowledged humans' dependence on natural resources and celebrated nature's intrinsic value (Oelschlaeger 1991). In "The Land Ethic," Leopold defines conservation as "a state of harmony between men and land (1949, 243). The word "harmony" implies a dynamic, mutually beneficial relationship among equals. Naess (1986, 508) argues that biological diversity is important because we in some degree *identify* with all forms of life. According to Naess, the broad majority of ordinary people believe that "every life form has its place in nature which we must respect."

The belief that nature has intrinsic value has been termed *deep ecology* by philosophers and historians (Oelschlaeger 1991, 301–3, and references therein). *Deep ecology* is a term that makes many ecologists uncomfortable, having overtones of mysticism and holism inconsistent with the scientific discipline of ecology. In addition to being somewhat anti-reductionist, and therefore antiscience, it is seen as politically naive. Presumably, only individuals in rich countries can afford to appreciate nature for its own sake. Persons in poorer countries, precisely those countries that are currently chewing up rain forest at an alarming rate, are perceived to lack the capacity (or the leisure) to appreciate nature on its own terms because the pressures to exploit resources are too great.

Nowhere is the split personality of conservation more apparent than in the practice of ecological restoration and all of the "re-" words that are associated with it—re-creation of wetlands, rehabilitation of endangered species habitat, reclamation of mine spoils, revegetation of degraded rangeland, reforestation of forest clear-cuts. Unlike preservation of relatively undisturbed habitat, restoration projects must have a goal in mind. Resolving to "make nature," we must consciously decide what to make and what sort of ecosystem to manage for. Should we rehabilitate ecosystem function for human use or attempt to mimic authentic communities as an academic and aesthetic exercise? Would anyone even consider restoring a damaged ecosystem that (1) contained no endangered species, (2) produced no useful products, (3) performed no significant ecosystem functions due to low primary productivity, and (4) never interested naturalists?

As an applied science that costs lots of money to do, restoration ecology is captive to forces outside the realm of biology. Social values translated into law, and the individuals and institutions that carry them out, determine whether restoration contributes to conservation and to what kind of conservation it contributes—a pre-agricultural landscape, species diversity, useful wild plants and game animals, "ecological functioning," grazing land, timber, or productive farmland. Prairie restoration, wetland mitigation, mine spoil reclamation, range and timber management, urban landscaping, and sustainable agroforestry projects in developing countries have different histories, motivations, sources of funding, and institutional constituencies. The potential for each restoration project to contribute to the conservation of biodiversity will depend on these factors, all of them unrelated to biological understanding or technique.

In this essay, I examine contrasting models of restoration that have taken place in this country, prairie restoration and rangeland improvement. Each has a limited ability to address the pressing need to conserve soil, biota, and people in developing countries, where the greatest number of species are being lost daily. Then I present a relatively new, evolving model of ecological restoration that takes advantage of and assists natural regeneration processes, operating on the scale of whole landscapes.

PRAIRIE RESTORATION—NATURE FOR ITS OWN SAKE

The Society of Ecological Restoration (SER) defines ecological restoration as "the process of intentionally altering a site to establish a defined, indigenous, historic ecosystem. The goal of this process is to emulate the structure, function, diversity and dynamics of the specified ecosystem" (SER 1991, 4). The organizational roots of the Society for Ecological Restoration are in the prairie—specifically, the fifty-year-old restored Curtis Prairie at the University of Wisconsin. The purpose of the Curtis Prairie, inspired by biologists John Curtis and Aldo Leopold, was to replicate authentic plant communities for preservation and research. The model of restoration defined by SER, with its emphasis on emulating the original ecosystem in form as well as function, springs from the idea of nature as having intrinsic value.

Prairies are most often restored for aesthetic and scientific purposes by state parks, private organizations, and colleges and universities (Kline and Howell 1987). Jordan et al. (1988), in an assessment of ecological restoration as a strategy for the conservation of biodiversity, point out that prairies are natural objects for ecological restoration because they lend themselves to agronomic techniques, they are species-rich and aesthetically pleasing, and satisfying results can be seen in a relatively short time as compared to forests. The popular appeal of prairie restoration is evident in the numerous restoration papers presented at the biannual North American Prairie Conference (e.g., Davis and Stanford 1986). The North Branch Prairie Project, involving scientists from all over the world and hundreds of volunteers, has successfully restored oak savanna to a suburban Chicago site. In the process, a new plant community was discovered that had all but disappeared (Packard 1988).

As a model and an ethos for conservation of biodiversity where it has been destroyed, the North American prairie restoration experience appears to be ideal. Careful attention to the authenticity of the product is the first consideration. As an added bonus, restoration projects educate citizens about the ecology of their region. Many workers feel the restoration process renews and celebrates their relationship with the earth.

There are, however, severe limitations on the total area that can be restored in this manner. First, land is expensive. Second, seeds, when they are available at all, can cost several hundred dollars per hectare. Transplants are even more expensive, and planting and watering them is labor-intensive. After planting, weed control is often crucial to keep agricultural weeds from dominating the restoration site. Even though many prairie plants would probably survive the first weedy year without weed control, good relations with neighbors—and in some states, noxious weed laws—make weeding mandatory. Because of these considerations, most restoration efforts are small, the largest so far being the 240-ha restoration within the oval formed by the nuclear accelerator at the state of Illinois' Fermi National Laboratory. This project has been implemented in small increments over thirteen years.

The utilitarian benefits of permanent grass cover are widely acknowledged, but this does not necessarily translate into prairie restoration. As public awareness of prairies has increased, state departments of natural resources and transportation have begun to use prairie grasses for road rights-of-way, utility corridors, parks, and nature preserves and to create wildlife habitat (Jordan et al. 1988). The Conservation Reserve Program (CRP) of the 1985 and 1990 Farm Bills pays farmers to plant a permanent vegetative cover on highly erodible land, but it is not clear to what extent mixtures of native species are being used. Domesticated species are usually cheaper and easier to work with, and true prairie restoration is probably uncommon. Although the greatest ecosystem function of prairies has been to build soil that now is mined for corn production, no one, to my knowledge, has restored a prairie in order to build soil for future generations of farmers.

RANGE IMPROVEMENT— CONSERVING A RESOURCE

Management of wildlands for livestock grazing has probably impacted more plant and animal communities in the United States, particularly in the western states, than any other land use. Range management is normally not mentioned in the same breath with ecological restoration, but the "golden age" of range improvement between 1945 and the mid-1960s (Young 1990) was a restoration movement of enormous proportions, and techniques first pioneered in range improvement were later transferred to mine spoil reclamation. Its successes and failures are worth remembering.

In the 1890s ranchers in the arid and semiarid Southwest began to notice a decline in the perennial bunchgrasses and an increase in trees and shrubs. The first range scientists around the turn of the century recognized the need to limit grazing and to revegetate large areas but had little success in the latter. Experiments with European pasture grasses worked only in wet mountain meadows that more closely approached the European climate. Seed from native species had to be collected by hand, had dispersal appendages that made mechanical sowing difficult, and germinated poorly (references in Young 1990; Call and Roundy 1990).

Meanwhile, the range continued to deteriorate. In 1922, Aldo Leopold concluded that the U.S. Forest Service's policy of grazing to control grass fires in what is now the Apache-Sitgreaves National Forest, Arizona, had caused the severe soil erosion he discovered in twenty-seven of the thirty mountain valleys surveyed (Flader 1990). The Dust Bowl of the 1930s ruined more rangeland and forced the abandonment of millions of acres of cultivated land throughout the Great Plains from North Dakota to Texas (Bennett 1939). Much of the farmland was repossessed by banks and later bought by the federal government and reseeded to perennial grass.

But what species could they use to revegetate damaged rangelands? An agronomist and forage grass breeder in Oklahoma tells me, "Only God can make a good stand of buffalograss, and even He fails two-thirds of the time" (C. Dewald, pers. com.). Range scientists did not have the skills or the technology to propagate native species like buffalograss (*Buchloe dactloides)*, and meanwhile the soil was blowing and washing away. In response, millions of acres of introduced crested wheatgrass (*Agropyron cristatum*) were seeded in the northern plains and the Great Basin. South African lovegrass species (*Eragrostis* spp.) and Old World bluestem (*Bothriochloa ischaemum*) were introduced to the southern plains with similar zeal.

At the time, Frederick Clements' (1928) theory of ecological succession was widely accepted, and ecologists and range scientists believed that these aggressive, weedy perennials would inevitably be superseded by the long-lived, native climax species that had evolved on this continent. For the most part they were wrong. Introduced grass populations have resisted the reinvasion of native grasses and shrubs (Hull and Klomp 1966). For example, Lehman's lovegrass (*Eragrostis lehmanniana*) continues to expand its range (Bock et al. 1986). In Oklahoma, weeping lovegrass (*Eragrostis curvula*) has escaped from seeded pastures and invaded overgrazed and drought-stricken native pastures, potentially replacing native species (Jackson, pers. obs.).

In retrospect, the seeding of millions of acres of grassland and forest to produce monocultures of palatable alien species for livestock use was a disaster from the point of view of species- and ecosystem-level biodiversity. However, it is easy to forget that much of the western range had been so drastically disturbed by a combination of overgrazing and drought that by the 1930s *any* plant cover was probably better than *no* plant cover at all. Devastating soil erosion was rapidly eliminating the option to carry out careful restoration of the pre-livestock flora; revegetation would at least buy time until the degradation could be reversed (Bennett 1939; Young 1990). Despite the need for stopgap measures, however, much of the range improvement was "seen as an end in itself" (Young 1990, 96).

Range "improvement" was then and is today grounded in the utilitarian view that rangeland is primarily a natural resource for the production of red meat.

What if range scientists in the "golden age of range improvement" had instead placed greater value on restoring a diverse, native ecosystem identical in composition, structure, and function to a pre-1890 condition? I believe that the substitution of domesticated or croplike alien species for the relatively intractable native species was inevitable. The central theme of Harlan's classic *Crops and Man* (1975) is that once humans begin to sow seed, selective forces are automatically set in motion that favor a "domestication syndrome": nonshattering habit, simultaneous ripening, naked or threshable seeds, and loss of seed dormancy mechanisms. Grasses used in range improvement appear to have been no exception.

Species that produce seeds in a harsh environment and disperse them via wind, animal fur, or other natural means have different adaptations than species that use humans as their dispersal agents (Harlan 1975). For instance, a perennial plant in an unpredictable environment will often produce flowers and seeds only when environmental conditions are very good, or it will have a phenology that extends flowering over a long period (Harper 1977). This behavior maximizes the chances of producing successful offspring while ensuring the parent's survival. It is exactly this behavior that also makes it difficult for humans to collect sufficient quantities of viable seed.

Seeds of wild plants often have dormancy mechanisms that help to ensure their germination under appropriate temperature and moisture conditions in the wild but that may be inappropriate in artificially created seedbeds. Awns, bristles, and other appendages protect, disperse, and position the seed in a natural setting but interfere with mechanical seed harvest and sowing (Dewald, Beisel, and Cowles 1983; Booth 1987). Thus the "first cut" in selecting grass species for reseeding is whether enough viable seeds can be collected from the wild and successfully propagated under controlled conditions, using existing or improvised technology to mechanize the process. Only species or ecotypes within species that pass this first basic requirement will undergo further testing for longevity, palatability, and so on. As a result, species that are and were available for large-scale revegetation are a restricted subset of wild species.

For this reason, I believe that any restoration on a large scale with a finite budget will be plagued by species selection and subsequent genotype selection within species, leading to a decrease in species richness and genetic diversity. Extensive use of transplants, as in current wetland restoration efforts, will likewise impose selection for those species and

genotypes that produce more stolons, offshoots, or other vegetative propagules. Basically, once we begin to manipulate plants by actively establishing them, they become "plant materials." Depending on the goals of
the project (ecosystem function vs. authentic reconstruction of a historical
plant community), this sacrifice may be acceptable. The dilemma suggests, however, that we pay more attention to natural regeneration processes as a tool in restoration.

Restoration in its many guises has added to our technical skill in
manipulating plant populations and even whole ecosystems. Enough of
this empirical knowledge has accumulated for us to assemble principles
of revegetation that will assist future projects (Bradshaw and Chadwick
1980; Allen 1988; Roundy and Call 1988; Call and Roundy 1990; Pyke
and Archer 1990). Restoration projects are beginning to be used as heuristic tools to investigate basic questions in ecology (Jordan et al. 1988).
How much can this new body of knowledge contribute to the conservation
of biodiversity in developing countries?

The prairie model of restoration, involving intensive transplanting
and reseeding of native species to achieve an authentic, historic community, is too expensive and resource-intensive to occur over a significant
area, even in an affluent country. Realistic cost estimates for authentic
restoration of communities are difficult to obtain, due to the extensive
use of volunteer labor (Jordan et al. 1988), but O'Malley (1991) reports
an average cost of $22,000 per 0.25- to 1-ha grove of ironwood (*Lyon-
othamnus floribundis*) on Santa Catalina Island, California. Large reclamation projects like those for degraded rangeland and mine spoils have
demonstrated a limited ability to reestablish diverse communities, both
because of ecological constraints on highly disturbed land (Allen 1988)
and because of their narrow, resource-oriented goals. We need a model
of restoration different from that given to us by the prairie restoration
movement and different from the model of range improvement.

A LANDSCAPE APPROACH TO ASSISTED NATURAL REGENERATION

A new model of restoration is taking shape that combines landscape
ecology; natural succession; management; and some strategically placed,
intensive revegetation. The best example is Dr. Daniel Janzen's efforts
to restore tropical dry forest in Costa Rica (Janzen 1988). Tropical dry
forest historically was favored for agriculture, and thus proportionately
less of it remains than the rain forests that attract most of our attention.
Janzen's strategy is to acquire degraded remnants of forest, as well as the

farm and ranch land surrounding them, and to encourage natural regeneration around these remnants via careful management. In the case of these tropical dry forests, protection from fire discourages pasture grasses and encourages natural forest regeneration. Cattle grazing is used as a "management tool" to suppress grass and to disperse some of the heavier-seeded tree species. In other ecosystems, fire and grazing might be detrimental to regeneration processes, and other tactics would be employed. This type of restoration requires a landscape perspective because disturbance and natural regeneration are sensitive to the heterogeneity of the landscape and land-use patterns (Forman and Godron 1986; Turner 1989). It requires attention to the human landscape as well as the biotic one, because such preserves and restoration areas must fit the needs and preferences of the indigenous people (Janzen 1988).

Another good example of landscape restoration linked to natural regeneration is a research project on the Tensas River basin, a tributary of the Mississippi in Louisiana (Gosselink et al. 1990). Large areas of lower Mississippi floodplain forests have been cleared in the last forty years for soybean, corn, cotton, and rice production, leaving only 15 percent of the original forested wetland in place. Gosselink et al. (1990) assessed the impacts of the incremental loss of forested wetlands over time in terms of species loss, water quality, and hydroperiod and developed a plan of wetland acquisition and restoration using biogeographic principles. Priority was placed on maximizing contiguous forest size and minimizing edge:interior dimensions of forest patches. Using a geographic information system, linkages were identified between adjacent forest patches that were feasible (i.e., did not cross a road) and significantly increased effective forest size. In this way restoring 400 ha of corridors would increase the size of the largest forest from 50,000 to 100,000 ha and create another 64,000-ha forest from many smaller forest patches.

Gosselink et al. (1990) suggest that any further clearing of bottomland hardwood forests could be regulated using a priority scale in which low-priority wetlands (small, isolated wetlands having little effect on water quality) received "light" protection. Mitigation where development was unavoidable could be targeted to the corridors between forests, where wetland restoration in a small area could leverage much larger gains in water quality and wildlife protection. In the absence of full-scale restoration, natural regeneration of wetland forests along target corridors could be assisted by the dissemination of large-seeded tree species, whose natural dispersal distances are short.

In these examples, large-scale restoration of native ecosystems or landscapes has leaned heavily upon the availability of numerous habitat fragments, or, as Janzen (1988) refers to them, "biotic debris," to provide the inoculum for natural regeneration. Indeed, even without human intervention, the simple availability of biotic debris can be enough to restore an area. Natural regeneration around biotic debris has occurred unaided in southern New England, where by 1900 up to 90 percent of the non-mountainous land was cleared (Marks and Smith 1989). Marks and Smith mapped the forest fragments remaining in 1900 and in 1980 for a portion of the Finger Lakes region of New York State. In 1900 only poorly drained woodlots and steep slopes adjacent to lakes, the vertical sides of gorges, and steep, inaccessible valleys remained wooded. After 1900, farming began to contract, and marginal fields were abandoned. By 1980, 50 percent of the landscape was once again wooded (Marks and Smith 1989; P. Marks, pers. com.).

Industrial Agriculture and Barriers to Natural Regeneration

New England had two advantages in its recovery by natural regeneration. First, its rough topography did not allow all forest to be cleared. Second and probably just as important, forest clearing occurred in the late 1800s, before agriculture became regionally specialized. Mixed farming required a diverse mixture of land types. Forest remnants were maintained for production of firewood, lumber, and maple syrup. Pigs foraged in the woods, and wild animals (deer, squirrels) and plants (e.g., wild ginger, berries) were economically important forest products. Animal husbandry required pastures and fences, and native trees and shrubs along fences were tolerated or encouraged. As a result, forest remnants were inter-spersed liberally throughout the farming region. Using a set of randomly located points falling within cleared fields, Marks and Smith determined that the median distance to a forest fragment in the important farming regions of Tompkins County, New York, was only 213 m in the year 1900 (range 0–629), despite the fact that only 10 percent of the landscape was still forested (Marks and Smith, unpublished data).

Where habitat fragments, including undisturbed soils, do not exist, the chances for large-scale restoration may be greatly reduced. For example, much of Arizona's lowland deserts were cleared *after* the age of mixed, small-scale farming, between 1935 and 1953. The lack of topographic barriers to clearing, and the regionally specialized and industrialized agricultural system, created a landscape radically different from that of New England, with vastly reduced potential for natural recovery. There is reason to believe that the methods of farming employed and the scale of agricultural development, combined with the special circumstances of a harsh Sonoran desert climate, have severely limited the possibility of large-scale restoration. Perhaps by learning from our own mistakes in agricultural development carried out only forty years ago, we can help to avert the same mistakes in other parts of the world.

As soon as new water-pumping technology made it possible to reach beyond shallow groundwater into deeper, fossil aquifers, farming in Arizona moved from the river bottoms onto higher ground (Smith 1940; Shapiro 1989). Diesel-powered machinery made quick work of the paper-flat, sparsely vegetated desert lowlands. Within twenty-five years, the 220,000-ha Santa Cruz valley in Arizona was almost completely cleared. Farming operations were specialized from the beginning, concentrating on cotton production (Shapiro 1989). Since the mid-1950s, up to 50 percent of the cleared land in some regions has been abandoned (L. Jackson, unpublished data). Changes in the cotton market and the high costs of pumping groundwater from increasing depth (Shapiro 1989) are just part of the complex reasons for farm abandonment.

Several factors may contribute to the speed and extent of natural regeneration in old fields in Arizona. Some factors, such as soil salinization and aridity, are specific to the desert, and others may be of more general importance. However, long distances to native seed sources is not specific to desert environments and is a formidable barrier to natural recovery in many environments. Recolonization depends on dispersal, and the colonization rates of native plant species in desert old fields are associated with their dispersal mode. The wind-dispersed weedy perennials burrowweed (*Isocoma tenuisecta*) and desert broom (*Baccharis sarothroides*) were once rare or infrequent components of the desert lowlands of Arizona (Shantz and Piemeisel 1924). They are now ubiquitous in fields abandoned five to forty years after cultivation (Karpiscak 1980). Velvet mesquite (*Prosopis velutina*), a subdominant of the original vegetation, has large, sweet pods dispersed by animals; it too has rapidly recolonized desert old fields. Creosotebush (*Larrea tridentata*) and desert saltbush (*Atriplex polycarpa*), the original dominant long-lived species, have heavy seeds that lack obvious adaptations for dispersal. These species appear to invade old fields very slowly.

To estimate the invasion rate of creosotebush into abandoned fields in the Santa Cruz Valley, Arizona, I located fields abandoned at different times that bordered natural creosotebush stands and measured the density of creosotebush as a function of distance from the seed source. (It was not possible to measure invasion rate of desert saltbush because so few remnant stands exist.) I determined the year of last cultivation by examining repeat aerial photographs and by quizzing neighbors. In all fifteen fields sampled (Figure 17.1; only three are shown), creosotebush exhibited a regular, negative exponential distribution as a function of distance from the seed source. In the oldest field, creosotebush had invaded 360 m in 37 years. In the younger field, creosotebush had invaded 100 to 200 m

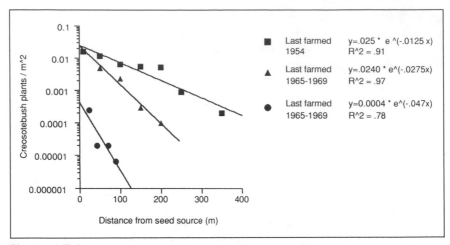

Figure 17.1
Density of creosotebush in three abandoned agricultural fields as a function of distance from the field edge. Each datum is the number of creosotebush per square meter along one transect 5 m to 20 m wide and 400 m to 800 m long. Note logarithmic scale on Y axis.

in 22 to 26 years. No firm estimate of creosotebush invasion rate is possible until more fields are censused, but if these data are representative, creosotebush may take 150 to 400 years to reach fields 1.6 km from the seed source[1]. The incidence of saltbush stands bordering old fields is so low that an estimate of their invasion rates is not possible.

Agriculture may alter the potential for natural recovery at the level of whole watersheds. To farm in the desert, the land must be leveled for gravity-fed irrigation. As a consequence, the original landscape pattern of dendritic ephemeral washes is converted to a cartesian grid of ditches. For a taste of the essentially permanent impact that irrigation has had on lowland desert landscapes, one need only consult a topographic map (Figure 17.2). A portion of the eastern edge of the Santa Cruz River Valley, Arizona, was selected in which leveled and unleveled land lie adjacent to one another. (In the center of the valley, not shown, essentially all of the land has been leveled, and so the contrast is not as evident). In this illustration, farmed land can be determined simply by the contour lines, which are perfectly straight. Abrupt dips in the contour lines reveal ditches along roads or between fields. The density and type of desert

---

[1]One mile, or 1.6 km, is the natural unit of measure for landscape studies in this area due to the grid of county roads placed every mile running north-south and east-west.

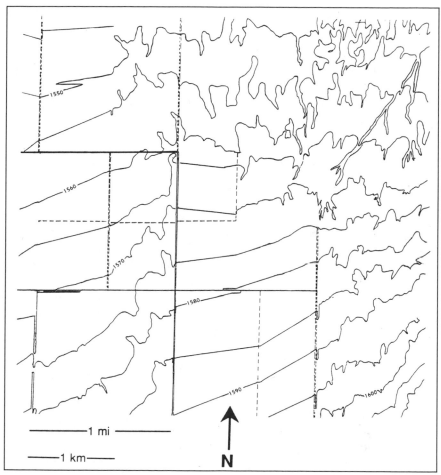

Figure 17.2
Topographic map of a portion of the Santa Cruz Valley in Pinal County,
Arizona, mapped in elevation increments of five feet (1.5 m). Where con-
tour lines are straight, land has been cleared and leveled for irrigation.
Deep horizontal and vertical diversions in the contour lines indicate the
presence of ditches.

vegetation, so closely tied to small differences in water availability due
to soils and topography, can never be the same where the land has been
leveled. In addition to channeling water movement, the grid of ditches,
roads, berms, and leveled fields also controls the movement of seeds. In
all likelihood, restoration of this area would require that at least some
of the original macro- and microtopography be re-created, possibly by
the removal of cement irrigation ditches and the closing of selected roads.

CONCLUSIONS

Our rather recent experience with agricultural development in the United States has taught us that agricultural acreage may very well contract after the initial primary conversion of all available land. We have learned that brute force restoration of farmland to pre-agricultural conditions is prohibitively expensive on a large scale, and that habitat fragments, however species-poor and degraded, are linchpins in the process of restoration at any scale. And we have learned that scale makes a difference, not just in natural regeneration and restoration, but in the quality of life of those who live on farms. In an essay about the spring dust storms in the northern Great Plains this year, Michael Melius, a farmer and writer in South Dakota, describes the incremental destruction of remnants of prairie during his lifetime. Unlike in New England, where eroding farms were abandoned and the forests began to take them over, the farm exodus in the Great Plains resulted in fewer and fewer people farming more and more acres:

> The less-wild, less-diverse, even industrial landscape that we are creating here is usually associated with an increased population density. It suggests that our land is becoming unlivable as it becomes un-lived in, in that one would like some wild companions, and something of a natural setting in which to work. . . .
>
> As these farms become virtual one-man factories, the landscape is becoming one only a "dirt farmer" could love, with little life but what the farmer allows. Dormant or dead (stubble) eight months of the year, it's planted to a monoculture of grains for the growing season. "Fencerow to fencerow" is passé; fences gone, it's now road ditch to road ditch. An industrial landscape, field after farmless field.
>
> It seems unfair that the depopulation of the plains has been accompanied not by the return of the prairie, but by its decrease. Sometimes it seems so dead out here, in terms of social life and natural history, especially when compared to the stories my older relatives tell. Thinking back only on my 30 years here, I realize that growing up here has also meant knowing so many people who are gone, and so many places. Whether for farmsteads razed and plowed under or the [native prairie] pastures around that I played in, there have been too many times I've been forced to leap almost bodily past grief into acceptance, for the places and times in my memory that were erased one morning and now are irrevocable. (Melius 1990, 13)

People and places, social life and natural history, farmsteads and native prairie. . . . In the history of the northern Great Plains, people and places were allies, not alternatives. In Melius's description of the emotional loss imposed by industrial agriculture, the ecologist gets a clear sense of the biotic loss as well. A land as sterile as this one is practically unredeemable. Janzen (1988) reminds us that the study and appreciation of the natural world is still the most intellectually stimulating activity in which humans can engage themselves. We study and enjoy nature not only to better manipulate natural resources, as Harper (1987) would have it, but also as an end in itself. Aldo Leopold's (1949) much-described hunting expeditions were for meat, for delight, and perhaps for worship. Marginal patches of nature are simultaneously refuges for study and enjoyment, and insurance for future biotic recovery. Restoration ecology's contribution to conservation will come from its defense and promotion of tiny, sometimes degraded fragments of habitat—not for what they are now, but for what they could become.

ACKNOWLEDGMENTS

This article was inspired by discussions with C.L. Jensen. I also wish to thank C.W. Bohlen, J. Poole, and B.A. Roundy for their comments and suggestions. J. Cole, J.R. McAuliffe, and G.P. Nabhan reviewed early versions of the manuscript and greatly improved it.

LITERATURE CITED

Allen, E.B. 1988. *The reconstruction of disturbed arid lands: An ecological approach.* American Association for the Advancement of Science Selected Symposium 109. Boulder, Colo.: Westview Press.

Bennett, H.H. 1939. *Soil conservation.* New York: McGraw-Hill.

Bock, C.E., J.H. Bock, K.L. Jepson, and J.C. Ortega. 1986. Ecological effects of planting African lovegrasses in Arizona. *Nat. Geogr. Res.* 2:456–63.

Booth, D.T. 1987. Diaspores of rangeland plants: Ecology and management. In *Seed and seedbed ecology of rangeland plants,* ed. G.W. Frasier and R.A. Evans, 202–11. Springfield, Va.: U.S.D.A. Agricultural Research Service, National Technical Information Service.

Bradshaw, A.D., and M.J. Chadwick. 1980. *The restoration of land: The ecology and reclamation of derelict and degraded land.* Oxford: Blackwell Scientific.

Call, C.A., and B.A. Roundy. 1990. New perspectives in rangeland revegetation. In *Perspectives and processes in rangeland revegetation,* comp. C.A. Call and B.A. Roundy, 1–33. Papers presented at the 43rd annual meeting of the Society for Range Management, Reno, Nevada, 11–16 February 1990.

Callicot, J.B. 1990. Standards of conservation: Then and now. *Cons. Biol.* 4:229–32.

Clements, F.E. 1928. *Plant succession and indicators.* New York: Hafner.

Davis, A., and G. Stanford, eds. 1986. *The prairie: Roots of our culture; foundation of our economy.* Proceedings of the 10th North American Prairie Conference. Dallas: The Native Prairies Association of Texas.

Dewald, C.L., A.E. Beisel, and S. Cowles. 1983. The Woodward chaffy seed conditioning system. In *Range and pasture seeding in the Southern Great Plains,* comp. H.T. Wiedemann and J.F. Cadenhead, 69–80. Vernon: Texas A&M University Agricultural Research and Extension Center.

Ehrenfeld, D. 1988. Why put a value on biodiversity? In *Biodiversity,* ed. E.O. Wilson, 212–16. Washington, D.C.: National Academy Press.

Ehrlich, P.R., and E.O. Wilson. 1991. Biodiversity studies: Science and policy. *Science* 253:758–62.

Flader, S.L. 1990. Let the fire devil have his due: Aldo Leopold and the conundrum of wilderness management. In *Managing America's enduring wilderness resource,* ed. D.W. Lime, 88–95. St. Paul: Minnesota Extension Service.

Forman, R.T.T., and M. Godron. 1986. *Landscape ecology.* New York: John Wiley & Sons.

Gosselink, J.G., G.P. Shaffer, L.C. Lee, D.M. Burdick, D.L. Childers, N.C. Leibowitz, S.C. Hamilton, R. Boumans, D. Cushman, S. Fields, M. Koch, and J.M. Visser. 1990. Landscape conservation in a forested wetland watershed: Can we manage the cumulative impacts? *BioScience* 40:588–600.

Harlan, J.R. 1975. *Crops and man.* Madison, Wis.: American Society of Agronomy and Crop Science Society of America.

Harper, J.L. 1977. *Population biology of plants.* London: Academic Press.

———. 1987. The heuristic value of ecological restoration. In *Restoration ecology: A synthetic approach to ecological research,* ed. W.R. Jordan III, M.E. Gilpin, and J.D. Aber. Cambridge: Cambridge University Press.

Hull, A.C., and G.J. Klomp. 1966. Longevity of crested wheatgrass in the sagebrush-grass type of southern Idaho. *J. Range Manage.* 19:5–11.

International Union for the Conservation of Nature and Natural Resources. 1980. *World conservation strategy.* Gland, Switzerland.

Janzen, D.H. 1988. Tropical ecological and biocultural restoration. *Science* 239:243–44.

Jordan, W.R., III, M.E. Gilpin, and J.D. Aber, eds. 1987. *Restoration ecology: A synthetic approach to ecological research.* Cambridge: Cambridge University Press.

Jordan, W.R., III, R.L. Peters II, and E.B. Allen. 1988. Ecological restoration as a strategy for conserving biological diversity. *Environ. Manage.* 12:55–72.

Karpiscak, M.M. 1980. Secondary succession of abandoned field vegetation in southern Arizona. Ph.D. dissertation, University of Arizona, Tucson.

Kline, V.M., and E.A. Howell. 1987. Prairies. In *Restoration ecology: A synthetic approach to ecological research,* ed. W.R. Jordan III, M.E. Gilpin, and J.D. Aber. Cambridge: Cambridge University Press.

Kusler, J.A., and M.E. Kentula, eds. 1990. *Wetland creation and restoration: The status of the science.* Washington, D.C.: Island Press.

Leopold, A. 1949. *A Sand County almanac.* London: Oxford University Press.

Marks, P.L., and B. Smith. 1989. Changes in the landscape: A 200 year history of forest clearing in Tomkins County, New York. *N. Y. State Food Life Sci. Quart.* 19:11–14.

Melius, M.M. 1990. Dakota dust: Denial, delusion, dishonesty. *High Country News* 11 March 1990.

Naess, A. 1986. Intrinsic value: Will the defenders of nature please rise? In *Conservation biology: The science of scarcity and diversity,* ed. M.E. Soulé, 504–16. Sunderland, Mass.: Sinauer.

Oelschlaeger, M. 1991. *The idea of wilderness: From prehistory to the age of ecology.* New Haven: Yale University Press.

O'Malley, P.G. 1991. In ironwood groves—restoration of Santa Catalina Island, California. *Restor. Manage. Notes* 9:7–15.

Packard, S. 1988. Chronicles of restoration: Restoration and the rediscovery of the tallgrass savanna. *Restor. Manage. Notes* 6:13–22.

Pyke, D.A., and S. Archer. 1990. Biotic interations affecting plant establishment and persistence on revegetated rangeland. In *Perspectives and processes in rangeland revegetation,* comp. C.A. Call and B.A. Roundy. Paper presented at the 43rd annual meeting of the Society for Range Management, Reno, Nevada, 11–16 February 1990.

Roundy, B.A., and C.A. Call. 1988. Revegetation of arid and semiarid rangelands. In *Vegetation science applications for rangeland analysis and management,* ed. P.T. Tueller, Dordrecht: Kluwer Academic Publishers.

Shantz, H.L., and R.L. Piemeisel. 1924. Indicator significance of the natural vegetation of the southwestern desert region. *J. Agric. Res.* 28:721–801.

Shapiro, E.A. 1989. Cotton in Arizona: A historical geography. Master's thesis, The University of Arizona, Tucson.

Smith, G.E.P. 1940. *The groundwater supply of the Eloy district in Pinal County, Arizona.* University of Arizona Agricultural Experiment Station Technical Bulletin no. 87, Tucson.

Society of Ecological Restoration. 1991. Society for Ecological Restoration program and abstracts, 3rd annual conference, Orlando, Florida, 18–23 May 1991.

Turner, M.G. 1989. Landscape ecology: The effect of pattern on process. *Ann. Rev. Ecol. Syst.* 20:171–97.

Worster, D. 1977. *Nature's economy: A history of ecological ideas.* Cambridge: Cambridge University Press.

Young, J.A. 1990. Revegetation technology. In Proceedings of Symposium, *Perspectives and processes in rangeland revegetation,* comp. C.A. Call and B.A. Roundy. Paper presented at the 43rd annual meeting of the Society for Range Management, Reno, Nevada, 11–16 February 1990.

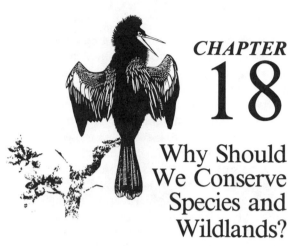

CHAPTER

18

Why Should We Conserve Species and Wildlands?

Essay by G. LEDYARD STEBBINS

WHY SHOULD WE CONSERVE . . . ?

As worldwide movements for conservation gain momentum, middle-class America is in the forefront of activity. A small but rapidly growing minority of people having a variety of professional and recreational interests are united in making serious efforts to preserve vignettes of our native landscape as well as rare species of plants and animals that were all but unknown a century ago. Some of these have always been rare, whereas others were formerly common but have become rare because of human exploitation during the past century. The voluntary associations that we support, such as the Sierra Club, Appalachian Club, The Nature Conservancy, California Native Plant Society, and similar organizations in other regions of the world, are growing in membership at a phenomenal pace. This growth has, in the United States, sent reverberations to state capitols and to federal government organizations like the U.S. Forest Service, Bureau of Land Management, and National Park Service. It not only has helped to elect conservation-minded members of the U.S. Congress, but also has motivated national administrators to appoint government officials who are sensitive to problems of the environment. A milestone in this progress was the decision of a major nationwide news magazine, in its annual year-end issue, to bypass a newly elected U.S. president in order to focus its attention on the Earth, the "Planet of the Year," and the serious, partly irreparable damage that it is suffering in the name of "progress."

Is all this conservation activity due to the misguided efforts of a few nature lovers who are distressed at seeing some of their favorite natural areas invaded by advancing civilization and prosperity? I admit that I, myself, during the past fifty years, have had uneasy feelings of this sort, providing me with one of several motives to slow this advance and give people a chance to learn how to promote prosperity in a more intelligent fashion. This essay is, in large part, an expression of these feelings. Yet if only selfish interests were involved, even the relatively large minority of active conservationists could do little against the great majority of those to whom exploitation for profit, or the subjugation of nature to human objectives, has for centuries been a motive and a basic ethic.

One of the natural areas that is dearest to me lies near the bottom of a deep canyon in California's Sierra Nevada, where a 30-m waterfall splashes into a deep, shaded pool surrounded by the fat stalks and umbrellalike leaves of Indian Rhubarb (*Darmera peltata*: Saxifragaceae) and beside which narrow ledges, intermittently sprinkled by spray from the falling water, support a fringe of leaf rosettes, slender stalks, and delicate pink flowers of the sawtooth-leaved bitter root (*Lewisia serrata*: Portulacaceae), which no botanist even knew existed until I found it there twenty years ago. Surely I would suffer a blow if human activity, such as the building of a new dam, should destroy this spot. Nevertheless, if that were the only issue, neither I nor anyone else should try to block well-planned projects that would benefit a large number of people in order to save this and a few other natural areas of this kind.

That, however, is not the main issue. The larger objective of conservation biologists should be to preserve as many of these beautiful natural areas as well as others that are by no means beautiful, such as the scrub-covered raised bench in the Sierra Nevada foothills south of Ione.

A perceptive visitor, driving from California's Central Valley near the cities of Stockton and Lodi toward the foothill town of Jackson, first crosses rolling hills covered by scattered blue oaks (*Quercus douglasii*) and introduced annual grasses, familiar and common sights throughout the foothills of central California. Soon, however, the scenery changes dramatically. The blue oaks and grass disappear, and their place is taken by scrubland or chaparral of a most unusual sort. The ravines and a few of the richer slopes support dwarf live oaks (*Quercus wislizenii*) and gray leaf manzanita (*Arctostaphylos viscida*), but most of the area that is not disturbed is covered by a low shrub bearing olive-green leaves: a species of manzanita (the Ione manzanita [*Arctostaphylos myrtifolia*]) that is found in only a few spots in this region and adjacent ones and is only distantly related to others of its kind. A few clearings on the hill slopes support a sparse stand of herbs, the most notable of which is the endemic Ione buckwheat (*Eriogonum apricum*), discovered only recently by J.T. Howell of the California Academy of Sciences. This diminutive plant likewise has no close relatives in the large genus to which it belongs and, according to James Reveal, monographer of the genus, is relatively primitive. Why does this little region of the Sierra Nevada support such a remarkable plant community? Part of the answer to this question lies in the combination of unusual ecological factors that prevail there. Its climate is extremely hot and dry during the five months of the rainless summer. The soil is a combination of sand and clay hardpan, the latter almost impervious to root penetration, and the soil is very low in mineral nutrients. According to measurements made by ecologist Jack Major, the soil is extremely acid (pH 3.5–4.0). Oaks, grasses, and other foothill plants cannot tolerate these extreme conditions and thus are confined to the few more favorable parts of the region.

The unusual soil of the Ione area is itself of great geological and evolutionary interest. Soil scientist Hans Jenny regards it as an extraordinary example of "fossilized" but unconsolidated soil that is tens of millions of years old. It was formed during the beginning of the Tertiary period when an arm of the Pacific Ocean extended to this area. The sand was originally a raised beach; the clay, a muddy shore. As the ocean retreated, the beach was uplifted and became a "raised beach," of which many examples having younger ages are scattered throughout California. Because different localized species of manzanita often inhabit more recent habitats of this kind, as on the Monterey Peninsula, one can speculate that the Ione manzanita is a relic of a mid-Tertiary raised-beach flora, twenty million or more years old.

Unfortunately, the unusual soil that makes Ione interesting to scientists also is of great value to exploiters, because both the clay and the sand are of exceptional value to industry. Quarries have been in operation for more than fifty years and are now expanding. Only a tiny corner of the area has been bought and is being conserved, and without large donations for this purpose, additional conservation will be very difficult to achieve.

Dozens of similar rare communities and species exist elsewhere in California, and thousands of comparable communities are known in developed nations throughout the world, where organizations exist or can be formed to save them. Many more thousands exist in developing nations, particularly in the tropics, but these raise even graver problems of a different nature, which I shall review later.

First, why and how shall we in the developed nations take care of our own problems of conservation? Do we need to establish an ethic or simply to recognize a responsibility? Most of us, I believe, share the latter opinion. We of the current generation in the developed world are stewards of prosperous countries, filled with pristine natural areas, most of which were discovered and admired by our predecessors who thought little about their conservation. As long as the land was sparsely populated and access to many wildlands was so difficult that only a favored few could visit and enjoy them, they were in little danger of destruction. As the human population has grown and the majority of us have been forced to live in or near crowded cities, the value of isolated, formerly remote natural areas as places to relax, unwind, explore, and prepare ourselves to resume city life in a refreshed state has grown continuously. At the same time, the advent first of the motor age, which replaced the horse and buggy, and after that the jet age, which made travel by railroad almost obsolete, has brought even the remote corners of our land within easy reach of all. Many so-called "modern" people heighten the problem by their desire to bring urban luxuries to previously unspoiled primitive areas.

What about the future? By itself, the rapid pace at which population size and ease of travel have simultaneously advanced should be warning signals, the canary in the mine shaft. If we are to be good stewards and transmit to coming generations the opportunities for relaxation and quiet enjoyment provided by the natural world that our ancestors and predecessors have bequeathed to us, we must do what we can, even at some sacrifice and reduction of short-term gains, to keep as much of our landscape as we can in its pristine condition. In doing so, we are living up to one of the most vital characteristics of the human species, called by the Polish philosopher Korzybski "conscious time binding." This is the ability, by conscious effort, to envisage and construct for the future a favorable order of family or community life based on past experience and logical deduction. Although limited, time binding, based on unlearned instincts or drives, is common in most species of higher animals; the far more expansive time binding of which conservation is a part not only is unique to humankind, but also has grown in strength in proportion to the advance of human culture. When we complain about the shortsightedness of modern men and women, particularly our leaders, we forget the fact that only a few centuries ago, and perhaps even today, the great majority of people thought about little except satisfying immediate needs and gratifying their own desires. Throughout the ages, the ability and motivation to develop a larger viewpoint, and to think intelligently about the needs of future generations, have been confined to a small, broad-minded elite and their immediate followers. Whatever desire exists for conserving nature represents only one step in the evolution of a worldwide culture dedicated to the preservation and future happiness of our species. Nevertheless, the activity of these few dedicated altruists has saved us from chaos in the past and can, if pursued with equal or greater enthusiasm and determination, guide us to a better future. We cannot afford to relax our efforts and let destructive forces continue unchecked.

On the other hand, these efforts will be to no avail unless they are directed in a realistic fashion. Although my own experience is limited, it has led to some guidelines for successes. As I have learned more by both reading and conversing with others about our successes and failures, the following guidelines assume prime importance.

1. *Whenever possible, avoid direct confrontation.* A successful conservationist must be both an idealist and a politician. If ideals are sacrificed, the motive for conservation disappears. Nevertheless, ultimate ideals must sometimes be temporarily put aside in order to get things done. The greatest mistake that any one of us can make is to assume that all of the people with whom we must deal either are motivated conservationists like ourselves or are so dominated by shortsightedness and personal greed that they must be regarded as enemies. Such a division of people into friends and enemies is par-

ticularly self-defeating when it is applied to governmental resource agencies. The people who guide natural resource policy are for the most part conscientiously trying to fulfill the goals of their organizations and the demands of their clients, among which conservationists are always a minority. Most of them listen to reason and are ready to compare various proposals that are put to them. Overstatements of a particular case, especially if they cause their proponents to be identified as "special interests," are as likely to produce a backlash as to bring about a desirable result. All strategies must be carefully considered. As I have become more acquainted with the people belonging to these resource agencies, I have found I can regard a large number of them as friends with whom I have interests in common.

2. *Be always on the lookout for allies*, particularly those who might have influence. Two personal incidents have convinced me of the value of this precept. The first concerns one of California's prime areas of natural beauty and scientific interest—the Monterey Peninsula. After becoming acquainted with it in about 1940 when I had just begun to teach a course in organismic evolution, I had given to a small wild area in its very center the sobriquet "Evolution Hill" because of the number of situations that its flora presented to one who was trying to explain plant evolution. Even the dominant trees, two species of pine, provided an instructive lesson. They belong to a group, the Pacific Coast closed-cone pines, that are confined to California and adjacent lower California and are notable for the local restriction of species and subspecies. Fossils of Tertiary age are so numerous that the first taxonomist, H.L. Mason, and more recently paleontologist D.I. Axelrod (Axelrod 1980) have worked out their evolutionary relationships throughout the Tertiary Period. A variety of soil types throughout the Monterey Peninsula support continuous forests of Monterey pine (*Pinus radiata*), but in this area only Evolution Hill supports a population of bishop pine (*P. muricata*), which is genetically distinct from and forms partly sterile hybrids with populations of *P. muricata* found elsewhere in California. Although hybrids between *P. radiata* and *P. muricata* have been grown under cultivation, they have never been found at Evolution Hill, although trees of the two species grow within a few meters of each other. An easy explanation for this absence is that in the Monterey area *P. radiata* sheds its pollen in February, whereas *P. muricata* does not shed pollen or open its female cones to receive pollen until March or even April. This pair exemplifies simultaneously three types of reproductive isolation that contribute to the origin of species: ecological, seasonal, and genetic. Moreover, good evidence supports the hypothesis that *P. radiata* arrived in Monterey from more southerly regions, where

early flowering would be favored by mild, sunny winters, whereas the Monterey population of *P. muricata* is derived from the much larger populations found in northern California where cool, rainy winters would render pollen shedding in February maladaptive.

The other tree species found on Evolution Hill are equally instructive. It is the gowen cypress (*Cupressus goveniana*), a pygmy species of which "trees" produce cones when only 30 cm tall. Only 4 km south of these dwarfs along the rocky shore grow the much larger, picturesque, and even rare trees of the Monterey cypress (*C. macrocarpa*). Garden experiments have shown that all species of cypress cross easily with each other and produce fully fertile hybrids. But in the only region in which two species of cypress grow near enough to each other so that pollen can easily be carried by the wind from one species to the other, no intermediates are found, in spite of the great differences between them in morphology and manner of growth. These two species emphasize the importance of short distances of spatial isolation plus ecological factors as ways that certain plant species maintain their identity over long periods.

Some shrubs found on Evolution Hill also provide valuable examples of evolutionary processes. Two species of the large manzanita genus *Arctostaphylos* (Ericaceae), *A. hookeri* and *A. tomentosa*, grow intermingled. *Arctostaphylos hookeri* has the diploid chromosome number of $2n = 26$. Its population is relatively constant, and in the Monterey area it is rarely found outside Evolution Hill. By contrast, *A. tomentosa* is tetraploid, is widespread on the Peninsula and elsewhere along the coast of central California, and forms intermediates with many other populations that by taxonomists are ranked as different species. This contrast deserves research by modern methods. A second pair of shrubby species, *Ceonothus dentatus* and *C. thyrsiflorus*, formed extensive and variable hybrid populations in 1940, soon after firebreaks had been cut through the area. Now, with the maturation of the vegetation, these hybrids have been largely crowded out, although the parental species still exist as they did formerly.

The ecology of the Evolution Hill area is as interesting as its vegetation. In contrast to the surrounding areas, it is underlain by a hardpan clay and is believed to be a raised beach, but more recent than that at Ione, perhaps not older than the Quaternary glacial-pluvial period. The evolutionary problems that it offers are of relatively recent origin.

For many years, I pondered over ways by which this area could be preserved. For thirty years it lay undisturbed. S.F.B. Morse, then the president of Del Monte Properties, which owns it, had rightly placed an economic value on it that is as high as or higher than that of any wildland in the United States, but he assured me that he wanted to see it preserved. With his passing, the crises came. I was suddenly telephoned by one of the homeowners whose property almost adjoins my Evolution Hill; I was who told that the new president felt the best use for the area would be quarrying for high-class sand. The wishes of homeowners who paid millions of dollars for a quiet retreat and the chance to enjoy undisturbed the beauty that lay behind their backyards were to be disregarded in the interest of immediate profit. I and other members of the California Native Plant Society did what we could to help, and in the end the supervisors of Monterey County set aside a part of Evolution Hill as a memorial to Mr. Morse. Eighteen years later, in 1988, I visited the S.F.B. Morse Botanical Area and had the satisfaction of realizing that at least some part of Evolution Hill is preserved and has become a favorite retreat for hikers and nature lovers.

The second example of helpful allies concerns the California rose mallow (*Hibiscus californicus*: Malvaceae), a splendid pink flower that is the emblem of the Sacramento Valley chapter of the California Native Plant Society. Formerly common in much of California's Central Valley, this species has become rare because its preferred habitat has been given over to cultivation of the rich, moist soil that it prefers. One hot summer afternoon I visited the largest area that it still occupies, in a part of the Sacramento Valley owned by private duck-hunting clubs. Before entering it, I asked for permission from the local guardian, explaining my mission. He asked "Does this flower make a lot of hard round seed?", to which I replied yes. Then he told me of successful efforts to propagate it because its seeds are most attractive to ducks. So, for the moment, California rose mallow will survive abundantly over several square miles. In case of future threats, we have allies.

3. *Look for weaknesses in your opposition* and capitalize on them as much as possible. Farsighted conservationists must recognize that their "get-rich-quick" opponents are likely to have their "Achilles heels" because of either their lack of knowledge or their reluctance, as people of action, to consider all aspects of a project. Another experience with the California Native Plant Society brought out this point clearly. A few years ago we were called upon to attend a hearing of the board of supervisors in Lake County because of a threat to a most remarkable, unique plant community. Because of wet winters and dry summers, small lakes, ponds, or "pools" that may consist of several acres during the winter but dry up completely in summer occur frequently in California. Most of them are in the lower valleys. Boggs Lake in Lake County is not only one of the largest, but it is

unique in that it is perched on a plateau 1,000 m above sea level and surrounded by pine-covered hills. The margins of the lake itself offer highly interesting problems in ecology and population genetics, on which only a little research has been done. During years of heavy rainfall, the water level reaches the upper margins of the basin, and during the months of May, June, and July it gradually recedes. During this recession there is a succession of flowering of the distinctive annual lake margin species, beginning with those that are the most drought-tolerant and ending with hydrophytic species that mature their seeds under fully moist conditions. Some species have a relatively narrow range of moisture tolerance, and therefore a relatively short period of flowering and fruiting, whereas for others it is longer. Based on transplants to the laboratory made by a graduate student many years ago, an endemic species (*Navarettia plieantha*: Polemoniaceae) may include genotypes that differ from each other with respect to moisture tolerance and time of flowering, possibly reflecting evolution by natural selection that has taken place only around Boggs Lake. Small wonder that I, as an evolutionary geneticist and ecologist, wished to see Boggs Lake preserved at all costs!

One spring a group of real estate developers from the San Francisco Bay area, flying over Boggs Lake in a small plane during the spring season, decided that, if properly developed, the shores of Boggs Lake could be converted into a prime summer resort. Our Society was resolved to block their efforts. Since the hearing took place in the middle of the summer, I drove from Davis to Lakeport via Boggs Lake so that I could present a factual account of the area at that season. As I expected, the "lake" had dried completely, leaving a few hundred acres of dusty flats with a small mud hole in the middle. The pine forest that bordered it consisted of small trees, densely packed together, so that several of them would have to be removed to make room for a single cabin. If by much effort a suitable area should be cleared, the surrounding forest in autumn, tinder dry after three completely rainless months, would pose a constant threat of an all-destroying fire. I could not see any possible source of drinking water for a housing development, and at the hearing members of the county board who knew the area verified the impossibility of supplying such water for the proposed development. Once the developers learned the facts, they gave up. Their threat made it possible to have Boggs Lake taken over by The Nature Conservancy so that it is now safe for the foreseeable future.

4. *If total victory appears unlikely, look for a compromise solution.* I have already mentioned examples of partial victory—the conservation of a vital part of an area that, if possible, should be entirely

conserved. Such compromises may more often be necessary in large, remote areas. One of the most annoying experiences that many of us had in the middle of California's desert areas has been the sight of a motorcycle churning up the sand of a pristine hillside for no other reason than the owner's desire to go up as steep a hill as the machine can climb. Here we are dealing with another minority group that is out for pleasure and has its rights just as we do. Moreover, they are backed by powerful commercial interests so that their banishment from all desert areas is impractical. Even if laws could be enacted prohibiting their activities entirely, these could not be enforced. A wiser course would be to restrict them to areas where damage would be minimal and to make these areas more easily accessible than areas that contain the most unusual of the fragile biotic communities of desert areas.

With a little knowledge, similar compromises can be made in areas that are nearer population centers for recreational facilities such as golf courses. The extent to which this can be done to our rivers, so that water can be made available for agricultural and urban communities and at the same time some stretches of wild river can be retained for boating and rafting, raises problems; solving these requires that we obtain additional information by means of careful research.

5. *As long as there is hope for conservation, never give up.* The effectiveness of persistence is illustrated by the history of the raised-beach scrub barrens of Ione that have already been mentioned. Some parts of its fossil soils are a fine source of clay for ceramics, particularly for irrigation tiles. Active quarries have been operating there for many years. Early efforts at conservation were met by a stone wall of opposition accompanied by colorful epithets and obscenities conveying the message that quarries would continue until the entire area had been exploited. Nevertheless, quiet persistence, particularly on the part of Jean Jenny, ardent conservationist and wife of Hans Jenny, eventually produced a partial victory. A sizable area in the center of the barrens that contains the rare plants is now under the control of The Nature Conservancy.

Compared to the enormous tasks that lie ahead, the examples of partial success that I have mentioned are of minor significance. Nevertheless, they show that enlightened self-interest can prove effective. Even more important, they show that knowledge and evidence combined with diplomacy can often overcome odds that at first sight appear to be insuperable. What is most important, action can be effective only if it is based on superior and accurate knowledge of each situation. That is why a book like this one is of inestimable value as a weapon for conservationists. The broad picture that must be understood is made up of many examples, small and large, including either a single species or many kinds of animals and plants. Anybody who wishes to be an effective conservationist must be fully familiar with the contents of this and similar volumes, such as those of Cox (1974), Owen (1980), and Diamond and Lovejoy (1985).

I should like to end my discussion with impressions of the larger picture of world conservation, using as a springboard the local California situation with which I am familiar. I regard this method as justified because with modifications, it can serve as a basis of effective action at broader levels.

Turning to worldwide conservation, we find ourselves faced with problems that have existed for millennia. One of the leaders in this effort, Garrett Hardin (1969), cites historical evidence in an extract from the *Critias* of Plato. This philosopher called attention to mountains in his native Greece that not long before his time had been covered with forests, but had been denuded to supply timber for building the cities of Attica. He had the perception to note that this deforestation caused supplies of water in springs to dry up, even though there had been no reduction in rainfall! Several authors have speculated that the floods that, according to archaeological records, periodically inundated civilizations that grew up in the Mesopotamian and other low valleys of the Middle East (and were the probable basis of the biblical legend of Noah) were caused chiefly by overgrazing and deforestation of the surrounding hills. The inhabitants of these countries were punished for their "sins"—not for sexual license or other social irregularities but rather for sins against the environment on which they depended for their existence. The cause of these inadvertent sins was, most probably, similar to that which has made conservation necessary in the United States during our century—that is, better nourishment that reduced infant mortality locally and, perhaps in combination with less intertribal conflict, caused a rapid increase in local population density. Then, as now, the effects of population pressure, if allowed to build beyond a point of no return, will inevitably lead to one or a succession of ecological catastrophes. What should or can we do about it?

This question, which is being asked by thinking people throughout the world, cannot be answered in a brief essay such as this one. Here I would like to point out only that the guidelines that I have just reviewed with respect to conservation in California are equally, and in some instances even more, relevant to the world picture.

Direct confrontation with exploiters and destroyers of world resources is even less desirable than it is within our democracy. The biotic communities and organisms that are in greatest danger of extinction live in foreign countries that resent incursions into their sovereignty and, with some justification, view the United States (the "colossus of the north") more as an enemy to be feared than as a friendly adviser. Moreover, most of us who rarely if ever have known what it is like to be hungry for days or weeks at a time, or to lack the fuel needed to keep one's dwelling warm during a cold winter, are unable to put ourselves in the position of those for whom hunger and cold are daily experiences or threats. During a trip many years ago to the Atlas Mountains of North Africa, I was appalled at the destructive erosion that many of the steep slopes there had suffered. At first I could only applaud the efforts of French forest rangers who at that time ruled over the Atlas Mountains, prohibiting people from cutting down the oak trees that held the soil in place and from digging up their stumps, and enforcing the prohibition with controls including severe fines, imprisonment, or both. As I thought further about the problem, however, I realized that oak wood was the inhabitants' only source of fuel for cooking and for keeping themselves warm during the cold winters that prevail in the mountains. How can situations like this be resolved in such a way that people living today can maintain themselves and at the same time hand down to their progeny an inhabitable world? I did not have an answer to this question then, and I do not have one now. Several long-term, indirect answers can be concocted, but they would not satisfy the needs of those who must, in the immediate present, maintain themselves with the resources that are at hand. This small example, one out of hundreds that exist everywhere within developing nations, represents a chronic situation that, if left alone, can only worsen as population pressures increase. The recent review in the news magazine *Time* presents, in Brazilian Amazonia, a large-scale counterpart of the dilemma that I saw in Algeria. Upon reading this account, a North American is immediately prompted to support urgent petitions to Brazilian officials to halt the destruction of Rondônia's forests. Should forest conversion be stopped, even if it means dashing the hopes of hundreds of previously impoverished settlers, whose government has promised them a better way of life if only they will colonize Rondônia and work for their own betterment?

Given the enormous problems involved and the great diversity of the people whose lives will be affected by any changes made, the need for alliances on a continental and worldwide scale should become clear to everyone. The private and public agencies that are helping nations like Costa Rica, which is doing more than others to solve its own conservation problems, should receive all possible support from conservationists everywhere. The logical organization that should be mobilized to lead a worldwide conservation movement is the United Nations, particularly

United Nations Educational, Scientific, and Cultural Organization (UNESCO). Unfortunately, political difficulties have prevented this from happening. Nevertheless, we in the developed world should become more aware of how these organizations are conducting themselves and be ready to support movements toward resuming U.S. membership in UNESCO should there arise hope that such membership would promote worldwide conservation.

Right now the increasing strength of cultural exchanges between the U.S. and the Soviet Union should provide a golden opportunity for joint leadership in conservation on the part of the two strongest and most influential superpowers. Whatever allies we seek, we in the developed world must inform ourselves of those allies' economic needs and points of view and make the most of the interests that we have in common. Worldwide conservation demands worldwide understanding of needs and issues.

Some events of history show that, when dealing with nature's resources, entrepreneurs can make serious mistakes, as damaging to themselves as they have been to others. One of the best-known examples is the history of the Amazonian rubber industry (Wolf and Wolf 1936). When the world's wheels started to roll on latex from *Hevea brasiliensis*, entrepreneurs who were exploiting the Amazon became suddenly rich on a bonanza that combined an abundant product with a labor supply of workers so needy that they accepted starvation wages plus constant abuse. The bubble burst when entrepreneurs obtained seeds of *Hevea* and planted them in Malaysia, where cultivation and labor conditions were better, and by various means created an industry based on the cultivated rubber plantations that drove out native rubber. Later, the Ford Motor Company tried to restore the industry in Amazonia by introducing improved cultivars. The company failed because it ignored a principle well known to plant pathologists: Cultivated plants are more successful when grown in regions away from that of their ancestors, because they can more easily be kept free from the disease and pest organisms that have coevolved the ability to attack them. This example shows clearly the value of a broad understanding of a problem's complexity and of the research that is needed to acquire this knowledge, and it exposes the weakness of exploitation based on greed and nearsightedness.

Equally important is the value of knowledge that tells us when compromises are necessary, as they will be many times in the future. Even the most optimistic conservationists cannot hope to save all the wildlands that are not threatened. Repeatedly, crises will arise that will demand from us quick choices. We cannot predict each individual crisis, but we can be sure that the wisdom of each decision will depend greatly on knowledge that has been accumulating during the years before the decision must be made. That is why one of the most immediate objectives of conservation movements at this time should be to expand the number of research institutions and teaching curricula that will provide as much of the necessary knowledge in the shortest possible time. At the same time, we must strengthen efforts to persuade research and educational administrators that there is a host of eager young men and women who are eager to devote their lives to the field of conservation and that, in every possible way, attractive and remunerative opportunities should be opened to them.

One compromise with nature that several biologists have recommended for saving rare species is to maintain them in domestication or cultivation. This method obviously has been highly successful for many species that are useful to humans: the horse, donkey, camel, domestic sheep and goat, and chicken survive only with human culture or as escapees from domesticated ancestors. Some of the world's finest trees and most beautiful flowers are also unknown in a completely wild state. In America, botanists recall the shrub named after our elder statesman (*Franklinia alatamaha*) while plant scientists everywhere respect the foresight of Oriental religious leaders who preserved and handed down to us the maidenhair tree (*Ginkgo biloba*), the only survivor of a plant order that flourished more than 50 to 100 million years ago.

Nevertheless, these domesticated and cultivated relics, however useful they may be in various ways, are no substitute for wild species that are preserved in their native habitat. Evolutionary change in nature is based on interaction between populations and the environment to which they have been or can become adapted via natural selection. The chief reason for preserving the majority of endangered species, those that lack any direct economic importance to humans and are so inconspicuous that they can give aesthetic pleasure only to nature lovers, lies in the information that they provide for our understanding of how organisms evolved. But to be truly valuable, this must include ecological information about the position of each species in the community of which it has been a part during tens of thousands or millions of years.

Finally, the adage "never give up: while there is life there is hope" has never been truer than it is today in the conservation movement. The literature of the twentieth century contains a series of milestones (see Soulé 1986) showing that battles can be lost or won, but, contrary to many predictions, the campaign goes on and has never ended in either complete victory or irretrievable defeat. Furthermore, these authors have established inextricable interconnections between conservation of resources, control of populations, and more equal distribution of opportunities for people to make a living and to grow old with dignity. For these reasons, I shall precede a discussion of the ethics of conservation with a summary of the highlights of the campaign for biotic conservation in the United States during our century.

I begin with the first presidential administration under which conservation received serious attention—that of Theodore Roosevelt. He was our first president who knew intimately the wildlands of our western states and the need for conserving as much of them as possible in pristine condition. He and his associates listened to voices like that of California's John Muir, who until then had literally been crying in the wilderness. Under Roosevelt's benign influence the first steps toward saving wildlands were taken. Roosevelt's close friend, Gifford Pinchot, who later became governor of Pennsylvania, wrote one of the first general books on the need for conservation (Pinchot 1910).

The first biologist to recognize that conservation is necessary to avoid a serious crisis was the Harvard geneticist Edward M. East, who in 1923 published *Mankind at the Crossroads*. His concern was with Malthus and overpopulation: The "crossroads" that he visualized was a choice between controlling the destiny of humankind by means of family planning or "of being tossed about until the end of time by the blind forces of the environment. . . ." Because his warning came at the beginning of the 1920s, the decade of prosperity, his warnings went unheeded. Later on, those who remembered him were also aware of his concepts of a genetic human elite and his fear that the white race would be outbred by people having other skin colors, which he thought would lead to disaster. His voice was therefore still unheeded, and his book is now only a period piece.

During the following decade, a combination of the harrowing experience of the Great Depression; the Dust Bowl that caused thousands of tons of prime agricultural topsoil to be blown away and poured into the ocean; plus the threat to human decency posed by Hitler, Nazism, and Fascism made thinking Americans ready to listen to well-considered warnings. The populist administration of Franklin Roosevelt provided opportunities to do something about those threats on a large scale. The prophet whose word provided a stimulus to both careful thought and effective action was ecologist Paul B. Sears. His book *Deserts on the March*, written in Oklahoma during the Dust Bowl crisis, was published in 1935 and followed by revisions in 1947, 1959, and 1980. It is both a chronicle and a guidebook for action in our field. In his first edition, Sears recognized the need to define the science of "ecology," a word that before 1935 was understood by only a few specialists. He stated that it dealt with "the relation between the individual living thing and the atmosphere and soil around it, and of course the relations which exist between and among living things" (p. 161). His characterization of the ecologist "who, when [s]he enters a forest or a meadow [s]he sees not merely what is there, but what is happening there" (p. 162) should be etched in the minds of all modern ecologists, lest they fall into the mistake of equating ecology either with picking up cans and bottles or with complex mathematical formulas that are helpful tools but not inspirations for intelligent action. In Sears' concept of ecology, conservation of agricultural soil, wildlands, and the animals and plants that live in them, and control of the growth of human populations are carefully interwoven. He pointed out that by the 1950s a "sea change" (p. 207) took place among spokespersons for natural resources. Those whose major concern had been depletion and disruption of natural resources and processes "began to say that no measures could be effective unless the increase of human numbers could be arrested" (p. 207). The foundation of his ethics of ecology and conservation he expressed as follows: "Discipline with a goal that makes ethical, aesthetic and scientific sense is thrice armed. There are no easy answers, no shotgun prescriptions for the deeply interwoven problems of human society and the planet whose unique fitness has made possible the origin and survival of mankind" (pp. 257–58).

A chronicle covering the critical postwar years 1948 to 1968, which saw the most serious challenge to traditional ethics and culture that America has ever experienced, is an anthology prepared by one of the most articulate spokesmen of his time, Garrett Hardin. The editor's comments that precede each group of excerpts are of prime value by themselves. Hardin's anthology is most valuable to the natural history conservationist because of the way in which it strengthens the bonds between conservation of natural resources and population control. Moreover, one can learn much from reading this volume to support the precept "never give up." The worst predictions made in 1965 to 1968 about what we would be facing in 1989 have not been realized. On the other hand, the events of the past twenty years have brought us no closer than we were to a solution, or even to a serious treatment of the causes of species extinction and environmental degradation rather than its symptoms. The symptoms, reduction in numbers of individuals and ecosystems supporting endangered species, have received more attention than ever before in the United States and are beginning to be considered by individuals and commissions in the more vulnerable regions of the tropics. On the other hand, the causes—increasing population pressure and the demand for short-term entrepreneurial profits—have hardly been affected either way during the forty-five years of peacetime prosperity that have followed World War II. Conservationists are, from time to time, winning minor skirmishes, but success in major campaigns has not yet been attained.

With respect to conservation ethics, the highlight of Hardin's anthology is the reprinting of his renowned essay "Tragedy of the Commons." The lesson of this essay is that conservation of any resource, whether it be pastures, forests, fuel resources, or endangered communities and species, that is of equal interest to many people is in the long run impossible if one assumes that everybody will always act according to selfish interest and short-term gain. Our conservation ethic must, therefore, include the principle that land, including natural communities and rare wild species, is not a commodity that people should always own and do with what they will, but a trust that has come down to us from our ancestors and forefathers and for which we have the duty to hand down to our successors in as good a condition as possible. The principle of "conscious time binding," including generosity and farsightedness, however often it has been violated in the past, has nevertheless been upheld and honored enough to make possible the rise of culture and to bring about the relatively comfortable life that many of us now enjoy. This principle must now be expanded as quickly as possible to encompass and promote better relationships between humanity and our natural environment throughout the world.

My own conservation ethic has been greatly influenced by reading the book that I consider to be the jewel in the crown of contemporary conservation literature: Aldo Leopold's *A Sand County Almanac* and the essays that were published with it (Leopold 1949). The final one of these, entitled "The Land Ethic," serves as the basis for my concluding remarks.

The principal guideline proposed by Leopold is this: "Quit thinking about decent land-use as solely an economic problem. Examine each question in terms of what is ethically and esthetically right, as well as what is economically expedient. A thing is right when it tends to preserve the integrity, stability, and beauty of the biotic community. It is wrong when it tends otherwise."

This proposal is valid only if it is supported by a reasonably large and well-informed part of our cultural community. Codes of ethics exist because they deter individuals from placing their own short-term desires above those of the community as a whole and the rights of coming generations. With respect to economic issues, one can always imagine a future compensation for present losses: When biotic communities, with their complex structure, or rare species that depend on these communities are at issue, no future compensation may be possible. Extinction is forever.

Even the most ardent conservationist must recognize that judgments of what is aesthetically right are highly subjective and that in a free society everyone has a right to his or her own opinion. This means that changes in the right direction must be brought about chiefly by persuasion: Laws and moral precepts are forceful only if they are widely supported. The way to increase support is intelligent instruction that teaches not only an accurate account of relevant facts, but also a broad appreciation of their significance. This is the purpose of volumes like this one. I believe that its content of information and collective viewpoints is such as to contribute much to the desired changes.

LITERATURE CITED

Axelrod, D.I. 1980. *History of the maritime closed-cone pines, Alta and Baja California.* University of California Publications in Geological Sciences 120. Berkeley: University of California Press.

Cox, G.W., ed. 1974. *Readings in conservation ecology,* 2nd ed. New York: Meredith Corp.

Diamond, A.W., and T.E. Lovejoy, eds. 1985. *Conservation of tropical forest birds.* Cambridge, England: International Council for Bird Preservation.

East, E.M. 1923. *Mankind at the crossroads.* New York: C. Scribners.

Hardin, G. 1969. *Population, evolution and birth control: A collage of controversial ideas,* 2d ed. San Francisco: W.H. Freeman.

Leopold, A. 1949. *A Sand County Almanac.* New York: Oxford University Press.

Owen, O.S. 1980. *Natural resource conservation: An ecological approach,* 3d ed. New York: Macmillan.

Pinchot, G. [1910] 1967. *The fight for conservation.* Seattle: University of Washington Press.

Sears, P.B. [1935] 1980. *Deserts on the march,* 4th ed. Norman: University of Oklahoma Press

Soulé, M.E. 1986. *Conservation Biology: The science of scarcity and diversity.* Sunderland, Mass.: Sinauer.

Wolf, H., and R. Wolf. 1936. *Rubber: A story of glory and greed.* New York: Covici, Friede.

EPILOGUE

EPILOGUE

The essential solutions entail dramatic and rapid changes in human attitudes, especially those relating to reproductive behavior, economic growth, technology, the environment, and resolution of conflicts.
—Ehrlich, Ehrlich, and Holdren 1977

We need ecosophy (wisdom of household) and not only ecology (knowledge of household).
—Arne Naess 1986

Intellectuals, as much as the religious, have delusions of grandeur.
—John Passmore 1974

In this volume our contributors have explored several central themes in conservation biology such as the patterns of species diversity, rates and causes of extinction, habitat losses resulting from fragmentation, community dynamics in islands or insular habitat patches, population genetics, demographic aspects of viable populations, design of nature reserves, and some other practical issues. Clearly, there remain many areas that should be covered, especially for a comprehensive text (e.g., agroecosystem theory and management, worldwide conservation programs for genetic resources in domesticated genera). It is equally cogent to point out that conservation involves much more than the biology of populations or communities; all of natural history, biosystematics, and other biological disciplines as well as social, economic, and political sciences dealing with the human use of natural resources are relevant areas. In fact, a recent forum on biodiversity (Wilson 1988) considers in some detail the various international programs of conservation that require changes in our perceptions and attitudes. For example, the international program developed by the International Union for the Conservation of Nature and Natural Resources involves eight different categories of protected lands that include cultural landscapes, retiring farmlands, and multiple-use areas, allowing humans to seek a fine-tuned balance of sustaining land use along with protection (Reid and Miller 1989). Several chapters (e.g., chapters 14 and 16) refer to such a broad view of nature conservation.

Let us now turn briefly to several major points of discussion and to certain areas of uncertainty for which biological research is a critical need. First, although we often focus attention initially on listed endangered species and their survival, minimum reserve size requirements, or evo- lutionary genetic fate under increased inbreeding, one must also deal with an equally urgent concern for communities, natural heritage landmarks, and habitat types (Lovejoy et al. 1986; Holland 1986) for which we have both botanical and zoological versions of species diversity models and reserve designs. We do not seem to have a good understanding, however, of how or where we could combine the species and the ecosystem levels of conservation efforts. There is an impressive body of community struc- ture and dynamics literature telling us of many advances about the po- tential nature or role of interspecies interactions; successional dynamics; invasibility; patch dynamics under disturbance; and ways to identify key- stone species, guilds, equivalent competitors or simply good indicator species in a community (Roughgarden, May, and Levin 1989). But we are still largely uncertain about the choices and juxtaposition of various spatial and temporal scales in dealing with the description of community types, boundaries, pattern regulation, or native uniqueness, let alone the resolution of best methods to protect, restore, or monitor them.

Second, and related to the point just made, we often find a serious lack of information exchange between the "ecosystem-ists" and the "pop- ulation-ists" (as noted emphatically by several contributors in the volume edited by Pomeroy and Alberts 1988). Population biologists emphasize conservation measures in relation to the maintenance of demographically viable populations with projections of long-term persistence and of ge- netic variation to avoid inbreeding effects as well as a loss of evolutionary flexibility. In contrast, energy, biomass, or nutrient flow variables are clearly the major themes in developing the so-called *ecological engi- neering* principles (Mitsch and Jorgensen 1989), but no particular species is necessarily identified, and thus population level studies become irrel- evant and overlooked. It has been repeatedly noted that populations, species, or communities in nature reserves require "normally" function- ing ecosystems, and, likewise, human impacts or utilitarian views of na- ture should require the ecosystem services to be maintained. Clearly, the abiotic environment becomes a major consideration in dealing with eco- system processes (e.g., mineral cycles, light canopies, watersheds). Perry et al. (1989) developed the idea that through close and positive inter- actions between plants, mycorrhizae, and soils, "some ecosystems are continually pulling themselves up by their own bootstraps," and this tends to correlate diversity with buffering against disturbance. Some re- cent developments (e.g., Long Term Ecological Research [LTER] funded by the National Science Foundation) have restored this emphasis on multilevel ecological research of great interest to conservation biologists.

Third, there are rather too few studies in which both genetics and demography have been sufficiently incorporated to allow a clear understanding for determining the criteria of population viability analyses (Lande 1988), to which the northern spotted owl (*Strix occidentalis caurina*) might become an exception (Gutierrez and Carey 1985; Marcot and Holthausen 1987). Inbreeding effects due to bottlenecks in population sizes of single versus few isolates are not equivocally known for many variables in judging (rather, prejudging) the potential adaptive role of diminished genetic variation or modified breeding system, enhanced mutability, and so on (Loveless and Hamrick 1984). Here, we certainly need extensive work on the genetic and evolutionary features of endemic species (Kruckeberg and Rabinowitz 1985). For example, Carson (1989) argues that many Hawaiian species may have evolved some tolerance to inbreeding so that many small reserves might work well. And finally, minimum viable population sizes determined on the basis of avoiding inbreeding effects will not necessarily be the same as those given by the demographic PVA criteria (Soulé 1987; Lande 1988). Clearly, we choose to play it safe by emphasizing an intuitive rule: All else being equal, the larger and more numerous these isolated populations are, the greater their chances for persistence as well as for evolutionary change. Management strategies designed to provide either natural gene flow or human-aided gene flow among the isolates sounds good in theory, but we need empirical results for assessing the positive and negative issues of mixing gene pools, especially if the programs of off-site breeding and captivity have modified them significantly. An important issue in population biology, the one of defining fitness in terms of both genetic features and demographic ones, is also relevant here. Hence, it is important to remind ourselves that several successful species recovery plans have used one or more of the following tactics: modeling life history, off-site breeding and reintroduction, habitat protection with "let-nature-alone" philosophy requiring minimal management, land acquisition, and ecosystem management. It follows that population biologists should seriously consider liaisons with the various environmental and habitat management–oriented efforts. We find it somewhat puzzling that the oldest questions in population ecology about the natural regulation of numbers (due to abiotic or biotic factors) and optimality of life histories have not been pursued directly in recent conservation biology writings. And even if genetic factors regulating population numbers are more difficult to establish, the links between the genetic and demographic fate of small populations still rely on some sort of bounded variation in population numbers (density-vague regulation) and in certain growth parameters.

The relative importance of genetic and demographic studies, and their interconnections, in developing conservation strategies requires focused and critical research. In reviewing literature on rare and endangered plant populations, we find many examples of genetic variation assays that show a weak correlation with the measures of population size or reproductive success. In fact, there are too many unknowns about the past history, recent habitat losses, or the current forces of evolutionary changes such that we cannot realistically predict either the patterns or the adaptive roles of variation. Furthermore, we need extensive information on mating system, local subdivision, gene flow, and fluctuations in size to estimate the effective population size. Almost all of the examples of variation studies at hand lack this information and frequently end with speculations about the potential threats from inbreeding depression, loss of adaptability, or both.

Most documented examples of extinct plant populations, in contrast, have clearly resulted from rapid habitat degradation or loss. In many situations, however, small remnant as well as undisturbed small populations could be used for joint observations on genetic variation, life history, and some fitness parameters. Only then would we get support for the speculations mentioned earlier. Another caveat in this approach must be kept in mind. It is not a self-evident or proven concept that small populations necessarily fail to adapt or evolve under changing environments. In developing models of interdeme selection, Sewell Wright and others have shown that drift, local selection, and founder events could in fact promote evolutionary advances. Here, too, ecology of dispersal, recruitment, and population growth must be considered together with the purely population genetic schema.

Fourth, some researchers are concerned about the distinction between conservation and restoration; the issues are of naturalness and management (e.g., see Miyawaki, Fujiwara, and Okuda 1987; Western and Pearl 1989). According to Soulé (1989), "soon, the distinctions between preservation, reintroduction, and restoration will vanish . . . the term natural will disappear from our working vocabulary." Following this kind of thought process, unfortunately, a lot of *in situ* conservation efforts could be replaced by off-site mitigation, reconstruction of habitats, the so-called "amenity areas," and largely off-site facilities such as botanical gardens or zoological parks. Perhaps we must accept the concept of natural stands or native communities/habitats on a sliding scale, with the goal of seeking "as natural as possible" in most cases. This approach is not simply romantic nostalgia but reflects a deep desire to conserve something beyond human artifacts. Restoration ecologists (or ecological engineers) will find situations varying from the one requiring little or no continuing intervention to almost highly managed systems allowing harvests and other consumptive uses of the biota. Whether the same theoretical ideas and protected areas would suffice for research on conservation versus restoration, that is, managing areas with disturbed or disrupted structure and function versus areas with clearly degraded habitats, remains to be investigated.

Besides the setting out of various kinds of protected areas for *in situ* conservation, there are many active and planned projects for restoring degraded, polluted, or simply use-impacted habitats. These include prairies, deserts, marshlands, lakes, local amenity parks, large regional ecosystems (e.g., riparian in California), and large reserves taken out of agricultural production, and emphases vary from a few dominant species to the overall energy and nutrient budgets of an ecosystem. Restoration ecology has become an important arena for defining ecological principles and experimental work with the dual objective of problem-solving applications as well as the testing of ecological theory of community structure, ecosystem function, succession, or niche evolution. In a recent series of books on applied population biology, several research groups have discovered that most so-called general models or theorems in ecology need to be redefined, expounded on, and validated with many case-by-case studies in applied areas of pest control, resource harvest optimization, or conservation. Adaptive learning thus requires simultaneous development of specific and useful theories and applications, and this is particularly so in the field of restoration ecology. The cumulative experience of restoring riparian streams or vernal pools, for example, will help us define the primary niche axes (e.g., water regimes and zonation), species equivalence, role of nitrogen-fixing taxa, and factors controlling germination/establishment success. Several prairie restoration projects are experimenting with the choices of species as well as genetic resources in initial seed mixtures. Monitoring succession in several wetland communities will help us learn about the dormant seed banks and their responses to water manipulation. Thus, restoration projects, in spite of some quibbling about the unknown predisturbance status or the highly empirical nature of such work, given proper spatial and temporal dimensions and statistical rigor, promise a great deal of both new ecological insights and conservation results.

Fifth, although agriculture and conservation have been treated in several ecological and resource science publications during the 1960s and the 1970s (Davidson and Lloyd 1977, for example), more recently this has become a pressing matter due to the changes desired in the high-input agricultural systems of the Western world (U.S. National Academy of Sciences 1989 report). Issues of chemical use, high energy use, environmental pollution, and increased soil erosion have raised ecological awareness about the sustainability of long-term resource conservation

(Gliessman 1990). Certain programs in the U.S. and U.K. have allowed the setting aside of land for conservation; here, research needs are widely recognized in terms of preserving traditional agroecosystems; using buffer areas, corridors, natural or seminatural recreation areas; and so on. For example, use of agroforestry practices on the one hand, and new ways of managing commercial forestry species with genetic conservation in mind on the other, rank high on new research and practical agendas. Landscape ecologists (Forman and Godron 1981) not only have to deal with the overall design of networks, corridors, roadways, cities, and farming areas but also have to consider species diversity, population dynamics of selected species, and ecosystem stability. A recent report on California biodiversity (Jensen, Torn, and Harte 1990), for example, provided a comprehensive review of the landscape-, community-, and species-level considerations.

Finally, a common observation and also a matter of concern is that there are rather very few general principles (or generalities in ecological or conservation biology theory) such that we often find uniqueness in each case history. In other words, no two communities, species, or populations share the same natural and human-impacted causes of extinction or the same demographic or evolutionary responses to such threats. There are just too many variables, so it appears, in life history, genetic systems, population size regulation, natural enemies, species interactions, and resource use. Many experimentalists find theoretical models either too complex or oversimplified, or even unrealistic, and tend to ignore them. Most researchers nowadays recognize that no single model can be all three things: general, precise, and realistic. Models help in clear thinking, as metaphors or step-by-step simulation tools that can serve well in some basic analytical tests of ideas, as well as in pedagogical matters (see Kareiva's chapter in Roughgarden et al. 1989). In a recent stimulating essay, Slobodkin (1988) succinctly pointed out that there is no dearth of intellectual challenges in applied ecology, and that even with its intractability or uncertainty in giving practical guidance, applied ecology is still the best hope. He further noted that models of population growth or persistence based on old logistic equations are seldom useful as they ignore complexities of age distributions, time lags, and nonlinear interactions, but "ecologists at their best remain to some degree naturalists, aided by modern technology and computational devices."

Conservation biologists deal with perhaps the most urgent and useful area of applied biology. Botsford and Jain (1991) reviewed the potential applications of population biology principles that are often perceived to be inadequate for the following reasons: (1) inadequate theory, (2) large inherent variability in populations, (3) a flawed approach to applications, and (4) various epistemological and sociopolitical factors. Several examples of applied population biology in plant and animal agriculture were discussed to illustrate these limitations. One may ask: Are there some unifying principles of conservation biology? As noted by Botsford and Jain (1991), certain population genetic ideas are almost broadly applicable as principles (e.g., inbreeding leading to certain genetic structure, natural selection in enhancing local population fitness, evolution of optimal migration rates), whereas certain ecological principles (such as limiting similarity of coexisting species, species diversity under intermediate disturbance, relay floristics, succession) are rightly still under further scrutiny. Island biogeography theory, too, is now no longer completely accepted as a general or unifying concept (see Shaffer 1991, and Saunders, Hobbs, and Margules 1991 for the most recent reviews). Botsford and Jain (1991) also concluded that we need a close collaboration among various disciplines, applying and developing basic ideas in an adaptive learning paradigm, working iteratively between theory and experiments, to understand the mechanisms case by case as well as to work toward generalities.

Various aspects of conservation biology in theory and practice will undoubtedly become the focus of graduate education and research just as much as the local and global actions will show some outstanding accomplishments. As a consensus there is a fair amount of optimism about these predictions for the next decade and even the following one. However, there also is a genuine cause for pessimism as well. The political and economic forces of rapid industrial growth, with their advertised and glamorized virtues of larger GNP, more white-collar jobs, increased food supply, and more goods of convenience or comfort are and always will remain with us. Many among us are less than fully convinced about the environmental agenda in which protection of endangered biota gets focused attention, and that occurs also at a cost of misplacing ecosystem-level arguments for conservation. We will probably always find conflicts between ecology and economics, but it is hoped that there will be numerous turnarounds in agriculture, forestry, and the recreation-related industry as well as limited energy-related production businesses.

Conservationists have often resorted to *utilitarian* (economic) arguments for conservation; that is, conservation will provide new and more resources for now and for the future, whether they be new plant products for pharmaceuticals or new medical advances using new model organisms for research. This view is further extended to put forth *ecological* reasons (viz., ecosystems provide "life-support" services—oxygen, a chemically healthy atmosphere, clean water). In contrast, some argue citing the *aesthetic* reasons such as enjoyment of wilderness or simply poetic imagination. What about *ethical* reasons (other organisms have a right to exist; it is morally wrong to kill or destroy, etc.)? Indeed, "a new

concern for developing a cogent and comprehensive environmental ethic
... is in the air" (Van DeVeer and Pierce 1986, ix). We need to argue
about wisdom as much as knowledge, ethics as much as technology, and
transpersonal or deep ecology as much as the current value-free ecological
sciences. Science is useful for discovering and knowing but not for choosing and making value judgments. We must also recognize, as noted in
an elegant book by Stone and Stone (1989, xix), that "Conservation biology is the combination of art and science, compromise and stubborn-
ness, judgment and serendipity." We may further add Darwinian instincts
of self-preservation as well as moral and societal values.

> Care follows naturally if the self is widened and deepened so that pro-
> tection of free Nature is felt and conceived as protection of ourselves.
> (Arne Naess, quoted by Warwick Fox, 1990, in *Toward a Transpersonal
> Ecology*, 95)

"The need is clear, but the task is enormous, and time is running out."
So concluded Ledig (1988, 471) in a very thoughtful essay on forestry
resources. We agree.

LITERATURE CITED

Botsford, L.W., and S.K. Jain. 1991. Population biology and its application to
practical problems. In *Applied population biology*, ed. S.K. Jain and L.W.
Botsford. Dordrecht: Kluever.

Carson, H.L. 1989. Gene pool conservation. In *Conservation biology in Hawai'i*,
ed. C.P. Stone and D.B. Stone, 118–24. Honolulu: University of Hawaii.

Davidson, J., and R. Lloyd, eds. 1977. *Conservation and agriculture*. Chichester:
John Wiley & Sons.

Ehrlich, P., A.H. Ehrlich, and J. Holdren. 1977. *Ecoscience: Population, resources
and environment*. San Francisco: Freeman.

Forman, R.T.T., and M. Godron. 1981. Patches and structural components for
a landscape ecology. *BioScience* 31:733–40.

Fox, W. 1990. *Toward a transpersonal ecology*. Boston: Shambhala.

Franklin, J.F., C.S. Bledsoe, and J.T. Callahan. 1990. Contributions of the long-
term ecological research program. *BioScience* 40:509–23.

Gliessman, S.R., ed. 1990. *Agroecology: Researching the ecological basis for sustainable agriculture*. New York: Springer-Verlag.

Gutierrez, R., and A. Carey, eds. 1985. *Ecology and management of the northern
spotted owl in the Pacific Northwest*. USDA Forest Service, Pacific Northwest
Forest and Range Experiment Station, Gen. Tech. Report PNW-185, Portland, Oregon.

Holland, R.F. 1986. *Preliminary descriptions of the terrestrial natural communities of California*. Nongame Heritage Program. California Deptartment of
Fish and Game. Sacramento, Calif.

Jensen, D.B., M. Torn, and J. Harte. 1990. *In our own hands: A strategy for conserving biological diversity in California.* Berkeley: California Policy Seminar, Research Report, University of California.

Kruckeberg, A.R., and D. Rabinowitz. 1985. Biological aspects of endemism in higher plants. *Ann. Rev. Ecol. Syst.* 16:447–79.

Lande, R. 1988. Genetics and demography in biological conservation. *Science* 241:1455–60.

Ledig, F.T. 1988. Conserving forest genetic resources. *BioScience* 38:471–79.

Likens, G.E., ed. 1989. *Long-term ecological research: Approaches and alternatives.* New York: Springer-Verlag.

Lovejoy, T.E., R.O. Bierregaard, Jr., A.B. Rylands, J.R. Malcolm, C.E. Quintela, L.H. Harper, K.S. Brown, Jr., A.H. Powell, G.V.N. Powell, H.O.R. Schubart, and M.B. Hays. 1986. Edge and other effects of isolation on Amazon forest fragments. In *Conservation biology: The science of scarcity and diversity*, ed. M.E. Soulé, 257–85. Sunderland, Mass.: Sinauer.

Loveless, M.D., and J.L. Hamrick. 1984. Ecological determinants of genetic structure in plant populations. *Ann. Rev. Ecol. Syst.* 15:65–95.

Marcot, B.G., and R. Holthausen. 1987. Analyzing population viability of the spotted owl in the Pacific Northwest. *Trans. N. Am. Wildl. Nat. Res. Conf.* 52:333–47.

Mitsch, W.J., and S.E. Jorgensen, eds. 1989. *Ecological engineering: An introduction to ecotechnology.* New York: Wiley-Interscience.

Miyawaki, A., K. Fujiwara, and S. Okuda. 1987. The status of nature and recreation of green environments in Japan. In *Vegetation ecology and creation of new environments*, ed. A. Miyawaki, A. Bogenrieder, S. Okada, and J. White, 357–76. Tokyo: Tokai University Press.

Naess, A. 1986. The deep ecological movement: Some philosophical aspects. *Philos. Inquiry* 8:10–31.

National Academy of Sciences. 1989. *Alternative Agriculture.* Washington, D.C.: National Academy Press.

Passmore, J. 1974. *Man's responsibility for nature: Ecological problems and western traditions.* London: Duckworth.

Perry, D.A., M.P. Amaranthus, J.G. Borchers, S.L. Borchers, and R.E. Brainerd. 1989. Bootstrapping in ecosystems. *BioScience* 39:230–36.

Pomeroy, L.R., and J.J. Alberts, eds. 1988. *Concepts of ecosystem ecology.* New York: Springer-Verlag.

Reid, W.V., and K.R. Miller. 1989. *Keeping options alive: The scientific basis for conserving biodiversity.* Washington, D.C.: World Resources Trust.

Roughgarden, J., R.M. May, and S.A. Levin, eds. 1989. *Perspectives in ecological theory.* Princeton, N.J.: Princeton University Press.

Saunders, D.A., R.J. Hobbs, and C.R. Margules. 1991. Biological consequences of ecosystem fragmentation: A review. *Cons. Biol.* 5:18–32.

Shaffer, C.L. 1991. *Nature reserves: Island theory and conservation practice.* Washington, D.C.: Smithsonian.

Slobodkin, L.B. 1988. Intellectual problems of applied ecology. *BioScience* 38:337–42.

Soulé, M.E., ed. 1987. *Viable populations for conservation.* Cambridge: Cambridge University Press.

Soulé, M.E. 1989. Conservation biology in the twenty-first century: Summary and outlook. In *Conservation for the twenty-first century,* ed. D. Western and M. Pearl, 297–303. Oxford: Oxford University Press.

Stone, C.P., and D.B. Stone, ed. 1989. *Conservation Biology in Hawai'i.* Honolulu: University of Hawaii.

Van DeVeer, D., and C. Pierce, eds. 1986. *People, penguins and plastic trees.* Belmont, Calif.: Wadsworth.

Western, D., and M.C. Pearl, eds. 1989. *Conservation for the twenty-first century.* New York: Oxford University Press.

Wilson, E.O., ed. 1988. *Biodiversity.* Washington, D.C.: National Academy Press.

GLOSSARY

• • •

GLOSSARY

Adaptation. Genetic or phenotypic response by individuals or populations to an environment so as to enhance fitness.

Age structure (see **Stage structure**). Structure of a population determined by the age (in any unit) of individuals, for example, one-year-olds, two-year-olds, and so forth.

Agroecology (= Agricultural ecology). The study of agricultural land uses and practices in relation to their impact on soil, water, and other resources, with emphasis on designing ecologically sound and sustainable agriculture.

Agroecosystem. An entire ecological system comprised of indigenous agricultural activities and the biological, genetic, and cultural processes that support it.

Allelopathy. The chemical inhibition of growth in one organism by another.

Allopatric. Referring to populations, species, or other taxa occupying different and disjunct geographical regions.

Biological diversity (= Biodiversity). Full range of variety and variability within and among living organisms, their associations, and habitat-oriented ecological complexes. Term encompasses ecosystem, species, and landscape as well as intraspecific (genetic) levels of diversity.

Biome. Classification of communities and ecosystems based on their convergent features due to similarities in environments.

Biosphere reserve (= Bioreserve). Nature reserve, wilderness park, or other kind of protected habitat that allows indigenous human cultures and societies to continue traditional resource use.

Biotype. Group of genetically identical individuals.

Bottleneck (= Population bottleneck). Episode of dramatic reduction in population size due either to environmental stress or a colonization event.

Breeding system. Mode, patterns, and extent to which individuals interbreed with others from the same or different taxa; outbreeding levels in plant species are estimated, for example, in terms of self- versus cross-pollination.

Cline. Gradient in genetic or phenotypic features of populations occurring along an environmental gradient or geographic transect.

Coevolution (= Coadaptation). Reciprocal (interactive) evolutionary and adaptive changes in two or more species or genetic units (races, gene pools) living in the same community or space.

Colonists. Individuals of various taxa that are the first to invade newly created habitats and establish viable populations.

Colonization. Successful invasion of a newly created habitat; successful recruitment in gaps or vacant niches following disturbance.

Community. Collection of organisms of different species that co-occur in the same habitat or region and that interact through trophic and spatial relationships.

Community disequilibrium. Condition of a community whereby the interactions of the member taxa do not permit steady-state conditions.

Community equilibrium. Condition of a community whereby the interactions of member taxa permit steady-state conditions.

Community modeling. Mathematical expressions of community processes, such as species coexistence, invasibility, and resilience in disturbed environments and in trophic relationships, among others.

Community structure. Physical and spatial arrangements of members of a community and patterns of species composition, as well as the dynamics of trophic interactions.

Community types. Distinct and definable units of the natural landscape, such as freshwater marsh, perennial bunchgrassland, thorn scrub, and so on.

Conservation. Judicious use and management of nature and natural resources for the benefit of human society and for ethical reasons.

Corridor. A more or less continuous connection between adjacent and similar habitats; examples in a landscape context include roads, hedgerows, streams, and irrigation ditches.

Demography. Study of birth and death rates and their consequences on the density or abundance of a population.

Deme. Discrete population of interbreeding individuals.

Deterministic. Resulting in one or a few outcomes predicted with absolute certainty (see **Stochastic**: models that incorporate the concepts of random processes and probabilistic predictions).

Disequilibrium. Condition where community structure is continually changing in terms of patchiness, patch occupancy, local extinction, and recolonization.

Disturbance. Any relatively discrete event in time that disrupts ecosystem, community, or population structure and changes resources, substrate availability, or the physical environment. Key descriptors are magnitude, frequency, size of area, and dispersion.

Diversity. Ecological measure based on the number of species and their relative abundance in a community; a low diversity refers to only a few species or unequal abundance, whereas a high diversity refers to many species or equal abundance.

Alpha diversity. Ecological measure of the intrinsic number of species within a community.

Beta diversity. Ecological measure of the turnover of species along an environmental gradient.

Gamma diversity. Ecological measure of the species turnover rate with distance between sites of similar habitat or with expanding geographic areas; rates at which additional species are encountered as geographic replacements within a habitat type in different localities.

Ecological gradient (see **Cline**). Change of environmental features along some environmental gradient, such as, the increase in temperature with increasing altitude, the decrease in species richness with decreasing latitude, and so on.

Ecosystem. Integrated set of biological components making up a biotic community and its abiotic environment; two primary axioms defining ecosystem structure and function are (1) recycling of essential elements, including biomass in different trophic levels following characteristic spatial and temporal patterns in each ecosystem type; and, (2) certain emergent properties such as homeostasis and self-regulation are definable and measurable in this highly aggregate unit of study.

Ecotone. Landscape boundary that exists between two (or more) adjacent communities or habitats; also known as edge.

Ecotype. Genetically differentiated subunit within species that represent ecological adaptation to certain local environments.

Edge-adapted species. Plant and animal taxa that spend some portion of the lives within ecotones.

Edge effects. Processes that characterize habitat fragmentation and the concomitant creation of edges.

Effective population size N_e. Average number of individuals in a population assumed to contribute genes equally to the succeeding generation; genetically effective population size.

Endangered species (taxon). Species that, for a variety of reasons, has been given formal protection status by state or federal (or both) statutes.

Endangered Species Act of 1973. Federal statute passed by the U.S. Congress in 1973 that provides for the formal protection and recovery (i.e., "delisting") of designated endangered and threatened species and their habitat; one of its purposes is to provide a means whereby the habitats on which endangered species depend may be conserved.

Endemic. Native or restricted to a limited or particular geographic region.

Enhancement. Management technique (using seeding, transplantation, fencing, watershed manipulations, etc.) that attempts to restore to predisturbance conditions those areas that are only partially disturbed by human influence.

Equilibrium community theory. Any theory concerning the structure of a community that suggests that species abundances remain constant over time.

Ethnobotany. Discipline within botany that examines the interactions between human societies and the plant species important to their culture.

Evolution. Any gradual, directional change; the cumulative change in the genotypic and phenotypic features that results in the modification of the populations or species over generations.

Exotic species. Nonnative species that have established viable populations within a community; species present within a community that did not exist there before the influence of human activities.

Ex situ conservation. Conservation of plant and animal taxa or their biological materials (e.g., seed embryo, DNA) away from their natural habitat, for example, zoological parks and botanic gardens.

Extinction. Process by which an individual, species, or population disappears from a given habitat or biota.

Extinction probability (EP). Probability of extinction calculated for multiple stochastic simulations using life history parameters; a working definition of an extinction threshold.

Extinction proneness (of small populations). Reproductive failures, synergistic genetic deterioration, and overall lack of adaptability over long time scales under the assumptions of both genetic consequences of drift and demographic/environmental stochasticity that increase a population's susceptibility to extinction.

Extinction rate. Number of species in a given habitat or region that become extinct per given unit of time.

Extinction vortex. Positive feedback loops of biological and environmental interactions that have further negative impacts on a population, most likely leading to extinction.

Extirpation. Process by which an individual, population, or species is totally destroyed.

Faunal collapse. Rapid or dramatic losses of species in fauna due to increased insularity (e.g., due to the disappearance of a corridor or bridge).

Faunal relaxation. Species loss on islands (habitat or oceanic) that can be approximated by an exponential function, such as ks^n, where k is the relation parameter and n, ranging from 1 to 4, is a constant for various fauna.

Fitness. In a broad evolutionary context, refers to some measure of specific or general adaptive outcome in relation to the environment; in population genetics, relative Darwinian fitness of a genotype is measured by its contribution to the next generation.

Founder effects. Nonselective changes in the genetic makeup of a colonizing population during its establishment by a few founding individuals.

Fragmentation. Process by which habitats are increasingly subdivided into smaller units, resulting in their increased insularity as well as losses of total habitat area.

Gap phase species. Taxa adapted to vegetation gaps created by small-scale disturbances, such as tree falls, particularly in tropical moist forests.

Gene pool. Total genetic material of a freely interbreeding population.

Generalist. Species that has a broad habitat range of food preference (also called habitat generalist).

Genetic conservation. Conservation measures emphasizing the maintenance of genetic systems and levels of genetic variation to provide for continued persistence and evolution; also considered generally under the broader term of biological conservation.

Genetic diversity. (also referred to as Genic diversity or Total genetic diversity $[H_T]$). Measures of genetic variation calculated from allelic frequencies and their combinations. Various comparisons among species allow their genetic diversity to be interpreted in terms of certain ecological and evolutionary features (e.g., gene flow among populations, rarity versus commonness over time). Total genetic diversity can be partitioned into within-population diversity and between-population diversity.

> **Within-population diversity (H_S).** Equivalent to the level of heterozygosity, that is, the probability that two alleles picked at random in a subpopulation will be different.
>
> **Between-population diversity (H_{ST}).** Genetic differentiation among populations, that is, the probability that two alleles picked at random from two subpopulations will be unlike.

Genetic drift. Genetic changes in the allelic composition or allelic frequencies in small or suddenly depauperate bottleneck populations. Genetic consequences of drift often may be harmful (e.g., loss of adaptability, inbreeding) but also occasionally useful in producing novel gene pools, interpopulation divergence, and new taxa.

Genetic load. Average number of lethal equivalents per individual in a population; a measure of deleterious genetic aspects of variation.

Habitat connectivity. A measure of connectedness in a habitat-oriented description of landscape elements.

Habitat degradation. Decline in habitat quality that accompanies non-natural forms of disturbance.

Habitat patchiness. Quality of many communities that exist in discrete units separated by other kinds of habitat.

Inbreeding. The mating of close relatives, that is, the mating of individuals likely to share some of their genes due to common ancestry; any mating between individuals more related than expected with the assumption of random mating.

Inbreeding coefficient (*F*). The probability that two alleles at a particular locus in an individual are identical by descent.

Inbreeding depression. Reduction of fitness and vigor by increased homozygosity as a result of inbreeding in a normally outbreeding population.

In situ conservation. Conservation of individuals, species, or habitats in their places of origin and natural occurrences.

Insular habitat. Patch of habitat that is completely surrounded by another habitat type, now isolated from other similar patches of habitat.

Integrated conservation strategy. Approach to preserve species through both *in situ* and *ex situ* conservation measures.

Interior species. Species limited to the center portions of large, undisturbed tracts of land.

Intermediate disturbance hypothesis. A hypothesis put forth by Joseph Connell proposing that communities living under an intermediate amount or severity of disturbance contain the greatest number of species. Communities that experience frequent or severe disturbances contain lower numbers of species due to the extreme nature of the disturbance; communities that rarely experience disturbance or experience only mild forms thereof also contain a smaller species complement due to the competitive exclusion by a few dominant species.

Introduction (of organisms, species; see **Reintroduction**). Management technique often used in restoration ecology or because of mitigation requirements involving the placement of a population (seed or individuals), typically of a legally endangered or threatened species, in an area in which the species was not formerly found.

Island biogeography. Theory developed by R.H. MacArthur and E.O. Wilson in 1967 that proposes that the number of species inhabiting an island is a function of island area and distance from the mainland and is determined by the relationship between the rates of species immigration and extinction.

Keystone resources. Resources, such as lipid-rich seeds, on which a large number of species within a given community depend for survival.

Keystone species. Species that serve as keystone resources.

Landrace. Variety or localized gene pool in an agricultural species that may have evolved a geographical or ecological identity; a locally adapted cultivar.

Landscape. Ecological mosaic of specific ecosystems; "a kilometer-wide area where a cluster of interacting stands or ecosystems is repeated in similar form" (Godron and Formann 1983, 13).

Landscape ecology. New branch of ecology dealing with the processes that determine pattern and function at a landscape scale.

Landscape management. Management of nature at a landscape scale that strives to maintain those functions and processes that characterize landscape features.

Leslie matrix. Schedules of age-dependent survivorship (dominant latent root or eigenvalue of a square nonnegative matrix) and fecundity (included in first row of matrix); use of matrix algebra allowing an analysis of population growth in terms of evolutionary changes in life histories.

Life history. Significant features of the life cycle through which an organism passes, particularly those influencing reproduction and survival.

Management. Manipulation of nature for a specific goal.

> **Proactive management.** Management of nature that seeks to avert a decline in habitat quality or quantity before an event is likely to occur.

> **Successional management.** Form of management, such as controlled burning, that manipulates the successional processes of a community for restoration or various multiple land uses.

Metapopulation. Series of populations (or population subdivisions described as local subpopulations) with dynamic patterns of local extinctions, recolonizations; gene flow or migration among subunits provides characteristic evolutionary and ecological features that help avoid extinction of the entire metapopulation; Sewell Wright's classic models of three-phase evolution had used these concepts long before the current popularization of the term.

Metapopulation dynamics. Demographic and evolutionary patterns and processes of the founding and extinction of subpopulations of a metapopulation.

Minimum viable population (MVP). As defined by Shaffer (1981), the smallest isolated population having a 99 percent chance of remaining extant for 1,000 years despite foreseeable effects of demographic, environmental, and genetic stochasticity as well as natural catastrophes; others subsequently have altered the time scale and survival probability (see Soulé 1987).

Mycorrhizae. Largely symbiotic relationship between large and taxonomically diverse groups of fungi and vascular plants that allows for the increased uptake of water and minerals by the vascular plant, and for the uptake of sugars and carbohydrates from the vascular plant by the associated fungus.

Nature reserve. Tract of land or habitat selected to be set aside for preservation in its natural condition; genetic reserves are designed specifically to conserve genetic resources in agricultural species.

Niche. Ecological role of a species in a community.

Nonequilibrium community theory. Any of several current theories concerning community organization describing any situation where species densities do not remain constant over time at each spatial location.

Nonnative species. See **Exotic species.**

Outbreeding depression. Disruption of coadapted genotypes through the mating of too distantly related individuals, sometimes resulting in the partial or complete infertility and inviability of offspring.

Parapatric. Referring to populations with contiguous but not overlapping geographic ranges.

Patch. Highly localized unit of population and community studies.

Patch dynamics. Continuous changes in community structure and its species abundances due to disturbance, creating shifting and mosaic patchiness.

Phenotype. Expression of genes or genotypes scored in terms of an individual's traits (e.g., form, function, behavior); in quantitative genetics, phenotypic variation is analyzed into its genetic and nongenetic components.

Phenotypic diversity. Expression of the physical variation that exists within a taxon or members thereof.

Phenotypic plasticity. Range of phenotypes produced by a genotype scored under different environments. This response can serve as an adaptive strategy under rapidly varying (fine-grained) environments.

Pioneer species. The first species to colonize or recolonize a barren or disturbed area, thereby beginning a new successional sequence.

Population biology. Study of biological populations where a population is defined as a collective group of organisms of a single species occupying a particular space and time; it incorporates subject areas from ecology, genetics, evolution, demography, behavior, and biostatistics and deals with the fundamental issues of structure and dynamics of populations.

Population genetics. The study of genetic and evolutionary properties of populations and species in terms of random drift; units of observation are gene or genotypic frequencies and the means and variances of morphological or physiological traits.

Population modeling. Analytical and computer simulation approaches to the study of population growth, persistence, evolution, and adaptedness.

Population structure. Distributions of individuals, potential mates, genotypes, or genes in space and time; furthermore, age-specific birth- and death-rate processes that determine the life history and reproductive strategy of a population or a series of populations in a species.

Population viability. Concept of a "viable" population number that represents a threshold for survival versus extinction.

Population viability analysis (PVA). Models and numerical estimation procedures to determine minimum viable population size (MVP) or area (MVA).

Preservation. Form of passive management that does not advocate any form of intervention or habitat manipulation by humans for a specific purpose.

Protected areas. Legally established areas under either public or private ownership where the habitat is managed to maintain a natural or seminatural state.

Rarity. Seldom occurring either in absolute number of individuals or in space; Rabinowitz (1981) described seven classes of rarity that may be defined by the combination of local or global abundance, habitat specificity, and geographic distribution.

Refuge. Area in which a prey individual or species may escape from or avoid a predator.

Refugium. Area that has escaped from major climatic changes that have occurred within the immediate region and that serves as a refuge for biota that was more widely distributed.

Reintroduction (of organisms, species; see **Introduction**). Placement of an individual, population, or species back into its former habitat after it had been extirpated from that habitat.

Relictual species. Persistent remnants of formerly widespread biotas, typically existing in specific isolated areas or habitats (also referred to as relict).

Resilience (of an ecosystem; see **Stability**). Degree to which an ecosystem's structure and long-standing composition can be disturbed and yet retain or return to its original features.

Restoration. Management of a disturbed and/or degraded habitat that results in the recovery of its original state.

Restoration ecology. The study of theoretical principles and applications in population and community ecology aimed to restore and rehabilitate highly disturbed or degraded ecosystems to their more natural states.

Scale. In ecology, referring to hierarchical units of measuring or modeling spatial processes (e.g., dispersal, niche divergence) or temporal processes (e.g., succession, species guild formation).

Secondary succession. Ecological change in the members of a community that results from the destruction of all or part of an earlier community.

Species area curve. Graph of the relationship $S = cA^z$, where number of species is represented by S and size of area by A, c is a constant, and z is a key parameter in island models.

Species richness (S). A simple measure of species diversity calculated as the total number of species in a habitat or community.

Stability (of an ecosystem). If used in the constancy sense, refers to the degree of invariance in species composition, relative abundances, and so forth that can be evaluated only in relation to well-defined environmental changes; term also is used in the persistence sense, that is, often regarded as the desirable outcome of diversity; resilient systems have low constancy since they are often subject to disturbance and change; spatial variability and the presence of multiple reserves or refugia promote both persistence and constancy.

Stage structure (see **Age structure**). Structure of a population of organisms determined by the stage (in any unit) of individuals, for example, seed, seedling, juvenile, adult, and so on.

Stochasticity. Chance or random variation.

Demographic stochasticity (DS). Chance variation in individual birth and death rates; in small populations this may lead to successful reintroduction or recruitment.

Environmental stochasticity (ES). Randomly varying environments causing fluctuations in the population-level demographic characteristics.

Stochastic population growth (see **Deterministic**). Dynamics of population numbers under randomly varying environments.

Succession. Process in which communities of plants and animal species in a habitat or region follow a time series of different and increasingly more complex and diverse communities.

Sympatric. Referring to populations, species, or other taxa that occur together within the same geographical area and with the potential of gene exchange and competition.

Translocation (of organisms, species; see **Introduction, Reintroduction**). Management technique often used in mitigation for endangered species protection whereby an individual, population, or species is removed from its habitat to be established in another area of similar or identical habitat.

Vicariance. Existence of closely related taxa or biota in different geographical regions that have been separated by the formation of a natural barrier to dispersal.

Weedy species. Species that is growing where it is not wanted; in a management context, an exotic and/or invasive undesirable species that often requires concerted effort (labor and economic) to remove it from its current location.

INDEX

INDEX

Conservation Biology
The Theory and Practice of Nature Conservation,
Preservation, and Management
Peggy L. Fiedler and Subodh K. Jain, editors
Foreword by John Harper

In the Foreword to *Conservation Biology: The Theory and Practice of Nature Conservation, Preservation, and Management,* John Harper tells us that nature conservation has shifted from an idealistic philosophy to a serious technology. Speaking to students and scientists of conservation biology, as well as those interested in resource management, this volume discusses why we have only recently begun to change the questions we ask about our world. Combining theory, lessons learned from the past, and current measures already underway, *Conservation Biology* furthers our understanding of the planet's resources.

The editors have brought together a diverse group of scientists and theorists to address just how and why this shift in attitude has occurred and what is to be done now about conserving Earth's biological resources. Personal viewpoints and experimental data are both used to explore options and ideas in this exciting, new field known as conservation biology.

Contents: Foreword; Prologue; *Part I: The Natural Order:* Species Richness in Plant Communities; Rarity in Vascular Plant Species; Peasant Farming Systems; Conservation Biology Above the Species Level; *Part II: Processes and Patterns of Change:* Reptilian Extinctions; Loss of Biodiversity in Aquatic Systems; Threats to Invertebrate Biodiversity; Forest Fragmentation; Issues of Scale in Conservation Biology; *Part III: Population Biology and Genetics:* Extinction in Plant Populations; Minimum Viable Population Sizes; Conservation of Asian Primates; *Part IV: The Practice of Conservation, Preservation, and Management:* Rare Plant Conservation; Ecological Management of Sensitive Natural Areas; Park Protection and Public Roads; Strategies for Protecting Plant Diversity; Ecological Restoration in Conservation; Why Should We Conserve Species and Wildlands?; Epilogue; Glossary; Bibliography.

Peggy L. Fiedler is in the Department of Biology at San Francisco State University. Subodh K. Jain is in the Department of Agronomy and Range Science at the University of California, Davis.

Also available from Chapman and Hall

Genetic and Ecological Diversity: The Sport of Nature
Lawrence M. Cook
Paperback (0 412 35620 1), 206 pages

Monitoring for Conservation and Ecology
F. B. Goldsmith, editor
Hardback (0 412 35590 6) Paperback (0 412 35600 7), 275 pages